DUMBARTON OAKS
MEDIEVAL LIBRARY

*Daniel Donoghue, General Editor*

MEDICAL WRITINGS FROM
EARLY MEDIEVAL ENGLAND

VOLUME I

DOML 81

# Medical Writings from Early Medieval England

## VOLUME I

## The Old English Herbal, Lacnunga, and Other Texts

Edited and Translated by

## JOHN D. NILES
## MARIA A. D'ARONCO

Dumbarton Oaks
Medieval Library

Harvard University Press
Cambridge, Massachusetts
London, England
2023

First Printing

*Library of Congress Cataloging-in-Publication Data*
Names: D'Aronco, M. A., editor ; translator. | Niles, John D., editor ; translator.
Title: Medical writings from early medieval England / edited and translated by John D. Niles, Maria A. D'Aronco.
Other titles: Old English herbal. (D'Aronco) | Dumbarton Oaks medieval library ; 81.
Description: Cambridge, Massachusetts : Harvard University Press, 2023– | Series: Dumbarton Oaks medieval library; DOML 81 | Includes bibliographical references and indexes. | Contents: v. 1. The Old English herbal, Lacnunga, and other texts. The Old English herbal / Maria A. D'Aronco — Old English remedies from animals / Maria A. D'Aronco — Lacnunga / John D. Niles — Peri didaxeon / John D. Niles — Miscellaneous remedies / John D. Niles. | Facing page translation with Old English on the versos and English on the rectos; includes some Latin and Greek.
Identifiers: LCCN 2022038047 | ISBN 9780674290822 (cloth)
Subjects: LCSH: Medicine, Medieval — Great Britain. | Anglo-Saxons — Medicine — Early works to 1800. | Herbals — Great Britain — Early works to 1800. | Materia medica, Vegetable — Great Britain — Early works to 1800. | Materia medica, Animal — Great Britain — Early works to 1800. | Medicine — Formulae, receipts, prescriptions — Early works to 1800.
Classification: LCC R128.6 .M435 2023 | DDC 615.3/210902 — dc23/eng/20221117
LC record available at https://lccn.loc.gov/2022038047

# Contents

# CONTENTS

vi

# Disclaimer

No statement in this book should be construed as medical advice. Anyone wanting to treat any illness mentioned in this collection of early medieval medical texts should consult a qualified professional before embarking on a course of treatment. While some of the remedies mentioned here might be benign and others innocuous, others could have deleterious effects, while certain of them involve the use of toxins that might be harmful or even fatal in their effects if administered as called for here.

# Introduction

One remarkable feature of the literature of early medieval England is the large number of medical texts that were produced in the English language at this early date. In other parts of Europe the vernacular language was not used in such a systematic manner for recording medical texts until a good deal later, after the rise of scholastic learning in university settings at Bologna, Paris, and other urban centers. While the corpus of Old English medical texts is worth knowing about for many reasons, one is that it represents a largely prescholastic body of knowledge with distinctly Insular traits.

"Medical texts" is perhaps too colorless a term to use with reference to the richly varied set of remedies and other medical writings recorded in English sources dating from the Anglo-Saxon period and its aftermath.[1] Recorded side-by-side with translations of some of the most prestigious late antique Latin medical treatises are cures that have no known sources or analogues in the Mediterranean tradition. While some of these texts are nondescript, others are as extravagant in their use of language as they are ambitious in the healing practices they prescribe. Alongside recipes for treating everyday ailments such as headaches or eyesores are unparalleled procedures for preventing infant mortality, restoring lost cattle, warding off elf-shot, or remedying the effects of flying venom. Granted, there is nothing startling about the great majority of Anglo-Saxon medical recipes, for most such remedies rest on a well-established tradition of Greek and Latin natural science. Some resemble cures still in use among herbalists. This can enhance their interest for present-day readers. At the same time, many Anglo-Saxon remedies rest on pseudoscientific beliefs about the nature and causes of disease. This too adds to their potential interest. Moreover, many rely on the power of the healer's voice. This is true of several unique remedies that rightly or wrongly have been called "shamanistic," and it is consistently true when Christian prayers or exorcisms are conducted in a liturgical style that evokes the doctrine of Christ as ultimate healer. There is much in this volume, then, to speak to the interests of specialists in the history of science, folklorists, social historians, literary scholars, students of popular religion, and medical practitioners alike.

The present contribution to the Dumbarton Oaks Medieval Library represents the first of a projected two-volume sequence on the medical literature of early medieval England. The second volume will treat the tenth-century vernacular treatise known as *Bald's Leechbook,* together with the work known as *Leechbook III* that accompanies it in London, British Library, MS Royal 12.D.xvii. The volume at hand covers the rest of the medical corpus, leaving aside a few texts that are readily available elsewhere. No comprehensive edition and translation of the earliest English medical texts has been undertaken since Thomas Oswald Cockayne's edition of 1864–1866, although the need for one has long been felt.

This volume is divided into four main parts. The first of these consists of an edition and translation of the major work *The Old English Herbal,* together with an edition and translation of the shorter work, titled here *Old English Remedies from Animals,* that regularly accompanies it in the manuscript tradition. The second part is an edition and translation of *Lacnunga,* a late tenth-century medical miscellany composed chiefly in the vernacular. The third part is an edition and translation of the medical treatise known as *Peri Didaxeon,* a post-Conquest work that has not been translated into modern English since Cockayne. The fourth part consists of a miscellany of Old English or Latin medical texts recorded in a variety of Insular manuscripts dating from the ninth century to the twelfth.

## SOURCES AND CONTEXTS OF ANGLO-SAXON MEDICAL KNOWLEDGE

The textual records with a bearing on medicine in early medieval England are firmly rooted in a classical and medieval Latin tradition of natural science and medical learning whose afterlife can be traced all over Europe, not just during the early medieval period but also in later centuries up to the present time.[2] This Latin tradition, in turn, was grounded in Greek, Egyptian, and Near Eastern precedents, as has been emphasized in standard sources on the history of science and medicine.[3] The special interest of the great bulk of the texts included in the present volume is that they are written in English: that is, they represent translations of Latin treatises, or of individual Latin remedies, into the language of daily use. Up to and through the time of the

Norman Conquest, English was increasingly becoming an important vehicle for scientific writings and religious translations as well as for polished examples of written literature composed in both verse and prose. The history of medicine in early medieval England, as far as can be discerned from the surviving textual evidence, is thus very much the story of how Latin learning was transmuted in innumerable slight ways as it was assimilated into English intellectual life and material practice.

Granted this overwhelming direction of influence from Italy and other parts of the Continent, there exist many medical texts in the Anglo-Saxon records that have no known sources or analogues in the Latinate tradition. While these remedies have sometimes been spoken of as pagan in inspiration or Germanic in origin, we prefer to refer to their sources as for the most part "unknown." More archaic versions of certain of these cures, with their reliance on beliefs in the powers of such creatures as elves, dwarves, witches, and mysterious night riders, appear to have been brought to Britain as part of the cultural heritage of Germanic-speaking settlers during the fifth- and sixth-century Age of Migrations. Other remedies may have had multiple antecedents—ancient ones, perhaps—among the indigenous Celtic-speaking groups that the incoming Angles and Saxons displaced, subdued, or intermarried with. Ongoing influences from Ireland, Scandinavia, and adjoining regions of northwest Europe are another factor to be taken into account, especially in the sphere of popular religion, given the close interactions of English-speaking people with the inhabitants of those regions during the early medieval period.

To judge from the available evidence, by the time that the texts presented in this volume were written down, they would have functioned primarily within the environment of Benedictine monasticism, which gave importance to the care of the sick and encouraged the production of books of medicine. The books in which these texts are written down would have been housed for the most part in monastic or cathedral libraries. To the extent that the remedies were put into practice as opposed to simply being recorded as part of a learned tradition, this would have happened in bilingual settings where the texts were at hand, where the requisite herbs were likely to be available in an herbarium, and where the words and rhythms of the Mass and other aspects of the liturgy were part of the daily tenor of life.[4] In addition, in an idealized scenario—from which the realities of life doubtless often fell short—each monastery would have had not just its own collection of books and its own well-stocked herbarium, but also its own infirmary. Moreover, each monastery would have had available to it a labor force of workers whose duties would have brought them to outlying gardens where healthful foods were grown, and to outbuildings where domesticated animals were sheltered, ale was brewed, and other commodities used in the recipes were produced. Extending beyond these confines were fields where barley, rye, and wheat were harvested and where animals were grazed, as well as ponds, streams, meadows, and woodlands where other plants or animals of potential medicinal value could be gathered or hunted. The texts that are edited here are best read with this plethora of resources in mind, from the cloister to outlying fields and woodlands.

Furthermore, each monastery would also have participated in well-defined channels of communication that linked English ecclesiastical centers with corresponding centers in regions where exotic plants such as frankincense and myrrh would have been available. Such interregional connections as these, within an overarching Church hierarchy, would have ensured that medicine as practiced in any tenth- or eleventh-century English monastery, even with its individual features, was in essential alignment with contemporary European norms.

This is not the sole historical context for these medical writings, however, for any English monastery would also have had connections with ordinary people living in surrounding communities. In nearby villages and estates, domestic medical practices may have persisted for years or centuries in patterns only peripherally affected by the Church. For the great majority of people, time-tested home cures would have been their initial recourse in cases of disease, discomfort, or injury. If fractures, for example, are barely mentioned in the medical collections, this may be in part for the reason that broken bones were either dealt with on the spot or were referred to lay bonesetters rather than to monastic physicians. Be that as it may, it is easy to imagine scenarios by which knowledge relating to domestic medical practices filtered into the monastic centers where manuscripts containing medical knowledge were made and housed. Priests serving either local communities or regional magnates might have served as intermediaries for such exchanges of knowledge, especially if they interacted with local people known for their medical knowledge and skills. Such people might have included herbalists and "cunning women," or women who were ascribed special healing powers.[5] As semilearned persons, parish priests may sometimes have been asked to serve as physicians themselves, as has not been unknown in later centuries in rural communities where regular physicians were few and far between. Certain English curates, or certain members of the literate laity, may have written down medical recipes for their own use in circumstances like these. It is natural to infer that the texts that are compiled in the present volume reflect these local influences alongside more clearly defined influences deriving from learned channels.

### The Old English Herbal and Old English Remedies from Animals

*The Old English Herbal* is a translation into English, dating from the end of the tenth century, of four Latin texts that in the Latin tradition appear as separate treatises. Its source is not one book: rather, it is a compilation based on these pre-existent medical texts. Their use to create this herbal demonstrates that the Anglo-Saxons had at their disposal almost all the important medical texts of the Carolingian period (ca. 750–900). These sources have been identified as (1) the *De herba vettonica liber* attributed to Antonius Musa; (2) the *Herbarius* of pseudo-Apuleius; (3) a version of the pseudo-Dioscorides treatise *Liber medicinae ex herbis femininis;* and (4) a version of the pseudo-Dioscorides treatise *Curae herbarum.* In addition, there are seven chapters without known sources.[6] In *The Old English Herbal,* these works, when run together, constitute 185 chapters of continuous text. The main part of the *Herbal* is preceded by a table of contents, where the name of a plant is followed by a list of the ailments for which it is a remedy.

*Old English Remedies from Animals* is a translation into Old

English of a Latin book of remedies, known today under the title *Medicina de quadrupedibus,* whose cures are mostly derived from four-footed animals rather than plants. This too is a composite work. Its fourteen unnumbered chapters are based on three distinct Latin works: (1) the A-version (or Christianized redaction) of *De taxone liber* or *Epistula Hipparchi de taxone,* on the medicinal virtues of the badger; (2) a short anonymous treatise on the medicinal virtues of the mulberry tree; and (3) the short version of the *Liber medicinae ex animalibus* attributed to Sextus Placitus Papiriensis.[7]

The translation into Old English of these Latin medical texts, and probably their compilation into a paired set of pharmacopoeias as well, was carried out in the cultural climate of the Anglo-Saxon Benedictine revival during approximately the last quarter of the tenth century. It is clearly the result of a well-defined translation project whose aim was to offer an English-speaking audience the most authoritative learning available in the field of pharmacy. Significantly, these texts require no special knowledge of physiology in order to be understood. Symptoms of disease are described without going far into details, and the entries generally address such common ailments as toothache, stomachache, headache, and digestive problems. There is nothing of the medical textbook about these pharmacopoeias, and in this regard they differ from *Bald's Leechbook,* where a number of more technical passages are introduced. The texts presented here are more like a first-aid manual, for they offer practical remedies that require no more than a minimum of experience on the part of the practitioner.

The importance of these two pharmacopoeias for the Anglo-Saxons is evident from the fact that this translation is preserved in four manuscript copies dating from the period 975 to 1200 and descending independently from the same original manuscript. Three of these are now housed in the British Library, London, while the fourth is in the Bodleian Library, Oxford, as is set forth in the Note on the Texts at the back of this volume. The most significant of these copies, London, British Library, MS Cotton Vitellius C.iii (abbreviated *V*), is a richly illustrated codex that was partially damaged by the fire that swept through Ashburnham House, London, on October 23, 1731. The presence in this manuscript of handsome illustrations of healing plants—and of venomous snakes, too, as their imagined adversaries—is a sign that care and expense went into its making. Viewed purely as a book, Cotton Vitellius C.iii would have been a splendid one to own, and it is probably for this reason that in the modern period it found its way into the library of Robert and Henry Cotton, whence it eventually came into the holdings of the British Library. The book evidences the wealth and skills of the monastic community that produced it, which was probably Christ Church, Canterbury.[8]

*The Old English Herbal,* together with *Old English Remedies from Animals,* represents a complex compendium of learning, organized plant by plant (or animal by animal) in an encyclopedia-like fashion rather than ailment by ailment, as in other treatises, including *Bald's Leechbook.* The plants that are described, and sometimes also the animals, such as the lion and the elephant, came from the Mediterranean basin and were therefore not necessarily familiar in England. The translator of the *Herbal* took care to make them recognizable to his readers, for failing to recognize a particular plant or, even worse, mistaking one species for another could have had life-threatening consequences. To aid this traditional matter of identification, the translator therefore furnished the text with a number of synonyms, systematically listing all the Latin names of the plants mentioned in individual chapters, adding next on some occasions the Greek synonym, and then, when possible, the corresponding English term. The translator refrained from coining new names *ad hoc,* so as to avoid the potentially dangerous consequences of mistaken identification. Thus, in *V* (as well as in Oxford, Bodleian Library, MS Hatton 76), when there was no corresponding vernacular name, a blank was left in its place (indicated in the present edition by an ellipsis framed by square brackets).

In sum, the making of *The Old English Herbal* and *Old English Remedies from Animals* was an initiative without precedents and without imitations. This bipartite pharmacopoeia is the result of exceptional competence in the subject matter treated, of a lucid capacity for synthesis, and of a strong awareness of one's own national identity, visible in the aim of providing English-speaking readers, regardless of their level of proficiency in Latin, with access to the greatest wealth of medical knowledge available at the time.

### Lacnunga, Peri Didaxeon, and Miscellaneous Remedies

The next text featured in the present volume is the treatise known, ever since Cockayne, as *Lacnunga,* an Old English term meaning "Remedies." Here medical procedures of very different character are juxtaposed with one another. There is a tendency all the same for remedies to be clustered according to the nature of the ailment to be treated, starting from the head and working down to the lower body, as is characteristic of medieval medical compendia.

*Lacnunga* is preserved in London, British Library, MS Harley 585, which can be dated on paleographical grounds to about the year 1000. The fact that Harley 585 also contains copies of *The Old English Herbal* and *Old English Remedies from Animals,* directly preceding *Lacnunga,* suggests that the compiler of this manuscript found these works to be of complementary benefit. As a collection, *Lacnunga* has no known single precedent, although certain of its individual remedies are paralleled elsewhere in the medical literature of early medieval England, and some of these parallel texts may have a common source. While most of its remedies are written out in the vernacular, some are written in Latin in whole or in part, especially at the end of the collection, where different hands can be discerned. Latin liturgical elements recur with some frequency, while an element of magic has a more prominent place here than in more scholarly works such as the *Leechbooks.* Like *The Old English Herbal,* this compendium appears to be the product of a milieu where bilingual literacy was taken for granted, whether or not expert competence in Latin was characteristic of the community where the book was made. In addition, an Irish element underlies certain invocations, though this is scarcely evidence that anyone using the manuscript was competent in Irish; the contrary is more likely to have been true.[9] At the end of the collection, a single remedy written out in the

Anglo-Norman dialect of French is one sign among many that the collection remained in active use during the second half of the eleventh century, and very likely longer.

Among the types of remedy included in *Lacnunga,* of special interest are the Old English metrical charms. These are the entries here numbered 76 ("The Nine Herbs Charm"), 86 (for dwarf-driven fever), 127 (for a stabbing pain), and 149 (for loss of cattle), plus three related remedies that address the problem of infant mortality (161, 162, and 163). In part because of their reputation as some of the most enigmatic texts in the corpus of Old English, the metrical charms have long been a magnet for scholars, particularly for their magical elements and for the evidence they provide as to the convergence of Christian and pre-Christian cultures.[10] Another item of special interest among the cures and prayers of *Lacnunga* is number 65, "The Lorica of Laidcenn," an elaborate poem composed in Latin that is attributed to the seventh-century Irish monastic scholar Laidcenn mac Buith Bannaig.[11] The copy of "The Lorica of Laidcenn" written out in British Library MS Harley 585 is accompanied by a word-for-word interlinear gloss into Old English. Although this gloss is of philological interest, the present edition features only the Latin poem, which is the primary document, set out with an accompanying modern English translation.[12]

The next work to be edited and translated here is the unique copy of the medical treatise known as *Peri Didaxeon* (On the Teachings). This is preserved in London, British Library, MS Harley 6258 B, a late twelfth-century manuscript, where it is written out just after an alphabetical reworking of *The Old English Herbal* and *Old English Remedies from Animals.* The treatise breaks off abruptly at entry 67, whether because some entries were lost or, conceivably, because the English text was never finished.

Although *Peri Didaxeon* has no single known source, what certain entries represent (or what certain parts of certain entries represent) are English-language paraphrases from the prior Latin work now known as *Tereoperica,* a text that was formerly known in the scholarly literature as *Practica Petrocelli Salernitani,* or *Petrocellus* for short.[13] This compilation of medical recipes is generally dated to the ninth century but may be of later date. It shows the influence of medieval scholastic modes of thought as developed first of all in Salerno and thereafter in other European centers of medical learning, including the theory of the four humors in its developed form. The English text that is based in part on this work was evidently produced as part of the same impulse to record medical literature in the vernacular that resulted in the making of *Bald's Leechbook, Leechbook III, The Old English Herbal, Old English Remedies from Animals,* and *Lacnunga.* Its language is later than that of these other texts, however: so late as to have led some scholars to regard it as pertaining to the corpus of Middle English literature. Others view it as a late record of an essentially Anglo-Saxon medical treatise. Regardless of these philological issues, and regardless also of the occasional substantive differences that distinguish these different medical compilations from one another, there is manifest continuity between *Peri Didaxeon* and the other works presented here.

The volume concludes with a miscellany of 101 short medical texts, many of which have little or nothing to do with the main contents of the manuscripts in which they are recorded. The language in which they are written reflects the bilingual culture from which they emerged: fifty-five of them are written out wholly or chiefly in Old English, forty-one are wholly or chiefly in Latin, and five use English and Latin in fairly equal measure. Their preservation can be ascribed in part to the availability of patches of unused parchment, sometimes in the margin of a page and sometimes in the space following another text. Despite the haphazard manner in which many of these remedies have been preserved, each one was thought to be of sufficient value for someone to want to record it for posterity, very likely in the belief that it recorded an efficacious cure for a serious medical condition.

For practical reasons, certain texts of a related kind are omitted from the present volume. Prominent among these are seven Old English metrical charms that are recorded elsewhere than in *Lacnunga;* these are the charms customarily numbered 1 and 7–12 in recent publications. Since Robert E. Bjork has edited and translated these in his DOML edition *Old English Shorter Poems,* their inclusion here would be redundant.[14] Also omitted here are five of the nine remedies written out on folio 15b of the Vitellius Psalter (London, British Library, MS Cotton Vitellius E.xviii). While these entries are not unlike the other four that are inscribed on that same folio and that are included in our edition, they are not medical in nature and so fall outside the scope of this volume.[15] For the same reason, just four of the five marginal texts that are written out at folios 129a and 140a of London, British Library, MS Cotton Caligula A.xv, are presented here. There are a few other omissions. A single gloss (a remedy for black boils) on folio 49r of London, British Library, MS Royal 12.D.xvii—the unique copy of *Bald's Leechbook*—is to be included in volume 2 of this projected two-volume DOML edition. Two remedies involving animal substances written out in London, British Library, MS Cotton Galba A.xiv are disregarded because the folios on which they are written out are so badly burned as to be virtually unreadable.[16] Finally, no attempt has been made to include Latin medical texts or treatises that happen to be recorded in English manuscripts of eleventh-, twelfth-, or thirteenth-century date and that show no conspicuously Insular features, for these texts pertain to the history of Western medicine more generally.[17] In like manner, medical treatises or recipe books that are composed in the Anglo-Norman dialect of French fall outside the scope of the present volume, as do texts composed in a dialect that, on philological grounds, is customarily classified as Middle English.

### TRANSLATING WORDS, TRANSLATING CULTURES

The translations offered here maintain close fidelity to the source texts within the idiom of modern English. While the cookbook-like style of many of the medical recipes has been maintained to a certain extent, pronoun reference has sometimes been clarified for the sake of easier understanding, while words or phrases that were left understood in the original text are sometimes supplied. The use of subjunctive constructions in the Old English texts ("let it be warmed"), as opposed to imperative ones ("warm it"), is sometimes mirrored in the translations, but at other times is ignored as being of no real consequence.

It should be understood that many of these remedies represent a kind of verbal shorthand: that is, they provide cues for a performance whose full nature could be known only by someone who was already a competent medical practitioner. As has rightly been remarked, analysis of these texts reveals that "tacit knowledge based on some kind of training must lie behind being able to use them effectively."[18] When a recipe calls for certain plants to be cooked and the Old English verb that is used for this operation is *wyllan* (to boil) for example, the practitioner would have had to know whether to boil the plants hard, as one might want to do with root vegetables such as beets or turnips, or to simmer them gently, as herbalists often take care to do when preparing a curative salve or drink. Either "boil" or "simmer" is therefore our preferred translation of *wyllan,* depending on what plants are being prepared, in what context.

No attempt has been made to translate nonsensical incantations or inscriptions. In our translations, though not on the text pages, conventional formulas are supplied to fill out abbreviated phrases that would have been familiar to educated readers during the early Middle Ages. At the end of entry 63 of *Lacnunga,* for example, the Latin word *per* (with a mark of abbreviation), meaning "by" or "through," is expanded in our translation to "by your holy name," with a note to account for this practice.

One of the most challenging tasks in translating the medical texts that date from this period is to determine the closest modern English equivalents to the Old English names for plants. Plant names were not necessarily used with the same referent from region to region or from writer to writer. Moreover, before the introduction of modern binominal nomenclature, a plant might have had more than one name, just as many plants have more than one name today. Nor was an Anglo-Saxon *æppel* the same as one of the large, glossy apples that one can buy in supermarkets today. The names offered in these translations are therefore sometimes approximations or, at times, no better than educated guesses. Detailed information about Old English botanical names —their variant spellings, their probable Latin equivalents, their apparent equivalents in modern scientific nomenclature, their various proposed translations into modern English—can be found in specialized resources on that topic.[19]

As for Old English terms for ailments and diseases, for the most part there is little problem in translating them, seeing that reflexes of many of these same ailments exist today. After all, a toothache is a toothache. Still, the meaning of certain terms is problematic. An example is the disease that in Old English is called *þeor,* and that we translate as "theor-disease." With his customary precision, M. L. Cameron concludes that the term *þeor* was used with reference to "a dry roughness externally of epithelial tissues, particularly the skin, such as might result from a vitamin deficiency or a reaction to an allergen and, by extension, a similar roughness of internal tissues."[20] Although one might be tempted to translate *þeor* as "dermatitis" or "psoriasis," the fact that the same word is used with reference to internal tissues as well as external ones leads us instead to fall back on the rather blank translation "theor-disease." Elsewhere, when faced with a problematic Old English term for disease, we have opted for a plausible modern English equivalent. For Old English *blegene,* for example, a term that others have translated "blains," we adopt the translation "boils," with the understanding that this term can refer to any skin swellings or sores, from fairly innocuous ones to the black boils *(blæce blegene)* that might have been symptoms of serious disease. For the disease named *dweorh,* a word whose root meaning is "dwarf," we opt for one or another periphrastic phrase. Although the use of this word to refer to a disease or affliction may reflect an archaic belief that high fevers, with their attendant delirium, were the result of malicious activity on the part of dwarves, this belief is likely to have been dormant in late Anglo-Saxon monastic settings.[21] To speak of "high fever" therefore seems an adequate equivalent. Likewise, the term "elf shot" might be a dead metaphor, except in entry 127 from *Lacnunga,* with its allusions to missiles hurled by elves, hags, or the numinous powers of ancient belief systems.[22] Although it is hard to say what was in the mind of a healer who spoke of a horse as being "shot," the reference is probably to a veterinary ailment—very likely the potentially life-threatening one now called colic—that was to be treated like any other, whether or not any thought of elves was in mind.

In general, if these remedies are to be understood, they should be read with an attempt at sympathetic understanding of the system of belief and practice within which they were once operative. Even though that system may be alien or repugnant to many people today, it made sense to men and women who were just as intelligent as ourselves and who were doing their best to preserve their well-being in a world where the consequences of disease could be both swift and deadly.[23]

These differences of mentality can sometimes lead to a sense of desperation on the part of the translator. The Old English noun *wyrm,* for example (plural *wyrmas*), has the basic sense "snake" or "serpent," and many of these remedies (like their Mediterranean counterparts) address the problem of snakebite. As readers of *Beowulf* know, however, a *wyrm* can also be a dragon, and we find an allusion to what appears to be a mythological serpent or dragon in entry 76 of *Lacnunga,* the "Nine Herbs Charm." Here Woden, once worshiped as a god of the pre-Christian northern pantheon, is said to have smitten a *wyrm* into nine pieces that apparently gave rise to nine venoms, or sources of infection. Elsewhere the same term *wyrmas* is used of innocent earthworms. It can also be used of intestinal worms, which were an affliction to humans and beasts alike. Moreover, in several of these cures reference is made to *wyrmas* that are tiny insects, including mites that can torment the flesh and cause scabies. Deciding how to translate *wyrm* in a particular instance can itself cause headaches. What remains clear is that the men and the women of this period conceived of many different ailments as resulting from the action of *wyrmas,* whether with or without accompanying venoms.

To generalize from these examples, the work of the Anglo-Saxon physician—as of all physicians—was devoted to warding off the effects of disease-causing agents by all means possible, whether this was through salves, preparations, steam baths, poultices, acts of ritual purification, surgical operations, the use of amulets and magical inscriptions, the making of the sign of the cross, or the singing of prayers, incantations, and spells. Joined to that work was the need to ward off the assaults of the devil, as well as the acts of witches, who were the devil's allies, in accord with the Christian doctrine that the ultimate cause of disease was the Evil One's assault on God's beneficent world order. Some remedies required complex procedures. Very many were of a simple and pragmatic kind, relying chiefly on knowledge of the beneficent qualities of certain herbs, many of which were readily available in herbaria or in the wild. The translations and notes presented here, it is hoped, will facilitate understanding of these texts, in their impressive variety, within the intellectual context of their time.

## Special Features of This Edition

References to the Bible are to the DOML edition *The Vulgate Bible,* in six volumes, as edited by Swift Edgar and Angela M. Kinney (2010–2013). Translations of biblical phrases are drawn from that same edition (that is, they represent the Douay-Rheims translation dating from between 1582 and 1610 with modernized capitalization, punctuation, and spelling). Correspondingly, psalms are cited by their Vulgate number.

Where the sign of the cross is a clearly visible part of the scribal text, it is reproduced in schematic form in these pages. Although the purpose of such inscribed crosses is unknown, there is no doubt that both the image of the cross and the gestural sign of the cross were thought to be potent means of warding off evil.[24] Certain of these marks may have been meant to cue the healer to make the sign of the cross at this point in the procedure. Alternatively, some such signs may be no more than marks of piety on the part of scribes.

In closing, the editors wish to express their gratitude to Jan Ziolkowski for his initial support of this project; to Daniel Donoghue for guiding it through to completion; to Nicole Eddy for her skilled editorial advice and assistance along the way; and to three specialist readers for the press—

Antonette diPaolo Healey, Peter S. Baker, and Diana Myers—with deep gratitude for their having spotted errors in the preliminary draft of the book and for making a number of perceptive suggestions for its improvement. Professor D'Aronco offers special thanks to Anne van Arsdall for allowing her to draw freely on Arsdall's previously published translation of *The Old English Herbal*. Professor Niles is specially indebted to his wife, Professor Carole Newlands, for help with troublesome Latin or pseudo-Latin passages.

## NOTES

1 Throughout this work, "Anglo-Saxon" is used in specific reference to the history and culture of the English-speaking people in what is now England from the sixth to the eleventh centuries CE.

2 For relevant discussion, see Maria Amalia D'Aronco, "Anglo-Saxon Plant Pharmacy and the Latin Medical Tradition," in *From Earth to Art: The Many Aspects of the Plant-World in Anglo-Saxon England,* ed. Carole P. Biggam (Amsterdam, 2003), 133–51; D'Aronco, "How 'English' Is Anglo-Saxon Medicine? The Latin Sources for Anglo-Saxon Medical Texts," in *Britannia Latina: Latin in the Culture of Great Britain from the Middle Ages to the Twentieth Century,* ed. Charles Burnett and Nicholas Mann (London and Turin, 2005), 27–41.

3 For additional information on this tradition and its influence in England, the reader is referred to Michael R. McVaugh, "Medicine in the Latin Middle Ages," in *Western Medicine: An Illustrated History,* ed. Irvine Loudon (Oxford, 1997), 54–65; and to C. H. Talbot's prior study *Medicine in Medieval England* (London, 1967).

4 On the availability of herbs, note D'Aronco, "Anglo-Saxon Plant Pharmacy," as well as her essays "The Benedictine Rule and the Care of the Sick: The Case of Anglo-Saxon England," in *The Medieval Hospital and Medical Practice,* ed. Barbara S. Bowers (Aldershot, 2007), 235–51, and "Gardens on Vellum: Plants and Herbs in Anglo-Saxon Manuscripts," in *Health and Healing from the Medieval Garden,* ed. Peter Dendle and Alain Touwaide (Woodbridge, 2008), 101–27.

5 This term has been used with reference to women who either claimed or were ascribed special healing powers. For archaeological evidence, see Tania Dickinson, "An Anglo-Saxon 'Cunning Woman' from Bidford-on-Avon," in *In Search of Cult: Archaeological Investigations in Honour of Philip Rahtz,* ed. Martin Carver (Woodbridge, 1993), 45–54.

6 For *De herba vettonica liber* attributed to Antonius Musa, see Ernst Howald and Heinrich E. Sigerist, eds., *Antonii Musae De herba vettonica liber, Pseudoapulei Herbarius, Anonymi De taxone liber, Sexti Placiti Liber medicinae ex animalibus etc.,* Corpus Medicorum Latinorum 4 (Leipzig, 1927), 1–11. This corresponds to chapter 1 of the present edition. For the *Herbarius* of pseudo-Apuleius, see Howald and Sigerist, *Antonii Musae De herba,* 13–225. This corresponds to chapters 2–132 of the present edition. For the pseudo-Dioscorides treatise *Liber medicinae ex herbis femininis,* see Heinrich F. Kästner, "Pseudo-Dioscoridis *De herbis femininis,*" *Hermes* 31 (1896): 578–636, and Kästner, "Addendum ad Pseudodioscoridis *De herbis femininis* ed. Hermae XXXI 578," *Hermes* 32 (1897): 160, corresponding to chapters 133, 135–36, 139, 141–44, 146–50, 152–57, 160–62, 165–70, 172–75, 181, and 184–85 of the present edition. For the pseudo-Dioscorides treatise *Curae herbarum,* see Sofia Mattei, "*Curae herbarum,*" (PhD diss., Università degli Studi di Macerata, 1995). The seven chapters without known sources are chapters 151, 159, 163–64, 171, 176, and 183 of the present edition. See Walter Hofstetter, "Zur lateinischen Quelle des altenglischen Pseudo-Dioskurides," *Anglia* 101 (1983): 315–60.

7 For the first of these three works, see Arsenio Ferraces Rodríguez, "Contaminación textual y censura ideológica en *La Epistula Hipparchi de taxone* (Montecassino, Biblioteca dell'Abbazia, v. 97, pp. 532b–533a)," *Medicina nei secoli* 29 (2017): 999–1032; the text is printed by Howald and Sigerist, *Antonii Musae De herba,* 227–32. For the second of the works, see Hubert Jan de Vriend, *The Old English "Herbarium" and "Medicina de quadripedibus,"* EETS o.s. 286 (Oxford, 1984), 238–40. For the third, see Howald and Sigerist, *Antonii Musae De herba,* 233–85.

8 The book's illustrations are reproduced to good advantage in Maria Amalia D'Aronco and M. L. Cameron, *The Old English Illustrated Pharmacopoeia: British Library Cotton Vitellius C.iii,* Early English Manuscripts in Facsimile 27 (Copenhagen, 1998). Oxford, Bodleian Library, MS Hatton 76, too, is an elegant copy of the Old English pharmacopeia that was prepared for being illustrated, though the illustrations were never added.

9 On this phenomenon see especially Howard Meroney, "Irish in the Old English Charms," *Speculum* 20 (1945): 172–82.

10 The Old English metrical charms recorded in *Lacnunga* and other manuscript sources are edited as a group by Elliott Van Kirk Dobbie in *The Anglo-Saxon Minor Poems,* ASPR 6 (New York, 1942), 119–24; and by Robert E. Bjork in *Old English Shorter Poems, Volume II: Wisdom and Lyric,* DOML 32 (Cambridge, MA, 2014), 194–213. For an annotated bibliography of scholarship on the metrical charms published up to 1990, see Russell Poole, *Old English Wisdom Poetry* (Cambridge, 1998), 154–90; and note also Jonathan Roper, *English Verbal Charms* (Helsinki, 2005), 29–36. Roper likewise reviews the scholarly literature on charms of the Middle English period (pp. 36–39) and the early modern period (pp. 40–51).

11 The Latin term *lorica* refers to a cuirass, corselet, or breastplate. According to a martial metaphor developed by Saint Paul in his sixth letter to the Ephesians, zealous Christians are urged to put on the breastplate of justice, the shield of faith, the helmet of salvation, and the sword of the spirit so as to join in spiritual combat against the devil's deceits. In early Christian Ireland, this martial metaphor led to the development of the literary genre of the lorica, a prayer addressed to God for protection. Some poems of the lorica genre included, as this one does, an enumeration of the parts of the body needing protection—hence their relation to the literature of healing. See Pierre-Yves Lambert, "Celtic *Loricae* and Ancient Magical Charms," in *Magical Practice in the Latin West,* ed. Richard L. Gordon and Francisco Marco Simón (Leiden, 2010), 629–48.

12 Readers are referred to Edward Pettit, ed., *Anglo-Saxon Remedies, Charms, and Prayers from British Library MS Harley 585: The Lacnunga* (Lewiston, 2001), vol. 1, pp. 40–56, and vol. 2, pp. 82–93, for an edition of the Old English gloss. "The Lorica of Laidcenn" survives in multiple copies. Michael W. Herren, *Hisperica Famina II: Related Poems* (Toronto, 1987), 23–31, 76–89, and 113–37, presents an edition of the Latin poem based on a collation of sources.

13 For a recent critical edition of the Latin text, see Laura López Figueroa, "Estudio y edición crítica de la compilación médica latina denominada *Tereoperica*" (PhD diss., Santiago de Compostela, 2011).

14 Bjork, *Old English Shorter Poems,* 186–93 and 214–27.

15 One of these is to prevent theft of bees; another is the drawing known as "Saint Columcille's Circle"; a third is for loss of property; a fourth is to

prevent mice from devouring grain; and the fifth is a blessing of the land. For edited texts see Thomas Oswald Cockayne, *Leechdoms, Wortcunning and Starcraft of Early England* (London, 1864–1866), vol. 1, pp. 388, 397, and vol. 3, p. 290; Godfrid Storms, *Anglo-Saxon Magic* (The Hague, 1948), nos. 50, 85, 86, and A2; and Karen L. Jolly, "Tapping the Power of the Cross: Who and for Whom?," in *The Place of the Cross in Anglo-Saxon England,* ed. Catherine E. Karkov, Sarah Larratt Keefer, and Karen Louise Jolly (Woodbridge, 2006), 77–79.

16 Those words and phrases that remain legible are transcribed by Bernard J. Muir, *A Pre-Conquest English Prayer-Book* (Woodbridge, 1988), 211. Muir's transcription is reprinted by Stephanie Hollis and Michael J. Wright, "The Remedies in British Library MS Cotton Galba A.xiv, fos. 139 and 136r," *Notes and Queries* 41 (1994): 146.

17 An example is Storms, *Anglo-Saxon Magic,* no. A1, which is an eleventh-century copy of an antique Latin address to the herbs. For discussion of the introduction of Latin medical texts, several of them of a scholastic nature, to England during the eleventh century, see Debby Banham, "England Joins the Medical Mainstream: New Texts in Eleventh-Century Manuscripts," in *Anglo-Saxon England and the Continent,* ed. Hans Sauer and Joanna Story (Tempe, 2011), 341–52. Lea Olsan traces continuities between orally based verbal remedies written in either Latin or English, taking into account both Old English and Middle English texts. See Olsan, "Latin Charms of Medieval England: Verbal Healing in a Christian Oral Tradition," *Oral Tradition* 7 (1992): 116–42.

18 Anne van Arsdall, "Medical Training in Anglo-Saxon England: An Evaluation of the Evidence," in *Form and Content of Instruction in Anglo-Saxon England in the Light of Contemporary Manuscript Evidence,* ed. Patrizia Lendinara, Loredana Lazzari, and Maria Amalia D'Aronco (Turnhout, 2007), 415.

19 For a helpful overview of this topic and its challenges, see Carole Biggam, "Introduction," in "Magic and Medicine: Early Medieval Plant-Name Studies," ed. Carole Biggam, special issue, *Leeds Studies in English* 44 (2013): 1–9. Readers are referred as well to the online Dictionary of Old English Plant Names (http://oldenglish-plantnames.org/); to Hans Sauer and Elisabeth Kubaschewski, *Planting the Seeds of Knowledge: An Inventory of Old English Plant Names* (Munich, 2018); to Peter Bierbaumer, *Der botanische*

*Wortschatz des Altenglischen,* 3 vols. (Bern, 1979); to Anne van Arsdall, *Medieval Herbal Remedies: The Old English Herbarium and Anglo-Saxon Medicine* (New York, 2002); and to Stephen Pollington, *Leechcraft: Old English Charms, Plantlore, and Healing* (Norfolk, 2001). In the first part of his book (pp. 93–176), Pollington includes a succinct plant-by-plant guide to the medical uses of plants and other curative substances known to the Anglo-Saxons.

20 M. L. Cameron, "On *Þeor* and *Þeoradl,*" *Anglia* 106 (1998): 129.

21 Note in this connection Philip Shaw, "The Manuscript Texts of *Against a Dwarf,*" in *Writing and Texts in Anglo-Saxon England,* ed. Alexander R. Rumble (Cambridge, 2006), 96–113; and Conan Doyle, "*Dweorg* in Old English: Aspects of Disease Terminology," *Quaestio Insularis* 9 (2008): 99–117.

22 For discussion of medical texts with a bearing on Anglo-Saxon and early Germanic beliefs about elves, see Alaric Hall, *Elves in Anglo-Saxon England: Matters of Belief, Health, Gender and Identity* (Woodbridge, 2007), 96–118.

23 On this topic see especially Audrey L. Meaney, "The Anglo-Saxon View of the Causes of Disease," in *Health, Disease, and Healing in Medieval Culture,* ed. Sheila Campbell, Bert Hall, and David Klausner (Toronto, 1992), 12–33; and Pollington, *Leechcraft,* appendix 3, "Causes of Disease" (pp. 449–63).

24 See Roy M. Liuzza, "Prayers and/or Charms Addressed to the Cross," in *Cross and Culture in Anglo-Saxon England: Studies in Honor of George Hardin Brown,* ed. Karen L. Jolly, Catherine E. Karkov, and Sarah Larratt Keefer (Morgantown, WV, 2007), 276–320, at 291–320, with reference to the use made of the sign of the cross in medical texts, cattle-theft charms, and personal prayers; and Karen L. Jolly, "Cross-Referencing Anglo-Saxon Liturgy and Remedies: The Sign of the Cross as Ritual Protection," in *The Liturgy of the Late Anglo-Saxon Church,* ed. Helen Gittos and M. Bradford Bedingford (Woodbridge, 2005), 228–36.

# THE OLD ENGLISH
# HERBAL

# Table of Contents

2

# Table of Contents

1.19. So that a person's food is easily digested

1.20. If a person cannot keep food down

1.21. For abdominal pain or if a person feels bloated

1.22. For having consumed poison

1.23. For snakebite

1.24. Again, for snakebite

1.25. For the bite of a mad dog

1.26. If a person's throat is sore or any part of the neck

1.27. For sore loins and if a person's thighs should ache

1.28. For a hot fever

1.29. For foot disease

2. The herb *arnoglossa;* that is, common plantain

2.1. For headache

2.2. For stomach pain

2.3. For abdominal pain

2.4. Again, if a person has a bloated stomach

2.5. If a person's anus is bleeding

2.6. If a person is wounded

2.7. If you want to soften a person's stomach

2.8. For snakebite

2.9. Again, for snakebite

2.10. For intestinal worms

2.11. If there is a hardening on the body

2.12. For the fever that comes every fourth day

2.13. For foot disease and for sore sinews

2.14. For the fever that comes every third day

2.15. For the fever that comes every second day

2.16. For inflamed wounds

2.17. If a person gets footsore on a journey

2.18. Wið þæt men weargebræde weaxe on þam nosum
oððe on þam hleore

2.19. Be æghwylcum uncuþum blædrum þe on mannes
nebbe sittað

2.20. Wið muðes wunde

2.21. Wið wedehundes slite

2.22. Wið ælces dæges mannes tyddernysse inneweardes

3. *Herba pentafilon;* þæt is, fifleafe

3.1. Wið þæt mannes lyþu acen oþþe ongeflogen sy

3.2. Wið wambe sare

3.3. Wiþ muþes ece, ond tungan, ond þrotan

3.4. Wiþ heafdes sare

3.5. Wiþ þæt men blod ut of nosum yrne to swyþe

3.6. Wiþ þæt mannes midrif ace

3.7. Wiþ nædran slite

3.8. Wiþ þæt man forbærned sy

3.9. Gyf þu wylle cancer ablendan

4. *Herba vermenaca;* þæt is, æscþrotu

4.1. Wið wunda ond deadspringas ond cyrnlu

4.2. Eft, wið cyrnlu

4.3. Wið þa þe habbað ætstandene ædran swa þæt þæt
blod ne mæg his gecyndlican ryne habban, ond heora
þygne gehealdan ne magon

4.4. Wið lifre sar

4.5. Wið þa untrumnysse þe stanas weaxeþ on blædran

4.6. Wið heafodsar

4.7. Wið nædran slite

4.8. Wið attorcoppan bite

4.9. Wið wedehundes slite

4.10. Wið niwe wundela

4.11. Wið nædran slite

2.18. If a sore should develop on the nose or on the cheek

2.19. For all strange pustules that are located on the face

2.20. For sores in the mouth

2.21. For the bite of a mad dog

2.22. For a person's daily internal weakness

3. The herb *pentaphyllon;* that is, cinquefoil

   3.1. If a person's joints ache or are attacked by a disease

   3.2. For stomach pain

   3.3. For an aching mouth, tongue, and throat

   3.4. For headache

   3.5. If a person has a severe nosebleed

   3.6. If a person has a pain in the midriff

   3.7. For snakebite

   3.8. If a person is badly burned

   3.9. If you want to heal an ulcerous sore

4. The herb *verbenaca;* that is, vervain

   4.1. For wounds and carbuncles and hard swellings

   4.2. Again, for hard swellings

   4.3. For those who have clogged veins so that blood cannot flow to the genitals, and who cannot keep their food down

   4.4. For pain in the liver

   4.5. For the infirmity when stones grow in the bladder

   4.6. For headache

   4.7. For snakebite

   4.8. For spider bite

   4.9. For the bite of a mad dog

   4.10. For fresh wounds

   4.11. For snakebite

5. *Herba simphoniaca;* þæt is, hennebelle
   5.1. Wiþ earena sar
   5.2. Wið cneowa geswell, oþþe sceancena, oððe swa hwær
   swa on lichaman geswell sy
   5.3. Wiþ toþa sare
   5.4. Wið þæra gewealda sar oþþe geswell
   5.5. Wiþ þæt wifes breost sare syn
   5.6. Wiþ fota sar
   5.7. Wiþ lungenadle

6. *Herba viperina;* þæt is, nædderwyrt
   6.1. Wiþ nædran slite

7. *Herba veneria;* þæt ys, beowyrt
   7.1. Wiþ þæt beon ne ætfleon
   7.2. Wiþ þæt man gemigan ne mege

8. *Herba pes leonis;* þæt is, leonfot
   8.1. Wiþ þæt man sy cis

9. *Herba scelerata;* þæt ys, clufþung
   9.1. Wið wundela ond deadspringas
   9.2. Wiþ swylas ond weartan

10. *Herba batracion;* þæt is, clufwyrt
    10.1. Wiþ monoðseoce
    10.2. Wiþ þa sweartan dolh

11. *Herba artemesia;* þæt is, mugcwyrt
    11.1. Wið innoðes sare
    11.2. Wið fota sare

12. *Herba artemesia tagantes;* þæt ys, oþres cynnes mucgwyrt
    12.1. Wið blædran sare
    12.2. Wiþ þeona sare
    12.3. Wiþ sina sare ond geswell
    12.4. Gyf hwa mid fotadle swyþe geswenced sy
    12.5. Gyf hwa sy mid feferum gedreht

5. The herb *symphoniaca;* that is, henbane
    5.1. For earache
    5.2. For swollen knees, or calves, or for wherever there is a swelling on the body
    5.3. For toothache
    5.4. For pain or swelling in the genitals
    5.5. If a woman's breasts are sore
    5.6. For sore feet
    5.7. For lung disease
6. The herb *viperina;* that is, adder's wort
    6.1. For snakebite
7. The herb *venerea;* that is, yellow iris
    7.1. So that bees do not swarm
    7.2. If a person cannot urinate
8. The herb *pes leonis;* that is, lion's foot
    8.1. If a person is fastidious
9. The herb *scelerata;* that is, celery-leaved crowfoot
    9.1. For wounds and carbuncles
    9.2. For swellings and warts
10. The herb *batrachion;* that is, buttercup
    10.1. For a person suffering from lunacy
    10.2. For black scars
11. The herb *artemisia;* that is, mugwort
    11.1. For abdominal pain
    11.2. For sore feet
12. The herb *artemisia tagantes;* that is, another kind of mugwort
    12.1. For bladder pain
    12.2. For sore thighs
    12.3. For sore sinews and for swelling
    12.4. If a person is severely tormented by foot disease
    12.5. If a person is suffering from fever

13. *Herba artemesia leptefilos;* þæt ys, þryddan cynnes mucgwyrt
   13.1. Wiþ þæs magan sare
   13.2. Wiþ þæra sina bifunge
14. *Herba lapatium;* þæt ys, docce
   14.1. Wið cyrnlu þe on gewealde weaxeþ
15. *Herba dracontea;* þæt ys, dracentse
   15.1. Wiþ ealra nædrena slite
   15.2. Wiþ banbryce
16. *Herba satyrion;* þæt ys, hrefnes leac
   16.1. Wið earfoðlice wundela
   16.2. Wiþ eagena sare
17. *Herba gentiana;* þæt ys, feldwyrt
   17.1. Wið nædran slite
18. *Herba orbicularis;* þæt ys, slite
   18.1. Wið þæt mannes fex fealle
   18.2. Wiþ innoðes styrunga
   18.3. Wiþ miltan sare
19. *Herba proserpinaca;* þæt ys, unfortredde
   19.1. Wiþ þæt man blod spiwe
   19.2. Wiþ sidan sare
   19.3. Wiþ breosta sare
   19.4. Wiþ eagena sare
   19.5. Wiþ earena sare
   19.6. Wiþ utsihte
20. *Herba aristolochia;* þæt ys, smerowyrt
   20.1. Wið attres strencðe
   20.2. Wiþ þa stiþustan feforas
   20.3. Wiþ næsþurla sare
   20.4. Wiþ þæt hwa mid cyle gewæht sy
   20.5. Wiþ naedran slite

20.6. Gif hwylc cyld ahwæned sy

20.7. Wiþ þæt wærhbræde on nosa wexe

21. *Herba nasturcium;* þæt is, cærse

21.1. Wiþ ðæt mannes fex fealle

21.2. Wiþ heafodsare; þæt ys, wið scurf ond gicþan

21.3. Wiþ lices sarnysse

21.4. Wiþ swylas

21.5. Wiþ weartan

22. *Herba hieribulbum;* þæt is, greate wyrt

22.1. Wiþ liþa sare

22.2. Gif nebcorn on wifmannes nebbe wexen

23. *Herba apollinaris;* þæt is, glofwyrt

23.1. Wið handa sare

24. *Herba camemelon;* þæt is, mageþe

24.1. Wið eagena sare

25. *Herba chamedris;* þæt is, heortclæfre

25.1. Gyf hwa tobrysed sy

25.2. Wið nædran slite

25.3. Wiþ fotadle

26. *Herba chameæleæ;* þæt is, wulfes camb

26.1. Wið liferseocnysse

26.2. Wiþ attres drenc

26.3. Wiþ wæterseocnysse

27. *Herba chamepithys;* þæt is, henep

27.1. Wiþ wundela

27.2. Wiþ innoþes sare

28. *Herba chamedafne;* þæt is, hrefnes fot

28.1. Wið innoþ to astyrigenne

29. *Herba ostriago;* þæt is, liðwyrt

29.1. Wiþ ealle þingc þe on men to sare innan acennede beoð

30. *Herba brittannica;* þæt is, hæwenhnydelu
    30.1. Wiþ muðes sare
    30.2. Eft, wið muþes sare
    30.3. Wiþ toþa sare
    30.4. Wiþ fæstne innoð to astyrigenne
    30.5. Wið sidan sare
31. *Herba lactuca silvatica;* þæt is, wudulectric
    31.1. Wiþ eagena dymnesse
    31.2. Eft, wiþ eagene dymnysse
32. *Herba argimonia;* þæt is, garclife
    32.1. Wið eagena sare
    32.2. Wið innoðes sare
    32.3. Wiþ cancor ond wið wundela
    32.4. Wiþ nædran slite
    32.5. Wiþ weartan
    32.6. Wið miltan sare
    32.7. Gyf þu hwilce þingc on þam lichoman ceorfan wille
    32.8. Wiþ slege isernes
33. *Herba astularegia;* þæt is, wudurofe
    33.1. Wið sceancena sare
    33.2. Wiþ lifre sare
34. *Herba lapatium;* þæt is, wududocce
    34.1. Gyf hwylc stiþnes on lichoman becume
35. *Herba centauria maior;* þæt is, curmelle seo mare
    35.1. Wið liferadle
    35.2. Wið wunda ond cancor
36. *Herba centauria minor;* þæt is, curmelle seo læsse
    36.1. Wið nædran slite
    36.2. Wið eagena sare
    36.3. Eft, wið þon ylcon
    36.4. Wiþ sina togunge

30. The herb *britannica;* that is, water dock
    30.1. For a mouth sore
    30.2. Again, for a mouth sore
    30.3. For toothache
    30.4. To move constipated bowels
    30.5. For a pain in the side
31. The herb *lactuca silvatica;* that is, prickly lettuce
    31.1. For dimness of the eyes
    31.2. Again, for dimness of the eyes
32. The herb *argemonia;* that is, agrimony
    32.1. For eye pain
    32.2. For abdominal pain
    32.3. For ulcerous sores and wounds
    32.4. For snakebite
    32.5. For warts
    32.6. For pain in the spleen
    32.7. If you want to cut anything from the body
    32.8. For a blow from a weapon
33. The herb *hastula regia;* that is, asphodel
    33.1. For pain in the shanks
    33.2. For pain in the liver
34. The herb *lapathum;* that is, common sorrel
    34.1. If a calloused spot should develop on the body
35. The herb *centaurea maior;* that is, yellowwort
    35.1. For liver disease
    35.2. For wounds and ulcerous sores
36. The herb *centaurea minor;* that is, lesser centaury
    36.1. For snakebite
    36.2. For eye pain
    36.3. Again, for the same
    36.4. For sinew spasm

36.5. Wiþ attres onbyrginge

36.6. Wiþ þæt wyrmas ymb nafolan derigen

37. *Herba personacia;* þæt is, bete

    37.1. Wið ealle wunda ond wiþ nædran slitas

    37.2. ond wiþ feforas

    37.3. Wið þæt cancor on wunde wexe

    37.4. Wiþ innoðes sare

    37.5. Wið wedehundes slite

    37.6. Wiþ niwe wunda

38. *Herba fraga;* þæt is, streawberge

    38.1. Wiþ miltan sare

    38.2. Wiþ nyrwyt

        Wiþ innoþes sare

39. *Herba hibiscus;* þæt is, merscmealwe

    39.1. Wið fotadle

    39.2. Wiþ ælce gegaderunga þe on þam lichoman acenned beoþ

40. *Herba ippirus;* þæt is, *æquiseia*

    40.1. Wiþ utsiht

    40.2. Wiþ þæt man blod swyþe hræce

41. *Herba malva erratica;* þæt is, hocleaf

    41.1. Wið blædran sare

    41.2. Wið sina sare

    41.3. Wið sidan sare

    41.4. Wið niwe wunda

42. *Herba buglossa;* þæt is, hundes tunge

    42.1. Gyf hwylcum men sy þæs þriddan dæges fefor oððe þæs feorþan

    42.2. Wiþ nyrwyt

43. *Herba bulbiscillitica;* þæt is, glædene

    43.1. Wið wæterseocnysse

43.2. Wiþ liða sare

43.3. Wiþ þa adle þe Grecas *paronichias* nemneð

43.4. Wiþ þæt man ne mæge wæterseoces mannes þurst gecelan

44. *Herba cotiledon;* þæt ys, *umbilicus Veneris*

44.1. Wið swylas

45. *Herba gallicrus;* þæt is, attorlaðe

45.1. Wið hundes slite

46. *Herba prassion;* þæt ys, harehune

46.1. Wið geposu ond wið þæt he hefelice hræce

46.2. Wið magan sare

46.3. Wið rengwyrmas abutan nafolan

46.4. Wiþ liþa sare ond wið geþind

46.5. Wið attres þigne

46.6. Wiþ sceb ond teter

46.7. Wið lungenadle

46.8. Wið ealle stiðnessa þæs lichoman

47. *Herba xifion;* þæt is, foxes fot

47.1. Wiþ uncuðe springas þe on lichoman acennede beoð

47.2. Wiþ heafodbryce ond ættrige ban

48. *Herba gallitricus;* þæt is, wæterwyrt

48.1. Gyf swylas fæmnum derien

48.2. Wiþ ðæt mannes fex fealle

49. *Herba temolus;* þæt is, singrene

49.1. Wið cwiðan sare

50. *Herba æliotrophus;* þæt is, sigelhweorfa

50.1. Wiþ ealle attru

50.2. Wið flewsan

51. *Herba gryas;* þæt is, mæderu

51.1. Wiþ banece ond wiþ banbryce

51.2. Wið ælc sar þe þam lichoman dereþ

52. *Herba politricus;* þæt is, hymele
    52.1. Wið innoðes sare ond wið þæt fex wexe
53. *Herba malochinagria;* þæt is, wudurofe
    53.1. Wiþ utsiht
    53.2. Wiþ innoðes flewsan
54. *Herba metoria;* þæt is, hwit popig
    54.1. Wið eagena sare
    54.2. Wiþ þunwonga sare
    54.3. Wið slæpleaste
55. *Herba oenantes*
    55.1. Wið þæt man gemigan ne mæge
    55.2. Gyf hwa swyþe hræce
56. *Herba narcisus;* þæt is, halswyrt
    56.1. Wiþ þa wunda þe on men beoð acenned
57. *Herba splenion;* þæt is, brunewyrt
    57.1. Wið miltan sare
58. *Herba polion*
    58.1. Wið monoðseoce
59. *Herba victoriola;* þæt is, cneowholen
    59.1. Wiþ ðone dropan ond þæs magan sare
60. *Herba confirma;* þæt is, galluc
    60.1. Wiþ wifa flewsan
    60.2. Gyf hwa innan toborsten sy
    60.3. Wið magan sare
61. *Herba asterion*
    61.1. Wið fylleseocnysse
62. *Herba leporis pes;* þæt is, haran hig
    62.1. Wið innoðes fæstnysse
63. *Herba dictamnus*
    63.1. Wið þæt wif hæbbe on hyre innoðe deadboren
    tuddur

52. The herb *polytrichum;* that is, common maidenhair
    52.1. For abdominal pain and to promote the growth of hair
53. The herb *moloche agria;* that is, asphodel
    53.1. For diarrhea
    53.2. For intestinal flux
54. The herb *meconium;* that is, white poppy
    54.1. For sore eyes
    54.2. For pain in the temples
    54.3. For sleeplessness
55. The herb *oenanthe*
    55.1. If a person cannot urinate
    55.2. If a person is coughing badly
56. The herb *narcissus;* that is, narcissus
    56.1. For sores that have grown on a person
57. The herb *splenion;* that is, spleenwort
    57.1. For pain in the spleen
58. The herb *polion*
    58.1. For a person suffering from lunacy
59. The herb *victoriola;* that is, butcher's broom
    59.1. For rheum and for stomachache
60. The herb *confirma;* that is, comfrey
    60.1. For women's excessive flow of menstrual blood
    60.2. If someone has an internal rupture
    60.3. For stomachache
61. The herb *asterion*
    61.1. For epilepsy
62. The herb *leporis pes;* that is, hare's-foot clover
    62.1. For constipation
63. The herb *dictamnus*
    63.1. If a woman is carrying a dead fetus in her womb

63.2. Wiþ wunda

63.3. Wiþ nædran slite

63.4. Wiþ attorþigene

63.5. Eft, wið niwe wunda

64. *Herba solago maior;* þæt is, *helioscorpion*

64.1. Eft, wið nædran slite

65. *Herba solago minor;* þæt is, *æliotropion*

65.1. Wið rengwyrmas abutan nafolan

66. *Herba peonia*

66.1. Wiþ monoðseocnysse

66.2. Wiþ hypebanece

67. *Herba peristereon;* þæt ys, *berbena*

67.1. Wiþ hundes beorc

67.2. Wið ealle attru

68. *Herba bryonia;* þæt is, hymele

68.1. Wið miltan sare

69. *Herba nymfete*

69.1. Wið utsiht

Eft, wið utsiht

69.2. Eft, wið innoþes sare

70. *Herba crision;* þæt is, clæfre

70.1. Wiþ gomena sare

71. *Herba isatis*

71.1. Wið næddran slite

72. *Herba scordea*

72.1. Eft, wið nædran slite

72.2. Wið sina sare

72.3. Wiþ fefor

73. *Herba verbascus;* þæt is, feltwyrt

Be þam þe Mercurius þas wyrte Ulixe sealde

73.1. Wið ealle yfele gencymas

73.2. Wið fotadle

74. *Herba heraclea*
    74.1. Wiþ þæt man wylle ofer langne weg feran ond him na sceaðan ondrædan
75. *Herba cælidonia;* þæt is, cyleþenie
    75.1. Wiþ eagena dymnysse ond sarnysse
    75.2. Eft, wið dymgendum eagum
    75.3. Wiþ cyrnlu
    75.4. Wiþ heafudece
    75.5. Wiþ þæt man gebærned sy
76. *Herba solata;* þæt is, solosece
    76.1. Wiþ geswel
    76.2. Wiþ earena sare
    76.3. Wiþ toðece
    76.4. Wiþ blodryne of nosum
77. *Herba senecio;* þæt is, grundeswylige
    77.1. Wið wunda, þeah hy ealde syn
    77.2. Wið isernes slege
    77.3. Wiþ fotadle
    77.4. Wiþ lendena sare
78. *Herba filix;* þæt is, fearn
    78.1. Wiþ wunda
    78.2. Wiþ þæt geong man healyde sy
79. *Herba gramen;* þæt is, cwice
    79.1. Wiþ miltan sare
80. *Herba gladiolum;* þæt is, glædene
    80.1. Wiþ blædran sare ond gemigan ne mæge
    80.2. Wiþ miltan sare
    80.3. Wiþ innoðes sare ond þæra breosta
81. *Herba rosmarinum;* þæt is, boðen
    81.1. Wið toðece
    81.2–3. Wið adligende, ond wið gicðan

81.4. Wiþ liferseocnysse ond þæs innoðes

81.5. Wiþ niwe wunda

82. *Herba pastinaca silvatica;* þæt is, feldmoru

   82.1. Wiþ þæt wifmen earfoðlice cennan

   82.2. Wiþ wifa afeormunge

83. *Herba perdicalis;* þæt is, dolhrune

   83.1. Wiþ fotadle ond wið cancor

84. *Herba mercurialis;* þæt is, cedelc

   84.1. Wið þæs innoðes heardnysse

   84.2. Wiþ eagena sare ond geswelle

   84.3. Gyf wæter on earan swyþe gesigen sy

85. *Herba radiola;* þæt is, eforfearn

   85.1. Wið heafodece

86. *Herba sparagiagrestis;* þæt is, wuducerville

   86.1. Wiþ blædran sare oþþe geswelle

   86.2. Wiþ toðece

   86.3. Wiþ æddrena sare

   86.4. Wiþ þæt yfel man þurh æfþancan oþerne begale

87. *Herba sabina;* þæt is, savine

   87.1. Wiþ togunga þæra sina ond wiþ fota geswell

   87.2. Wiþ heafodece

   87.3. Wiþ deadspringas

88. *Herba canis caput;* þæt is, hundes heafod

   88.1. Wiþ eagena sare ond geswel

89. *Herba erusti;* þæt is, bremel

   89.1. Wiþ earena sare

   89.2. Wiþ wifes flewsan

   89.3. Wiþ heortece

81.4. For liver and abdominal disease

81.5. For fresh wounds

82. The herb *pastinaca silvatica;* that is, wild carrot

82.1. If women have difficulty in giving birth

82.2. For women's cleansing

83. The herb *perdicalis;* that is, pellitory-of-the-wall

83.1. For foot disease and for ulcerous sores

84. The herb *mercurialis;* that is, dog's mercury

84.1. For hardening of the abdomen

84.2. For pain and swelling of the eyes

84.3. If water has penetrated into the ears

85. The herb *radiolum;* that is, polypody

85.1. For headache

86. The herb *asparagus agrestis;* that is, asparagus

86.1. For pain or swelling of the bladder

86.2. For toothache

86.3. For sore veins

86.4. If any wicked person enchants another person out of malice

87. The herb *sabina;* that is, savine

87.1. For contraction of the sinews and swelling of the feet

87.2. For headache

87.3. For carbuncles

88. The herb *canis caput;* that is, snapdragon

88.1. For soreness and swelling of the eyes

89. The herb *erusti,* that is, bramble

89.1. For earache

89.2. For a woman's excessive flow of menstrual blood

89.3. For heart pain

89.4. Wiþ niwe wunda

89.5. Wiþ liþa sare

89.6. Wiþ nædran slite

90. *Herba millefolium;* þæt is, gearwe

   90.1. Wiþ isernes slege, ond þæt Achilles þas wyrte funde

   90.2. Wiþ toðece

   90.3. Wiþ wunda

   90.4. Wiþ geswell

   90.5. Wiþ þæt man earfoðlice gemigan mæge

   90.6. Gyf wund on men acolod sy

   90.7. Gyf men þæt heafod berste oððe uncuð swyle on
gesytte

   90.8. Eft, wiþ þam ylcan

   90.9. Gyf hwylcum men ædran aheardode syn oþþe his
mete gemyltan nylle

   90.10. Wiþ þæra þearma ece ond þæs innoðes

   90.11. Wið þæt men sogoða eglige

   90.12. Wiþ heafodece

   90.13. Wiþ þam næddercynne þe man *spalangius* hateð

   90.14. Eft, wið nædran slite

   90.15. Wiþ wedehundes slite

   90.16. Wiþ næddran slite

91. *Herba ruta;* þæt ys, rude

   91.1. Wið þæt blod of nosum flowe

   91.2. Wið toþundennesse

   91.3. Wið þæs magan sare

   91.4. Wið eagena sare ond geswelle

   91.5. Wið ofergitulnesse

   91.6. Wið eagena dymnesse

   91.7. Wið heafodece

92. *Herba mentastrus;* þæt is, minte
    92.1. Wiþ earena sare
    92.2. Wiþ hreoflan
93. *Herba ebulus;* þæt is, wealwyrt
    93.1. Wiþ þæt stanas on blædran wexen
    93.2. Wiþ nædran slite
    93.3. Wiþ wæterseocnysse
94. *Herba pollegion;* þæt is, dweorgedwosle
    94.1. Wið þæs innoþes sare
    94.2. Wiþ þæs magan sare
    94.3. Wiþ gicþan þæra sceapa
    94.4. Eft, wið þæs innoðes sare
    94.5. Wiþ þam fefore þe þy þriddan dæge egleþ
    94.6. Gif deadboren cild sy on wifes innoðe
    94.7. Gif hwa on scipe wlættan þolige
    94.8. Wiþ blædran sare, ond þæt stanas þæron wexen
    94.9. Gyf hwa onbutan his heortan oððe on his breostan
    sar þolige
    94.10. Gyf hwylcum men hramma derie
    94.11. Wiþ ðæs magan aðundenysse ond þæs innoþes
    94.12. Wiþ miltan sare
    94.13. Wiþ lendenece ond wið þeona sare
95. *Herba nepitamon;* þæt is, nepte
    95.1. Wiþ nædran slite
96. *Herba peucedana;* þæt is, cammoc
    96.1. Eft, wið nædran slite
    96.2. Wiþ gewitlæste þæs modes
97. *Herba hinnula campana;* þæt ys, sperewyrt
    97.1. Wiþ blædran sare
    97.2. Wiþ toþa sare ond wagunge
    97.3. Wiþ rengwyrmas ymb þone nafolan

92. The herb *mentastrum;* that is, horsemint
  92.1. For earache
  92.2. For a severe skin disease
93. The herb *ebulus;* that is, dwarf elder
  93.1. If stones should grow in the bladder
  93.2. For snakebite
  93.3. For dropsy
94. The herb *pulegium;* that is, pennyroyal
  94.1. For abdominal pain
  94.2. For stomachache
  94.3. For itching genitals
  94.4. Again, for abdominal pain
  94.5. For the fever that afflicts one every third day
  94.6. If a stillborn child should remain inside a woman
  94.7. If a person suffers nausea on board ship
  94.8. For bladder pain, and if stones should grow there
  94.9. If a person suffers pain around the heart or in the chest
  94.10. If a person suffers from a spasm
  94.11. For swelling of the stomach and the abdomen
  94.12. For pain in the spleen
  94.13. For pain in the loins and for sore thighs
95. The herb *nepeta;* that is, catmint
  95.1. For snakebite
96. The herb *peucedanum;* that is, hog's fennel
  96.1. Again, for snakebite
  96.2. For loss of one's wits
97. The herb *inula Campana;* that is, elecampane
  97.1. For bladder pain
  97.2. For toothache and loose teeth
  97.3. For harmful worms around the navel

98. *Herba cynoglossa;* þæt is, ribbe

    98.1. Wiþ nædran slite

    98.2. Wiþ þam fefore þe þy feorþan dæge on man be-
cymeþ

    98.3. Wiþ þæt man well gehyran ne mæge

99. *Herba saxifragiam;* þæt is, sundcorn

    99.1. Wiþ þæt stanas on blædran wexen

100. *Herba hedera nigra;* þæt is, eorðifig

    100.1. Eft, wið þæt stanas on blædran wexen

    100.2. Wiþ heafodece

    100.3. Wiþ miltan sare

    100.4. Wiþ þæra wyrma slite þe man *spalangiones* nemneþ

    100.5. Eft, wiþ þara wunda lacnunge

    100.6. Wiþ þæt næsðyrlu yfele stincen

    100.7. Wiþ þæt man ne mæge wel gehyran

    100.8. Wiþ þæt heafod ne ace for sunnan hætan

101. *Herba serpillus;* þæt is, organe

    101.1. Wiþ heafdes sare

    101.2. Eft, wið heafodece

    101.3. Gyf hwa forbærned sy

102. *Herba absinthius;* þæt is, wermod

    102.1. Wiþ læla ond wið oþre sar

    102.2. Wiþ rengwyrmas

103. *Herba salvia*

    103.1. Wiþ gicþan þæra gesceapa

    103.2. Eft, wið gicþan þæs setles

104. *Herba coliandra;* þæt is, celendre

    104.1. Wið rengwyrmas

    104.2. Wiþ þæt wif hrædlice cennan mæge

98. The herb *cynoglossum;* that is, ribwort
    98.1. For snakebite
    98.2. For the fever that comes on a person every fourth day
    98.3. If a person cannot hear well
99. The herb *saxifraga;* that is, saxifrage
    99.1. If stones should grow in the bladder
100. The herb *hedera nigra;* that is, ground ivy
    100.1. Again, if stones should grow in the bladder
    100.2. For headache
    100.3. For pain in the spleen
    100.4. For bite of those venomous creatures that are called *spalangiones*
    100.5. Again, for the healing of those wounds
    100.6. If the nostrils emit a bad smell
    100.7. If a person cannot hear well
    100.8. So that the head will not ache because of the sun's heat
101. The herb *serpyllum;* that is, wild thyme
    101.1. For headache
    101.2. Again, for headache
    101.3. If a person is burned
102. The herb *absinthium;* that is, wormwood
    102.1. For bruises and other sores
    102.2. For harmful worms
103. The herb *salvia*
    103.1. For itching genitals
    103.2. Again, for itching anus
104. The herb *coriandrum,* that is, coriander
    104.1. For harmful worms
    104.2. So that a woman may give birth quickly

105. *Herba porclaca*
    105.1. Wiþ swyþlicne flewsan þæs sædes
106. *Herba cerefolia;* þæt is, cerfille
    106.1. Wiþ þæs magan sare
107. *Herba sisimbrius*
    107.1. Wiþ blædran sare, ond ne mæge gemigan
108. *Herba olisatra*
    108.1. Eft, wiþ blædran sare ond þæs micgan
109. *Herba lilium;* þæt is, lilie
    109.1. Wið nædran slite
    109.2. Wiþ geswell
110. *Herba tytymallus calatites;* þæt ys, *lacterida*
    110.1. Wiþ þæra innoþa sare
    110.2. Wiþ weartan
    110.3. Wiþ hreoflan
111. *Herba carduus silvaticus;* þæt is, wuduþistel
    111.1. Wiþ þæs magan sare
    111.2. Wiþ þæt þu nane yfele gencymas þe ne ondræde
112. *Herba lupinum montanum*
    112.1. Wiþ þæt wyrmas ymb þone nafolan dergen
    112.2. Wiþ þæt cildum þæt sylfe derige
113. *Herba lactyrida;* þæt is, giþcorn
    113.1. Wiþ þæs innoðes heardnysse
114. *Herba lactuca leporina;* þæt is, *lactuca*
    114.1. Wið feforgende
115. *Herba cucumeris silvatica;* þæt is, hwerhwette
    115.1. Wiþ þæra sina sare ond fotadle
    115.2. Gyf cild misboren sy
116. *Herba cannave silvatica*
    116.1. Wiþ þæra breosta sare
    116.2. Wiþ cile bærnettes

105. The herb *porcilaca*
    105.1. For excessive flow of semen
106. The herb *chaerephyllum;* that is, chervil
    106.1. For stomachache
107. The herb *sisymbrium*
    107.1. For bladder pain, and if a person cannot urinate
108. The herb *holus atrum*
    108.1. Again, for bladder pain and when urinating
109. The herb *lilium;* that is, lily
    109.1. For snakebite
    109.2. For swellings
110. The herb *titymallos calatites;* that is, caper spurge
    110.1. For abdominal pain
    110.2. For warts
    110.3. For a severe skin disease
111. The herb *carduus silvaticus;* that is, sow thistle
    111.1. For stomachache
    111.2. So that you do not fear any evil encounters
112. The herb *lupinus montanus*
    112.1. If worms are doing harm around the navel
    112.2. If the same disease is troubling children
113. The herb *latirida;* that is, spurge laurel
    113.1. For hardening of the abdomen
114. The herb *lactuca leporina;* that is, lettuce
    114.1. For a person who is feverish
115. The herb *cucumis silvaticus;* that is, squirting cucumber
    115.1. For sore joints and for foot disease
    115.2. If a child is born prematurely
116. The herb *cannabis silvatica*
    116.1. For sore breasts
    116.2. For frostbite

117. *Herba ruta montana;* þæt is, rude
   117.1. Wiþ eagena dymnysse
   117.2. Eft, wið breosta sare
   117.3. Wiþ lifersare
   117.4. Wiþ þæt man gemigan ne mæge
   117.5. Wiþ nædran slite
118. *Herba eptafilon;* þæt is, seofonleafe
   118.1. Wiþ fotadle
119. *Herba ocimus;* þæt is, mistel
   119.1. Wiþ heafodece
   119.2. Eft, wið eagena sare ond geswelle
   119.3. Wiþ ædrena sare
120. *Herba apium;* þæt is, merce
   120.1. Wiþ eagena sare ond geswelle
121. *Herba hedera crysocantes;* þæt is, ifig
   121.1. Wiþ wæterseocnysse
122. *Herba menta;* þæt ys, minte
   122.1. Wiþ teter ond wiþ pypylgende lic
   122.2. Wiþ yfele dolh ond wiþ wunda
123. *Herba anetum;* þæt is, dile
   123.1. Wiþ gicþan ond wið sar þæra gesceapa
   123.2. Gyf þonne wifmen hwæt swilces derige
   123.3. Wið heafodece
124. *Herba origanum;* þæt is, organe
   124.1. Wiþ þone dropan ond liferadle ond nyrwytte
   124.2. Wiþ gebræceo
125. *Herba sempervivus;* þæt ys, sinfulle
   125.1. Wiþ ealle gegaderunga þæs yfelan wætan
126. *Herba fenuculus;* þæt ys, finul
   126.1. Wiþ gebræceo ond wið nyrwyt
   126.2. Wiþ blædran sare

117. The herb *ruta montana;* that is, wild rue
   117.1. For dimness of the eyes
   117.2. Again, for a sore chest
   117.3. For liver pain
   117.4. If a person cannot urinate
   117.5. For snakebite
118. The herb *heptaphyllum;* that is, tormentil
   118.1. For foot disease
119. The herb *ocimum;* that is, basil
   119.1. For headache
   119.2. Again, for pain and swelling of the eyes
   119.3. For kidney pain
120. The herb *appium;* that is, wild celery
   120.1. For pain and swelling of the eyes
121. The herb *hedera crisocantes;* that is, ivy
   121.1. For dropsy
122. The herb *menta;* that is, mint
   122.1. For impetigo and a pimply body
   122.2. For bad open sores and wounds
123. The herb *anethum;* that is, dill
   123.1. For itching and pain in the genitals
   123.2. If anything is troubling a woman
   123.3. For headache
124. The herb *origanum;* that is, marjoram
   124.1. For rheum and liver sickness and shortness of breath
   124.2. For coughs
125. The herb *sempervivum;* that is, houseleek
   125.1. For all accumulations of corrupted matter
126. The herb *feniculum;* that is, fennel
   126.1. For coughs and shortness of breath
   126.2. For bladder pain

127. *Herba erifion;* þæt is, lyþwyrt
   127.1. Wiþ lungenadle
128. *Herba sinfitus albus*
   128.1. Wiþ wifes flewsan
129. *Herba petroselinum;* þæt is, petersilie
   129.1. Wiþ nædran slite
   129.2. Wiþ þæra sina sare
130. *Herba brassica silvatica;* þæt is, caul
   130.1. Wiþ ealle geswell
   130.2. Wiþ sidan sare
   130.3. Wiþ fotadle
131. *Herba basilisca;* þæt is, nædderwyrt
   131.1. Wiþ eall næddercyn
132. *Herba mandragora*
   132.1. Wiþ heafodece
   132.2. Wiþ þæra earena sare
   132.3. Wið fotadle
   132.4. Wiþ gewitleaste
   132.5. Eft, wiþ sina sare
   132.6. Gyf hwa hwylce hefige yfelnysse on his hofe geseo
133. *Herba lychanis stephanice;* þæt is, læcewyrt
   133.1. Wiþ eal næddercyn
134. *Herba action*
   134.1. Wiþ þæt man blod ond worsm gemang hræce
   134.2. Wiþ þæra liða sare
135. *Herba abrotanus;* þæt is, suþernewuda
   135.1. Wyþ nyrwyt ond wið banece ond wið þæt man ear-
   foþlice gemigan mæge
   135.2. Wiþ sidan sare
   135.3. Wiþ attru, ond wið nædrena slite

127. The herb *eriphion;* that is, rue
    127.1. For lung disease
128. The herb *symphytum album*
    128.1. For women's excessive flow of menstrual blood
129. The herb *petroselinum;* that is, parsley
    129.1. For snakebite
    129.2. For sore sinews
130. The herb *brassica silvatica;* that is, cabbage
    130.1. For all swellings
    130.2. For pain in the side
    130.3. For foot disease
131. The herb *basilica;* that is, sweet basil
    131.1. For all kinds of snakes
132. The herb *mandragoras*
    132.1. For headache
    132.2. For earache
    132.3. For foot disease
    132.4. For insanity
    132.5. Again, for sore sinews
    132.6. If one should see any grievous evil in his home
133. The herb *lychnis stephanomatica;* that is, rose campion
    133.1. For all kinds of snakes
134. The herb *arcion*
    134.1. If a person coughs up blood and foul matter
        together
    134.2. For pain in the joints
135. The herb *habrotonum;* that is, southernwood
    135.1. For shortness of breath, for pain in the thigh, and
        for difficulty urinating
    135.2. For pain in the side
    135.3. For poison and for snakebite

135.4. Eft, wið nædrena slite

135.5. Wiþ eagena sare

136. *Herba sion;* þæt ys, laber

136.1. Wiþ þæt stanas on blædran wexen

136.2. Wiþ utsiht ond innoðes styrungæ

137. *Herba eliotropus;* þæt ys, sigilhweorfa

137.1. Wiþ ealra næddercynna slitas

137.2. Wiþ þæt wyrmas ymb þone nafolan derigen

137.3. Wiþ weartan

138. *Herba spreritis*

138.1. Wiþ þone colan fefor

138.2. Wiþ wedehundes slite

138.3. Wiþ miltan sare

139. *Herba aizos minor*

139.1. Wiþ homan ond eagena sare ond fotadle

139.2. Wiþ heafodece

139.3. Wiþ þæra wyrma slite þe man *spalangiones* hateþ

139.4. Wiþ utsiht ond wiþ innoþes flewsan, ond wiþ wyrmas þe on þam innoðe deriaþ

139.5. Eft, wiþ gehwylce untrumnysse þæra eagena

140. *Herba elleborus albus;* þæt is, tunsingwyrt

140.1. Be þysse wyrte mægenum

140.2. Wiþ utsiht

140.3. Wiþ adla ond wið ealle yfelu

141. *Herba buoptalmon*

141.1. Wiþ gehwylce yfele springas

141.2. Wiþ æwyrdlan þæs lichoman

142. *Herba tribulus;* þæt is, gorst

142.1. Wiþ mycele hætan þæs lichaman

142.2. Wiþ þæs muðes ond þæra gomena fulnysse ond forrotudnysse

142.3. Wiþ þæt stanas on blædran wexen

142.4. Wiþ nædran slite

142.5. Wiþ attres drinc

142.6. Wið flean

143. *Herba coniza*

143.1. Wiþ nædran slite ond afligennysse, ond wið gnættas ond micgeas ond wið flean, ond wunda

143.2–3. Wiþ wifes cwiþan to feormienne ond wið þæt wif cennan ne mæge

143.4. Wiþ þa colan feforas

143.5. Wiþ heafodece

144. *Herba tricnos manicos;* þæt is, foxes glofa

144.1. Wiþ homan

144.2. Wiþ pypelgende lic

144.3. Wiþ heafodes sare ond þæs magan hætan, ond wið cyrnlu

144.4. Wiþ earena sare

145. *Herba glycyrida*

145.1. Wiþ þone drigean fefor

145.2. Wiþ breosta sare ond þære lifre ond þære blædran

145.3. Wiþ leahtras þæs muþes

146. *Herba strutius*

146.1. Wiþ þæt man gemigan ne mæge

146.2. Wiþ liferseocnysse ond nyrwytte, ond wiþ swyðlicne hracan ond innoþes togotennysse

146.3. Wiþ þæt stanas on blædran wexen

146.4. Wiþ hreoflan

146.5. Wiþ yfele gegadirunge

147. *Herba aizon*

147.1. Wiþ toborsten lic ond forrotudnysse, ond wið eagena sare ond hætan ond forbærnednysse

147.2. Wiþ heafodece

147.3. Wiþ nædran slite

147.4. Wiþ utsiht ond wið wyrmas on innoþe, ond wiþ
swyðlicne cyle

148. *Herba samsuchon;* þæt ys, ellen

148.1. Wiþ wæterseocnysse ond unmihtilicnysse þæs
migðan ond innoþa astyrunge

148.2. Wiþ springas, ond wið toborsten lic

148.3. Wiþ *scorpiones* stincg

148.4. Wiþ mycele hætan ond geswel þæra eagena

149. *Herba stecas*

149.1. Wiþ þæra breosta sare

150. *Herba thyaspis*

150.1. Wiþ ealle yfele gegaderunga þæs innoþes, ond wið
wifa monoðlican

151. *Herba polios;* þæt is, *omnimorbia*

151.1. Wiþ nædran slite

151.2. Wiþ wæterseocnysse

151.3. Wiþ miltan sare, ond wið nædran to afligenne, ond
wið niwe wunda

152. *Herba hypericon;* þæt ys, *corion*

152.1. Wiþ migþan ond monoðlican astyringe

152.2. Wiþ fefor þe þy feorþan dæge egleþ

152.3. Wiþ þæra sceancena geswel ond ece

153. *Herba acantaleuca*

153.1. Wiþ þæt man blode hræce, ond þæs magan sare

153.2. Wiþ þæs migðan astyrunge

153.3. Wiþ þæra toða sare ond yfele læla

153.4. Wiþ hramman ond nædran slite

154. *Herba acanton;* þæt is, beowyrt

154.1. Wiþ innoþes astyrunge ond þæs migðan

154.2. Wiþ lungenadle, ond gehwylce yfelu

147.3. For snakebite

147.4. For diarrhea and intestinal worms, and for a severe chill

148. The herb *sampsuchum;* that is, sweet marjoram

148.1. For dropsy and for difficulty urinating and for movement of the bowels

148.2. For carbuncles, and for a ruptured body

148.3. For a scorpion's sting

148.4. For severe inflammation and swelling of the eyes

149. The herb *stoechas*

149.1. For chest pain

150. The herb *thlaspis*

150.1. For all accumulations of harmful matter in the abdomen, and for women's menses

151. The herb *polium;* that is, wood sage

151.1. For snakebite

151.2. For dropsy

151.3. For pain in the spleen, to put snakes to flight, and for fresh wounds

152. The herb *hypericum;* that is, Saint John's wort

152.1. To stimulate the urine and women's menses

152.2. For the fever that comes every fourth day

152.3. For swelling and pain of the legs

153. The herb *acantha leuca*

153.1. If a person should cough up blood, and for a stomachache

153.2. To stimulate the urine

153.3. For toothache and bad bruises

153.4. For cramps and snakebite

154. The herb *acantion;* that is, Scotch thistle

154.1. To stimulate the bowels and the urine

154.2. For lung disease, and for any disease

155. *Herba quiminon;* þæt is, cymen
   155.1. Wiþ þæs magan sare
   155.2. Wiþ nyrwyt, ond nædran slite
   155.3. Wiþ innoða toðundennysse ond hætan
   155.4. Wiþ blodryne of næsþyrlon
156. *Herba camelleon alba;* þæt is, wulfes tæsl
   156.1. Wiþ þæt wyrmas on þam innoðe ymb þone naflan
     dergen
   156.2. Wiþ wæterseocnysse ond þæs micðan earfoðlic-
     nysse
157. *Herba scolymbos*
   157.1. Wiþ fulne stenc þæra oxna ond ealles þæs lichoman
   157.2. Wiþ fulstincendne migðan
158. *Herba iris Yllyrica*
   158.1. Wiþ mycelne hracan ond innoða astyrunge
   158.2. Wiþ nædran slite
     Wiþ wifa monoðlican to astyrigenne
   158.3. Wiþ cyrnlu ond ealle yfelu cumlu
   158.4. Wiþ heafdes sare
159. *Herba elleborus albus*
   159.1. Wiþ liferseocnysse ond ealle attru
160. *Herba delfinion*
   160.1. Wiþ þam fefore þe þy feorþan dæge on man be-
     cymeþ
161. *Herba acios*
   161.1. Wiþ nædrena slitas, ond lendena sare
162. *Herba centimorbia*
   162.1. Wiþ þæt hors on hrycge oððe on þam bogum awyrd
     sy ond hyt open sy
163. *Herba scordios*
   163.1–2. Wiþ þæs migðan astyrunge ond wið nædrena sli-
     tas ond ealle attru, ond magan sare

155. The herb *cuminum;* that is, cumin
    155.1. For stomachache
    155.2. For shortness of breath and snakebite
    155.3. For bloating and inflammation in the abdomen
    155.4. For nosebleed
156. The herb *chamaeleon albus;* that is, carline thistle
    156.1. For intestinal worms that do harm around the navel
    156.2. For dropsy and for difficulty urinating
157. The herb *scolymos*
    157.1. For a foul smell from the armpits or any other part
       of the body
    157.2. For foul-smelling urine
158. The herb *iris Illyrica*
    158.1. For a bad cough, and for movements of the bowels
    158.2. For snakebites
       To stimulate woman's menses
    158.3. For hard swellings and all harmful lumps
    158.4. For headache
159. The herb *elleborus albus*
    159.1. For liver disease and for all poisons
160. The herb *delphinion*
    160.1. For the fever that comes on a person every fourth
       day
161. The herb *echios*
    161.1. For snakebites, and for pain in the loins
162. The herb *centimorbia*
    162.1. For a horse injured on the back or the shoulder and
       the wound is open
163. The herb *scordion*
    163.1–2. To stimulate the urine, and for snakebite and all
       poisons, and for stomachache

163.3. Wiþ þa gerynnincge þæs wormses ymb þa breost

163.4. Wiþ fotadle

163.5. Wiþ niwe wunda

164. *Herba ami;* þæt is, *milvium*

164.1. Wiþ þæs innoðes astyrunge ond earfoðlicnysse þæs migðan, ond wildeora slitas

Wiþ wommas þæs lichoman

164.2. Wiþ æblæcnysse ond æhiwnysse þæs lichoman

165. *Herba viola;* þæt ys, banwyrt

165.1. Wiþ þæs cwiðan sare ond wið þone hætan

165.2. Wiþ misenlice leahtras þæs bæcþearmes

165.3. Wiþ cancor þæra toða

165.4. Wiþ þa monoðlican to astyrigenne

165.5. Wiþ miltan sare

166. *Herba viola purpurea*

166.1. Wið niwe wundela ond eac wið ealde

166.2. Wið þæs magan heardnysse

167. *Herba zamalentition*

167.1. Wiþ ealle wundela

167.2. Wiþ wunda cancor

168. *Herba ancusa*

168.1. Wiþ forbærnednysse

169. *Herba psillios*

169.1. Wiþ cyrnlu ond ealle yfelu gegaderunga

169.2. Wiþ heafodes sare

170. *Herba cynosbatus*

170.1. Wiþ miltan sare

171. *Herba aglaofotis*

171.1. Wiþ þone fefor þe þy þriddan dæge ond þy feorþan on man becymeð

171.2. Gif hwa hreohnysse on rewytte þolige

171.3. Wiþ hramman ond wiþ bifunge

172. *Herba capparis;* þæt is, wudubend

172.1. Wiþ miltan sare

173. *Herba eryngius*

173.1. Wiþ þæs migðan astyrunge, ond wið þa monoðlican
ond þæs innoþes astyrunge

173.2. Wið mænigfealde leahtras þæs innoþes

173.3. Wiþ þæra breosta geswell

173.4. Wiþ *scorpiones* styng ond ealra næddercynna slitas,
ond wið wedehundes slite

Wiþ oman ond wið fotadle

174. *Herba philantropos*

174.1. Wiþ nædrena slitas, ond wið þæra wyrma þe man
*spalangiones* hateþ

174.2. Wiþ earena sare

175. *Herba achillea*

175.1. Wiþ niwe wunda

175.2. Gif wif of ðam gecyndelican limon þone flewsan
þæs wætan ðolige

175.3. Wið utsiht

176. *Herba ricinus*

176.1. Wiþ hagol ond wið hreohnysse to awendenne

177. *Herba polloten;* þæt ys, *porrum nigrum*

177.1. Wiþ hundes slite

177.2. Wiþ wunda

178. *Herba urtica;* þæt is, netele

178.1. Wiþ forcillede wunda

178.2. Wið geswell

178.3. Gyf ænig dæl þæs lichoman geslegen sy

178.4. Wiþ lyþa sare

178.5. Wiþ fule wunde ond forrotude

178.6. Wiþ wifes flewsan

178.7. Wiþ þæt ðu cile ne þolige

179. *Herba priapisci;* þæt is, *vicapervica*

    179.1. Wið deofulseocnyssa, ond wið nædran, ond wið wildeor, ond wiþ attru, ond wið gehwylce behatu, ond wið andan, ond wið ogan, ond þæt þu gife hæbbe, ond wið þæt þu gesælig beo ond gecweme

180. *Herba litosperimon*

    180.1. Wið þæt stanas on blæddran wexen

181. *Herba stavisagria*

    181.1. Wiþ þone yfelan wætan þæs lichoman

    181.2. Wiþ scurf ond wið sceab

    181.3. Wið toða sare ond toðreomena

182. *Herba gorgonion*

    182.1. Wiþ gehwylce yfele fotswaðu

183. *Herba milotis*

    183.1. Wiþ eagena dymnysse

    183.2. Wiþ sina togunge

184. *Herba bulbus*

    184.1. Wiþ geswel, ond wið fotadle, ond wið gehwylce gedrecednesse

    184.2. Wið wæterseocnesse

        Wiþ hunda slitas, ond wið þæt man swæte, ond wið þæs magan sare

    184.3. Wiþ wundela ond scurfe ond nebcorne

    184.4. Wiþ þæra innoþa toðundennysse ond toborstennysse

185. *Herba colocynthisagria;* þæt is, *cucurbita*

    185.1. Wið innoþes astyrunge

178.6. For a woman's excessive flow of menstrual blood

178.7. So that you do not suffer from the cold

179. The herb *priapiscus;* that is, greater periwinkle

179.1. For possession by demons, and for snakes, and for wild animals, and for poisons, and for any threats, and for envy, and for terror, and that you may have good luck, and so that you may be happy and agreeable to others

180. The herb *lithospermon*

180.1. If stones should grow in the bladder

181. The herb *staphis agria*

181.1. For the corrupt humor in the body

181.2. For scurf and scab

181.3. For toothache and sore gums

182. The herb *gorgonion*

182.1. For all evil footsteps

183. The herb *melilotus*

183.1. For dimness of the eyes

183.2. For sinew spasm

184. The herb *bulbus*

184.1. For swellings, and for foot disease, and for any injury

184.2. For dropsy

For dog bite, and for sweating, and for stomachache

184.3. For sores and scurf and spots

184.4. For swelling and rupture in the abdomen

185. The herb *colocynthis agria;* that is, bitter cucumber

185.1. To stimulate the bowels

# I

## *Betonica*

1.1. Ðeos wyrt, þe man betonican nemneð, heo biþ cenned on mædum ond on clænum dunlandum ond on gefriþedum stowum. Seo deah gehwæþer ge þæs mannes sawle ge his lichoman; hio hyne scyldeþ wið unhyrum nihtgengum, ond wið egeslicum gesihðum ond swefnum. Ond seo wyrt byþ swyþe haligu, ond þus þu hi scealt niman, on Agustes monðe, butan iserne. Ond þonne þu hi genumene hæbbe, ahryse þa moldan of þæt hyre nanwiht on ne clyfie, ond þonne drig hi on sceade swyþe þearle. Ond mid wyrttruman mid ealle gewyrc to duste; bruc hyre þonne ond hyre byrig þonne ðu beþurfe.

1.2. Gif mannes heafod tobrocen sy, genim þa ylcan wyrte betonican; scearfa hy þonne ond gnid swyþe smale to duste; genim þonne twega trymessa wæge, þige hit þonne on hatum beore. Þonne halað þæt heafod swyðe hraðe æfter þam drince.

1.3. Wið eagena sar: genim þære ylcan wyrte wyrttruman; seoð on wætere to þriddan dæle, ond of þam wætere beþa þa eagan. Ond genim þæræ sylfan wyrfe leaf ond bryt hy, ond lege ofer þa eagan on þone andwlatan.

# Chapter 1

# Betony

1.1. This plant, which is called betony, grows in meadows and on cleared hilly land and in sheltered places. It is good both for the soul and the body; it gives protection against dreadful nightmares, and against terrifying visions and dreams. And this plant is very holy, and this is how you must gather it: during the month of August, without using an iron tool. When you have gathered it, shake off the dirt so that none sticks to it, and then dry it very thoroughly in the shade. Pound to dust the whole plant with its roots; then use it and taste it when you have need.

1.2. If a person's head is fractured, take the same plant betony; shred it and pound it into very thin powder; then take the weight of two tremisses and have him drink it in hot beer. After that drink, the head will heal very quickly.

1.3. For eye pain: take the roots of this same plant; simmer in water reducing it by two thirds, and bathe the eyes with this liquid. And take the leaves of this same plant and break them up, and lay them on the forehead over the eyes.

1.4. Wið earena sar: genim þære ylcan wyrte leaf þonne heo grenost beo; wyl on wætere ond wring þæt wos; ond siþþan hyt gestanden beo, do hit eft wearm ond þurh wulle drype on þæt eare.

1.5. Wið eagena dymnesse: genim þære ylcan wyrte betonican anre tremesse wæge ond wyl on wætere, ond syle drincan fæstendum. Þonne gewanað hit þone dæl þæs blodes ðe seo dymnys of cymð.

1.6. Wiþ tyrende eagan: genim þa ylcan wyrte betonican ond syle þigccean. Heo gegodað ond onliht þæra eagena scearpnysse.

1.7. Wiþ swyþlicne blodryne of nosum: genim þa ylcan wyrte betonican, ond cnuca hy, ond gemeng þærto sumne dæl sealtes; ond genim þonne swa mycel swa þu mæge mid twam fingrum geniman, wyrc hit sinewealt, ond do on þa næsþyrlu.

1.8. Wiþ toðece: genim þa ylcan wyrte betonican ond wyl on ealdan wine oþþe on ecede to þriddan dæle. Hit hælþ wundurlice þæra toða sar ond geswell.

1.9. Wiþ sidan sare: genim þære ylcan wyrte þreora trymessa wæge; seoð on ealdum wine, ond gnid þærto seofon ond twentig piporcorna. Gedrinc his þonne on niht nistig þreo full fulle.

1.10. Wiþ lændenbrædena sare: genim þare ylcan betonican þreora trymessa wæge, seofontiene piporcorn; gnid tosomne; wyll on ealdum wine. Syle hym swa wearm on niht nistig þreo full fulle.

1.4. For earache: take the leaves of this same plant when they are at their greenest; boil them in water and press the juice out; and when it has stood for a while, warm it again and drip it through some wool into the ear.

1.5. For dimness of the eyes: take one tremiss-weight of this same plant betony and boil it in water, and give it to the patient to drink while fasting. Then it will dilute that part of the blood from which the dimness originates.

1.6. For watery eyes: take the same plant betony and give it to the patient to eat. It will improve and sharpen a person's eyesight.

1.7. For excessive flow of blood from the nose: Take the same plant betony and pound it, and add some salt; and then take as much as you can pinch with two fingers, work it into a ball, and put it into the nostrils.

1.8. For toothache: take the same plant betony and boil it in old wine or vinegar, reducing it by two-thirds. It will heal the toothache and the swelling remarkably well.

1.9. For pain in the side: take three tremiss-weights of the same plant; simmer it in old wine, and pound twenty-seven peppercorns into it. Then drink three cupfuls of it, having fasted for a night.

1.10. For pain in the loins: take three tremiss-weights of the same plant betony, plus seventeen peppercorns; pound them together; boil them in old wine. Give the person three cupfuls, warm, after fasting for a night.

1.11. Wið wambe sare: genim þære ylcan wyrte twega trymessa wæge; wyl on wætere; syle hyt þonne him wearm drincan. Ðonne bið þæs innoðes sar settende ond liðigende þæt
hit sona nænig lað ne bið.

1.12. Gif mannes innoð to fæst sy: anbyrge þas ylcan wyrte on
wearmum wætere on niht nistig. Þonne bið se man hal on
þreora nihte fyrste.

1.13. Wiþ þon ðe men blod upp wealle þurh his muð: genim
þære ylcan wyrte þreora trymessa wæge ond cole gate meolc
þreo ful fulle. Ðonne bið he swyþe hraðe hal.

1.14. Gif man nelle beon druncen: nime þonne ærest, onbyrge betonican ðære wyrte.

1.15. Gif men wylle spring on gesittan: genime þonne anes
trymeses gewæge; cnucige wið eald smeoru; lecge on ðone
stede þe se spring on gesittan wolde. Þonne byþ hit sona hal.

1.16. Gif mon sy innan gebrocen oþþe him se lichoma sar sy,
genime þonne betonican þære wyrte feower trymessan gewæge; wyll on wine swyþe; drince þonne on niht nistig.
Þonne leohtað him se lichoma.

1.17. Gif mon on mycelre rade oþþe on miclum gangum
weorðe geteorad, nime þonne betonican þære wyrte ane
trymessan fulle; seoð on geswettum wine; drince þonne on
niht nihstig þreo full fulle. Þonne bið he sona unwerig.

1.18. Gif man sy innan unhal oþþe hyne wlatige, þonne genim
ðu betonican þære wyrte twa trymessan gewæge ond huniges anre yndsan gewæge; wylle þonne on beore swyþe þearle;
drince ðreo ful fulle on niht nistig. Þonne rumað him sona se
innað.

1.11. For stomach pain: take two tremiss-weights of this same plant; boil in water; then have the patient drink it warm. Then the pain in the stomach will diminish and be relieved so that soon there will be no discomfort.

1.12. If a person is constipated: let him take this same plant in warm water after fasting for a night. The person will be healthy in three days' time.

1.13. If a person spits blood: take three tremiss-weights of this same plant and three cupfuls of cold goat's milk. Then he will very quickly be well.

1.14. If a person does not want to get drunk: let him first of all take and eat the plant betony.

1.15. If a boil is developing on a person: take one tremiss-weight; pound in old grease; lay it on the place where the boil was starting to grow. Then it will soon be healed.

1.16. In case a person is ruptured internally or if the body is sore, then take four tremiss-weights of this same plant betony; boil this well in wine; then let him drink it after fasting for a night. Then the body will be relieved.

1.17. If a person becomes tired from much riding or much walking, then take one tremiss-weight of this same plant; simmer it in sweetened wine; then let him drink three cupfuls after fasting for a night. He will soon be refreshed.

1.18. If a person is feeling poorly inside or is nauseous, then take two tremiss-weights of this same plant betony and one ounce of honey; then boil them very thoroughly in beer; let him drink three cupfuls after fasting for a night. Then his insides will soon be freed of discomfort.

1.19. Gif þu ðonne wylle þæt ðin mete eaðelice gemylte, genim þonne betonican þære wyrte þreo trymessan gewæge ond huniges ane yndsan; seoð þonne þa wyrte oðþæt heo heardige; drinc hy þonne on wætere twa full fulle.

1.20. Wiþ ðon þe man ne mæge his mete gehabban ond he spiwe ðonne he hyne geðigedne hæbbe: genim þonne betonican þære wyrte feower trymesan gewæge ond awylled hunig; wyrc þonne lytle poslingas feower þærof. Ete þonne ænne ond ænne on hatum wætere ond on wine tosomne; geðicge ðonne þæs wætan þreo full fulle.

1.21. Wið innoþes sare, oððe gif he aþunden sy: genim betonican þa wyrt; gnid on wine swyðe smale; lege þonne abutan þa wambe, ond þyge hy. Þonne eac hraðe cymeþ þæt to bote.

1.22. Gif þonne hwylc man attor geþycge, genime ðonne þære ylcan wyrte þreo trymessan gewæge ond feower ful fulle wines; wylle tosomne ond drince. Þonne aspiweð he þæt attor.

1.23. Gif hwylcne man nædre toslite, genime þære wyrte feower trymesan gewæge; wyll on wine, ond gnid swyþe smale. Do þonne gehwæþer: ge on ða wunde lege, ond eac drinc swyþe þearle. Ðonne meaht ðu æghwylcere nædran slite swa gehælan.

1.24. Eft, wið nædran slite: genim þære ylcan wyrte ane trymesan gewæge; gecnid on read win; gedo þonne ðæt þæs wines syn þreo ful fulle. Smyre ðonne mid þam wyrtum ða wunde ond mid þy wine. Þonne byð hio sona hal.

1.19. If you want to digest your food easily, then take three tremiss-weights of this same plant betony and one ounce of honey; simmer the plant until it thickens; then drink it in two cupfuls of water.

1.20. If a person cannot keep food down and he vomits what he has eaten: take four tremiss-weights of this same plant betony, plus boiled honey; then make four little pills of it. Then give him one to eat and put one in a mixture of wine and hot water; then have him drink three cupfuls of this liquid.

1.21. For abdominal pain or if a person feels bloated: take this plant betony; crush it very fine in wine; then apply it around the stomach and have him eat some. This too will quickly bring about a cure.

1.22. If a person should ingest poison, then have him take three tremiss-weights of this same plant and four cupfuls of wine; boil them together and have him drink it. Then he will vomit up the poison.

1.23. If a snake bites someone, take four tremiss-weights of this plant; boil it in wine, and crush it very fine. Then do both these things: lay this on the wound, and also have him drink a great deal of it. You can heal the bite of any snake in this way.

1.24. Again, for snakebite: take one tremiss-weight of this plant; pound it in red wine; see to it that there are three cupfuls of wine. Smear the wound with the plants and with the wine. The wound will soon be healed.

1.25. Wið wedehundes slite: genim betonican ða wyrte; gecnuca hy swyþe smale, ond lege on þa wunde.

1.26. Gif þe ðin þrotu sar sy oððe þines swyran hwylc dæl, genim þa ilcan wyrte ond gecnuca swyðe smale; wyrc to clyþan; lege on þone swyran. Ðonne clænsað heo hit æghwær, ge innan ge utan.

1.27. Wiþ lendena sare ond gif men his ðeoh acen: genim þare ylcan wyrte twega trymessa gewæge; wyll on beore; syle him drincan.

1.28. Gif he ðonne sy febrig ond he sy mycelre hætan ðrowiende, syle ðonne þa wyrte on wearmum wætere, nalæs on beore. Ðonne godiað þæra lendena sar ond þæra ðeona swyðe hræðe.

1.29. Wiþ fotadle: genim þa ylcan wyrte; seoð on wætere oþðæt þæs wæteres sy ðriddan dæl on besoden. Cnuca ðonne þa wyrte ond lege on þa fet ond smire þærmid, ond drinc þæt wos. Þonne findest ðu þæræt bote ond ælteowe hælo.

1.25. For the bite of a mad dog: take the plant betony; pound it very fine and lay it on the wound.

1.26. If your throat or any part of your neck is sore, take this same plant and pound it very fine; make a poultice; lay it on the neck. It will clear up the soreness everywhere, both internally and externally.

1.27. For sore loins and if a person's thighs should ache: take two tremiss-weights of this plant; boil it in beer; let him drink it.

1.28. If the person is feverish and suffering from a high temperature, then give him this plant in warm water, by no means in beer. The pain in the loins and the thighs will get better very quickly.

1.29. For foot disease: take this same plant; simmer it in water until it has been reduced by one third. Then pound the plant and apply it to the feet and rub it on, and drink the juice. By this means you will find recovery and excellent health.

## 2

# Wægbræde

2.1. Gif mannes heafod ace oððe sar sy: genime wegbrædan wyrtwalan ond binde him on swyran. Ðonne gewiteð þæt sar of þam heafde.

2.2. Gif men his wamb sar sy: genime wegbrædan seaw ðære wyrte; gedo þæt hio blacu sy, ond þyge hy. Ðonne mid mycelre wlatunge gewiteþ þæt sar onweg. Gif hyt þonne sy þæt sio wamb sy aþundeno, scearfa ðonne þa wyrte ond lege on þa wambe. Ðonne fordwineð heo sona.

2.3. Wið þæs innoðes sare: genim wegbrædan seaw; do on sumes cynnes ealo, ond þycge hyt swyðe. Þonne bataþ he inneweard, ond clænsað þone magan ond þa smælþearmas swyþe wundrum well.

2.4. Eft, wið þon þe man on wambe forweaxen sy: seoð þonne þa wegbrædan swyþe, ond ete þonne swyþe. Ðonne dwineþ seo wamb sona.

2.5. Eft wið þon þe man þurh hys argang blode ut yrne: genim wegbrædan seaw; syle him drincan. Þonne bið hit sona oðstilled.

2.6. Gif man gewundud sy: genim wegbrædan sæd; gnid to duste, ond scead on þa wunde. Heo bið sona hal. Gif se lichoma hwær mid hefiglicre hæto sy gebysgod, gecnuca ða

# Chapter 2

# Common plantain

2.1. If a person's head aches or is sore: take the roots of common plantain and bind them round the person's neck. Then the soreness will leave the head.

2.2. If a person's stomach is aching: take the juice of the same plant plantain; work it until it becomes clear, and have him eat it. Then with much vomiting the soreness will go away. Then if it happens that the stomach is bloated, shred the plant and lay it on the stomach. The swelling will quickly disappear.

2.3. For abdominal pain: take the juice of the plantain; put it in some kind of ale, and have him drink a lot of it. Then his insides will feel better, and it will purge the stomach and the intestines marvelously well.

2.4. Again, if someone should have a bloated stomach: simmer the plantain for a long time and have him eat a lot of it. Then the swelling of the stomach will soon subside.

2.5. Again, if a person's anus is bleeding: take the juice of the plantain; give it to him to drink. The bleeding will quickly stop.

2.6. If a person is wounded: take plantain seeds; pound them into a powder and sprinkle them over the wound. It will quickly heal. If the body is troubled anywhere by high tem-

sylfan wyrte ond lege þæron. Ðonne colað se lichoma ond halað.

2.7. Gif ðu þonne wylle mannes wambe þwænan, þonne nim ðu þa wyrte; wyll on ecede. Do þonne þæt wos ond þa wyrte, swa awyllede, on win; drince þonne on niht nihstig, symle an ful to fylles.

2.8. Wið nædran slite: genim wegbrædan ða wyrt; gnid on wine, ond ete hy.

2.9. Wiþ *scorpiones* slite: genim wegbrædan wyrtwalan; bind on þone man. Þonne ys to gelyfenne þæt hyt cume him to godre are.

2.10. Gif men innan wyrmas eglen: genim wegbrædan seaw; cnuca ond wring, ond syle him supan. Ond nim ða sylfan wyrte; gecnuca; lege on þone naflan, ond wrið þærto swyðe fæste.

2.11. Gif hwylces mannes lichoma sy aheardod: nim þonne wegbrædan þa wyrte ond gecnuca wið smeru, butan sealte, ond wyrc swa to clame; lege þonne on þær hit heardige. Hnescaþ hyt sona ond bataþ.

2.12. Gif hwylcum men sy þæs feorðan dæges fefer getenge: genim ðonne þære wyrte seaw; cnid on wætere; syle him drincan twam tidum ær he hym þæs feferes wene. Þonne ys wen þæt hyt him cume to mycelre freme.

2.13. Wið fotadle ond wið sina sare: genim þonne wægbrædan leaf; gnid wið sealt; sete ðonne on þa fet ond on þa syna. Þonne ys þæt gewisslice læcedom.

2.14. Wið þam fefore þe ðy þriddan dæge on man becymeð:

perature, pound this same plant and apply it to the area. Then the body will cool down and heal.

2.7. If you want to soften a person's stomach, then take this plant; boil it in vinegar. Then pour the juice and the plant, thus boiled, into wine; have him drink it after fasting for a night, always one cup as the full dose.

2.8. For snakebite: take the plant plantain; crush it in wine, and eat it.

2.9. For the bite of scorpions: take roots of plantain; bandage them onto the person. It is believed that this will be a big help to him.

2.10. If intestinal worms trouble a person: take plantain juice. Pound and strain, and give it to him to drink. And take this same plant; pound it, lay it on the navel, and fasten it there very tightly.

2.11. If there is a hardening on the body: take the plant plantain and pound it in grease, without salt, and work it into a poultice in this way. Apply it to where there is a hardening. It will quickly soften and get better.

2.12. If the fever that comes on a person every fourth day affects a person: take the juice of this plant; crush it in water; give it to him to drink two hours before he expects the fever to arrive. Then there is hope that it will do him much good.

2.13. For foot disease and sore sinews: take the leaves of the plantain; crush them with salt; put this on the feet and on the sinews. This is a surefire remedy.

2.14. For the fever that comes on a person every third day:

genim wegbrædan þry cyðas; cnid on wætere oþþe on wine; syle him drincan ær þon se fefor him to cume on niht nihstig.

2.15. Wiþ ðy fefore þe ðy æftran dæge to cymeð: gecnuca þas ylcan wyrte swyþe smale; syle him on ealoð drincan. Þæt ys to gelyfenne þæt hit dyge.

2.16. Wið wunda hatunge: nim þonne wegbrædan þa wyrt; cnuca on smerwe butan sealte; lege on þa wunde. Þonne bið he sona hal.

2.17. Gif mannes fet on syþe tydrien: genim þonne wegbrædan ða wyrt; gnid on ecede; beþe ða fet þærmid, ond smyre. Ðonne þwineþ hy sona.

2.18. Gif hwylcum weargbræde weaxe on þam nosum oððe on þam hleore, genim ðonne wegbrædan seaw; wring on hnesce wulle; lege þæron; læt licgan nigon niht. Þonne halaþ hyt hraðe æfter ðam.

2.19. Be æghwylcum uncuþum blædrum ðe on mannes nebbe sittað: nim wegbrædan sæd; drig to duste ond gnid; meng wið smeoru; do lytel sealtes to; wes mid wine; smyre þæt neb mid. Þonne smeþað hyt ond halað.

2.20. Wiþ muþes wunde: genim wegbrædan leaf ond hyre seaw; gnid tosomne; hafa ðonne swiþe lange on þinum muðe, ond et ðone wyrtwalan.

2.21. Gif wedehund man toslite, genim þas ylcan wyrte ond gegnid ond lege on. Ðonne bið hit sona hal.

2.22. Wiþ ælces dæges mannes tyddernysse inneweardes:

take three roots of the plantain; crush them in water or in wine; give him this to drink before the fever arrives, after fasting for a night.

2.15. For the fever that comes every second day: crush this same plant very fine; give it to him to drink in ale. It is believed that this will help.

2.16. For inflamed wounds: take this plant plantain; pound in grease without salt; apply to the wound. He will soon be well.

2.17. If a person gets footsore on a journey, then take the plant plantain; crush it in vinegar; bathe the feet with it, and rub it in. They will quickly become less swollen.

2.18. If a sore should grow on the nose or cheek, take the juice of the plantain; wring it onto soft wool; lay it on the sore; let it stay there for nine days. After that the sore will quickly heal.

2.19. For any strange pustules that are located on someone's face: take the seeds of plantain; dry them and pound them into a fine powder; mix with grease; add a little salt: soak in wine; smear the face with it. The face will become smooth and will heal.

2.20. For sores in the mouth: take plantain leaves and its juice; crush them together; keep them in your mouth as long as possible, and eat the root.

2.21. If a mad dog bites a person, take this same plant and pound it and apply it to the bite. It will quickly heal.

2.22. For a person's daily internal weakness: take plantain,

nime þonne wegbrædan, do on win, ond sup þæt wos ond et
ða wegbrædan. Ðonne deah hit wið æghwylcre innancundre
unhælo.

# 3

# Fifleafe

3.1. Gif men his leoðu acen oððe ongeflogen sy: genim fifleafe ða wyrt; cnuca on smeorwe swyþe smale; lege ðæron
butan sealte. Ðonne halað hyt sona.

3.2. Wiþ wambe sare: genim fifleafan seaw þære wyrte; gewring twegen cuculeras fulle; syle him supan. Þonne clæn-
saþ hit onweg þæt sar eall.

3.3. Wiþ muðes ece, ond wið tungan, ond wið þrotan: genim
fifleafan wyrtwalan; wyll on wætere; syle him supan. Þonne
clænsað hit ðone muð innan, ond bið se ece litliende.

3.4. Wiþ heafdes sare: genim fifleafan ða wyrt; bewrit þriwa
mid þam læstan fingre ond mid þam ðuman; ahefe þonne
upp of ðære eorþan ond gegnid swyþe smale; ond bind on
þæt heafod. Ðonne biþ se ece lytliende.

3.5. Gif men blod ut of nosum yrne to swiðe: syle him drincan fifleafan on wine, ond smyre þæt heafud mid þam.
Ðonne oðstandeþ se blodgyte sona.

put it in wine, and drink the juice and eat the plantain. This will be good for all kinds of internal ailment.

# Chapter 3

# Cinquefoil

3.1. If a person's joints ache or are attacked by a disease: take the plant cinquefoil; pound it in very thin grease; apply this to the place, using no salt. It will quickly heal.

3.2. For stomach pain: take the juice of the same plant cinquefoil; press out two spoonfuls; give it to the patient to drink. It will clear away all the soreness.

3.3. For aching in the mouth, tongue, and throat: take roots of cinquefoil; boil in water; give it to him to eat. This will cleanse the inside of the mouth, and the pain will diminish.

3.4. For headache: take the plant cinquefoil. Scratch a line around it three times with your little finger and your thumb, then pull it up from the ground and crush it very fine. Bind it on the head. Then the pain will diminish.

3.5. If a person has a severe nosebleed: let him drink cinquefoil in wine, and smear his head with it. Then the bleeding will quickly stop.

3.6. Gif mannes midrife ace: genime fifleafan seaw; mencg to wine, ond drince ðonne þreo ful fulle þry morgenas ond on niht nihstig.

3.7. Wiþ nædran slite: genim fifleafan þa wyrte; gnid on wine, ond drince swiðe. Ðonne cymeð him þæt to bote.

3.8. Gif man forbærned sy: genime fifleafan þa wyrt; bere on him. Ðonne cweþað cræftige men þæt him þæt to gode cume.

3.9. Gif ðu wille cancer ablendan, genim ðonne fifleafan ða wyrte; seoð on wine ond on ealdes bearges rysle, butan sealte; mencg eall tosomne; wyrc to clyðan, ond lege ðonne on þa wunde. Þonne halað heo sona. Ðu scealt ðonne eac gewyrcean þa wyrt on Agustus monðe.

# 4

# Æscþrote

Ðeos wyrt, þe man *vermenacam* ond oðrum naman æscþrote nemneð, bið cenned gehwær on smeþum landum ond on wætum.

4.1. Wiþ wunda, ond wið deadspringas, ond wið cyrnlu: genim þære ylcan wyrte wyrtwalan ond gewrið abutan ðone swyran. Þonne fremað hit healice.

3.6. If a person has pain in the midriff: take the juice of cinquefoil, mix with wine, and have him drink three cupfuls for three mornings and while fasting overnight.

3.7. For snakebite: take the plant cinquefoil; crush it into wine, and have him drink a good deal of it. This will be his cure.

3.8. If someone is badly burned: take the plant cinquefoil; have him carry it on his person. Learned people say that this will do him good.

3.9. If you want to heal an ulcerous sore, then take the plant cinquefoil; simmer it in wine and in the fat of an old barrow pig, using no salt; mix it all together; make this into a poultice, and then apply it to the wound. This will soon heal it. In addition, you must prepare the plant in the month of August.

# Chapter 4

# Vervain

This plant, which is called *vermena* and by another name vervain, grows everywhere in flat areas and in damp ones.

4.1. For wounds and carbuncles and hard swellings: take the roots of this plant and tie them around the neck. This will do a great deal of good.

4.2. Eft, wið cyrnlu: genim ða sylfan wyrte *vermenacam;* ge-cnuca hy ond lege ðærto. Heo hælð wundorlice.

4.3. Wiþ ða þe habbað ætstandene ædran swa þæt þæt blod ne mæg hys gecyndelican ryne habban, ond heora þigne ge-healdan ne magon: nim þære ylcan wyrte seaw ond syle drin-can. Ond syððan genim win ond hunig ond wæter, mencg tosomne, ond hyt sona hælð þa untrumnysse.

4.4. Wið lifre sar: genim on Middesumeres Dæg þa ylcan wyrte ond gegnid to duste. Nim þonne fif cuculeras fulle ðæs dustes ond þry scenceas godes wines; mencg tosomne; syle drincan. Hyt fremað miclum; eac swa same manegum oðrum untrumnyssum.

4.5. Wiþ þa untrumnysse þe stanas weaxað on blædran: genim þære ylcan wyrte wyrtwalan ond cnuca hy; wyll þonne on hatan wine; syle drincan. Hyt hælð þa untrumnysse wun-dorlicum gemete, ond na þæt an, eac swa hwæt swa þone migðan gelet hyt hrædlice gerymð ond forð gelædeþ.

4.6. Wið heafodsar: genim þa ylcan wyrte ond gebind to þam heafde. Ond heo gewanað þæt sar ðæs heafdes.

4.7. Wið nædran slite: swa hwylc man swa þas wyrt *vermena-cam,* mid hyre leafum ond wyrttrumum, on him hæfð, wið eallum nædrum he bið trum.

4.8. Wiþ attorcoppan bite: genim þære ylcan wyrte leaf; seoð on wine, gecnucode. Gif hyt mid geswelle on forboren byð, gelege þærto; seo wund sceal sona beon geopenud. Ond syððan heo geopenud beo, þonne gecnuca þa wyrt mid hu-nige ond lege þærto oþðæt hyt hal sy. Þæt bið swiðe hrædlice.

4.2. Again, for hard swellings: take this same plant vervain; pound it and apply it to the swellings. This will heal them wonderfully.

4.3. For people who have clogged veins so that blood cannot flow to the genitals, and who cannot keep their food down: take the juice of this plant and give it to the patient to drink. And afterward take wine and honey and water, mix them together, and this will soon heal the infirmity.

4.4. For pain in the liver: gather this same plant on Midsummer's Day and crush it into a powder. Then take five spoonfuls of the powder and three cupfuls of good wine; mix them together; give it to the patient to drink. It will be of great benefit; likewise it will help cure many other ailments.

4.5. For the infirmity when stones grow in the bladder: take the roots of this same plant and pound them; then simmer them in hot wine; give it to the patient to drink. This will heal the infirmity in a wonderful manner, and not just that one ailment, but it will also quickly dissolve and remove anything that hinders the flow of urine.

4.6. For headache: take the same plant and bind it to the head. It will lessen the head pain.

4.7. For snakebite: anyone who keeps on his person this plant vervain, including its leaves and roots, will be safe from all snakes.

4.8. For spider bite: take the leaves of this same plant; simmer in wine, pounded. If the poison is retained in the swelling, apply them there; the wound will quickly open up. And after it has opened up, pound the plant in honey and apply it to the spot until it heals. That will be very quickly.

4.9. Wiþ wedehundes slite: genim þa ylcan wyrte *vermena-cam* ond hwætene corn, swa gehale, ond lege to þære wunde swa oþþæt ða corn þurh þone wætan gehnehsode syn ond swa toðundene. Nim þonne ða corn ond gewurp to sumum henfugule. Gif he hy þonne etan nelle, ðonne nim ðu oþre corn ond mencg to þære wyrte þam gemete þe þu ær dydest, ond lege to ðære wunde swa oþðæt þu ongite þæt seo frecnys of anumen sy ond ut atogen.

4.10. Wiþ niwe wundela: genim þa ylcan wyrte ond cnuca mid buteran, ond lege to þære wunde.

4.11. Wið nædran slite: genim þære ylcan wyrte twigu ond seoð on wine, ond cnuca syþþan. Gyf se slyte blind bið ond mid þam geswelle ungeheafdud, þonne lege ðu þa wyrte þærto; sona hyt sceal openian. Ond syððan hyt geopenud beo, þonne nim ðu ða ylcan wyrte, ungesodene, ond cnuca mid hunige; lege to þære wunde oðþæt heo hal sy. Þæt is swyþe hrædlice gyf man hy þyssum gemete þærto alegð.

# 5

# Hennebelle

Ðeos wyrt, þe man *symphoniacam* nemneð ond oðrum naman *belone,* ond eac sume men hennebelle hataþ, wihst on beganum landum ond on sandigum landum ond on

4.9. For the bite of a mad dog: take this same plant vervain and some grains of wheat, intact, and apply them to the bite until the grains are softened by the moisture and become swollen. Then take the grains and throw them to a hen. If it refuses to eat them, then take some more grains and mix them with the plant in the same way as you did before, and apply this to the wound until you feel that the danger has been eliminated and the poison has been drawn out.

4.10. For fresh wounds: take this same plant and pound it in butter, and apply it to the wound.

4.11. For snakebite: take twigs of this same plant and simmer them in wine, and then pound them. If the bite is closed and the swelling has not come to a head, then apply the plant to it; soon the wound will open. And as soon as it has opened, then take this same plant, without boiling it, and pound it in honey; lay it on the wound until it heals. That will be very quickly if one applies the plant in this way.

# Chapter 5

# Henbane

This plant, which is called *symphoniaca* and by another name *bellonaria,* and some people also call it henbane, grows in cultivated places and in sandy soils and in gardens. There is

wyrtunum. Þonne ys oðer þisse ylcan wyrte, sweart on hiwe ond stiðran leafum ond eac ætrigum; þonne ys seo aerre hwitre, ond heo hæfð þas mægnu.

5.1. Wið earena sar: genim þysse ylcan wyrte seaw ond wyrm hit; drype on þæt eare. Hyt wundorlicum gemete þæra earena sar afligð; ond eac swa same, þeah þær wyrmas on beon, hyt hy acwelleð.

5.2. Wið cneowa geswell, oððe sceancena, oððe swa hwær swa on lichoman geswell sy: nim þa ylcan wyrte symphonia-can ond cnuca hy; lege þærto. Þæt geswell heo of animeð.

5.3. Wiþ toða sare: genim þære ylcan wyrte wyrtwalan; seoð on strangum wine; supe hit swa wearm, ond healde on his muðe. Sona hit gehælð þara toða sar.

5.4. Wið þæra gewealda sar oððe geswell: genim þære ylcan wyrte wyrtwalan ond gewrið to ðam þeo. Ge þæt sar ge þæt geswell þara gewalda hio of animeð.

5.5. Gif wifes breost sare sien: genim ðonne þære ylcan wyrte seaw; wyrc to drence ond syle hyre drincan, ond smyre ða breost þærmid. Þonne byð hyre sona þe sel.

5.6. Wið fota sar: genim þa ylcan wyrte mid hyre wyrtruman ond cnuca tosomne; lege ofer ða fet ond þærto gebind. Hyt hælþ wundurlice ond þæt geswell of animð.

5.7. Wiþ lungenadle: genim þære sylfan wyrte seaw; syle drin-can. Mid healicre wundrunge he bið gehæled.

another variety of this same plant, dark in color and with stiffer leaves, ones that are poisonous, too. The former type is whiter, and it has the following powers.

5.1. For earache: take the juice of this same plant and warm it; drip it into the ear. It will expel the earache in a wonderful way; and likewise, if there are worms, it will kill them.

5.2. For swollen knees, or calves, or for wherever there is a swelling on the body: take this same plant henbane and pound it; apply it to the place. It will take away the swelling.

5.3. For toothache: take the roots of this same plant; simmer in strong wine; have the patient eat this while it is warm, and have him keep it in his mouth. This will soon cure the toothache.

5.4. For pain or swelling in the genitals: take the roots of this same plant and tie them to the thigh. It will take away both the genital pain and the swelling.

5.5. If a woman's breasts are sore: take the juice of this same plant; make it into a beverage and have her drink it, and smear her breasts with it. She will soon be better.

5.6. For sore feet: take this same plant, including its roots, and pound them together; place this on top of the feet and bind it on. It will provide a wonderful cure and will take away the swelling.

5.7. For lung disease: take the juice of this same plant; give to drink. The person will be healed, to people's great astonishment.

# 6

# Nædrewyrt

Ðeos wyrt, þe man *viperinam* ond oðrum naman nædderwyrt nemneð, bið cenned on wætere ond on æcerum. Heo bið hnesceum leafum ond bitterre on byrgingce.

6.1. Wiþ næddran slite: genim ðas sylfan *viperinam;* cnuca hy; mengc mid wine; syle drincan. Heo hælð wundorlice þone slyte ond þæt attor todrifð. Ond þas wyrte ðu scealt niman on ðam monðe þe man Aprelis nemneð.

# 7

# Beowyrt

Ðeos wyrt, þe man on Leden *veneriam* ond on ure geþeode beowyrt nemneð, heo bið cenned on beganum stowum ond on wyrtbeddum ond on mædum. Ond þas wyrte þu scealt niman on þam monðe þe man Augustum nemneð.

7.1. Wiþ ðæt beon ne ætfleon: genim þas ylcan wyrte þe we *veneriam* nemdon ond gehoh hy to ðære hyfe; þonne beoð hy wungynde ond næfre ne swicað, ac hym gelicað. Þeos wyrt

# Chapter 6

# Adder's wort

This plant, which is called *viperina* and by another name adder's wort, grows in water and in open fields. It has soft leaves and is bitter to the taste.

6.1. For snakebite: take this same plant adder's wort; pound it; mix it with wine; give to drink. It will heal the bite wonderfully and will expel the poison. And you must gather this plant in the month called April.

# Chapter 7

# Yellow iris

This plant, which is called *venerea* in Latin and yellow iris in our language, grows in cultivated places and in garden plots and in meadows. And you must gather this plant in the month called August.

7.1. So that bees do not swarm: take this same plant that we have called yellow iris and suspend it from the hive; then they will remain in place and will never depart; on the

byð seldon funden, ne hy man gecnawan ne mæg buton
ðonne heo grewð ond blewð.

7.2. Gif hwa ne mæge gemigan ond se micgða ætstanden sy,
nime þysse ylcan wyrte wyrtwalan ond seoþe on wætere to
þriddan dæle; sylle drincan. Þonne binnan þrym dagum he
mæg þone migþan forð asendan. Hyt hælð wundorlice þa
untrumnysse.

# 8

# Leonfot

Ðeos wyrt, þe man *pedem leonis* ond oðrum naman leonfot
nemneð, heo bið cenned on feldon ond on dicon ond on
hreodbeddon.

8.1. Gyf hwa on þære untrumnysse sy þæt he sy cis, þonne
meaht ðu hine unbindan. Genim þysse wyrte þe we leonfot
nemdon, fif ðyfelas, butan wyrttruman; seoð on wætere on
wanwægendum monan, ond ðweah hine þærmid. Ond læd
ut of þam huse onforan nihte, ond ster hyne mid þære wyrte
þe man *aristolochiam* nemneð; ond þonne he ut ga, ne beseo
he hyne na onbæc. Þus ðu hine meaht of þære untrumnysse
unbindan.

contrary, it will please them. This plant is seldom found, and it cannot be recognized except when it is putting forth foliage and is in bloom.

7.2. If a person cannot urinate and the flow of urine is stopped, take the roots of this same plant and simmer them in water, reducing the amount of fluid by two-thirds; have him drink this. Then within three days he will be able to pass urine. It heals the infirmity wonderfully.

# Chapter 8

# Lion's foot

This plant, which is called *pes leonis* and by another name lion's foot, grows in fields and in ditches and in reedbeds.

8.1. If a person is suffering from an infirmity that makes him fastidious, then you can free him from it. Take this plant we have called lion's foot, five bushy shoots, leaving the roots aside; boil them in water when the moon is waning, and wash him with this. And lead him out of the house before nightfall, and fumigate him with the plant called *aristolochia;* and when he goes outside, he must not look back. In this manner you can free him from the infirmity.

# 9

# Clufþunge

Ðeos wyrt, þe man *sceleratam* ond oðrum naman clufþunge nemneð, heo bið cenned on fuhtum ond on wæteregum stowum. Swa hwylc man swa þas wyrte fæstende þigð, hlih-hende he ðæt lif forlæteð.

9.1. Wið wundela ond wið deadspringas: genim þas ylcan wyrte ond gecnuca hy mid smeruwe butan sealte; lege to þære wunde. Ðonne yt heo, ond feormað gyf ðær hwæt horwes on bið. Ac ne geþafa þæt heo lengc þæræt licge þonne hyt þearf sy þy læs heo þone halan lichoman fornime. Gyf þonne mid orþance þisses ðinges fandian wille, gecnuca ða wyrte ond wrið hy to þinre halan handa. Sona heo yt þone lichaman.

9.2. Wið swylas ond wið weartan: genim þa sylfan wyrte ond gecnuca hy mid swinenum gore; lege to þam swylum ond to þam weartum. Binnan feagum tidum heo drifð þæt yfel ond þæt worsm ut atyhð.

# Chapter 9

# Celery-leaved crowfoot

This plant, which is called *scelerata* and by another name celery-leaved crowfoot, grows in damp and watery places. Whoever eats it while fasting, he will take leave of his life laughing.

9.1. For wounds and for carbuncles: take this same plant and pound it in unsalted grease; apply it to the wound. Then it will eat away the wound, and it will cleanse it if any pus is present. But do not let it stay there longer than necessary, or else it will eat away the healthy flesh. If you wish to test this with an experiment, pound the plant and bind it to your healthy hand. It will soon eat into the flesh.

9.2. For swellings and for warts: take this same plant and pound it together with pig's dung; apply this to the swellings and the warts. Within a few hours it will drive out the malignancy and will draw out the pus.

## 10

# Clufwyrt

Ðeos wyrt, þe man *batracion* ond oþrum naman clufwyrt nemneð, bið cenned on sandigum stowum ond on feldum. Heo bið feawum leafum ond þynnum.

10.1. Wið monoðseoce: genim þas wyrte ond gewrið mid anum readum þræde onbutan þæs monnes swyran, on wanwegendum monan on þam monþe ðe man Aprelis nemneð ond on Octobre foreweardum. Sona he bið gehæled.

10.2. Wiþ ða sweartan dolh: genim þas ylcan wyrte myd hyre wyrtwalan ond gecnuca hy; mengc eced þærto; lege to ðam dolchum. Sona hyt fornimð hy ond gedeð þam oþrum lice gelice.

## 11

# Mugcwyrt

Ðeos wyrt, þe man *artemesiam* ond oðrum naman mucgwyrt nemneð, bið cenned on stanigum stowum ond on sandigum. Þonne hwa siðfæt onginnan wille, ðonne genime he him on

# Chapter 10

# Buttercup

This plant, which is called *batrachion* and by another name buttercup, grows in sandy soils and in fields. It has few leaves and they are thin.

10.1. For a person suffering from lunacy: take this plant and bind it around the person's neck with a red thread, when the moon is waning in the month called April and at the beginning of October. He will soon be healed.

10.2. For black scars: take this same plant with its roots and pound it; mix in some vinegar; apply to the scars. This will quickly remove them and will make the area resemble the rest of the body.

# Chapter 11

# Mugwort

This plant, which is called *artemisia* and by another name mugwort, grows in rocky and sandy soils. When a person wants to begin a journey, then he should take this plant

hand þas wyrte *artemesiam* ond hæbbe mid him; ðonne ne ongyt he na mycel to geswynce þæs siðes. Ond eac heo afligð deofulseocnyssa, ond on þam huse þe he hy inne hæfð, heo forbyt yfele lacnunga, ond eac heo awendeð yfelra manna eagan.

11.1. Wiþ innoðes sar: genim þas ylcan wyrte ond gecnuca hy to duste ond gemengc hy wið niwe beor; syle drincan. Sona heo geliðegað þæs innoþes sar.

11.2. Wiþ fota sar: genim þas ylcan wyrte ond gecnuca hy mid smeruwe; lege to þam fotum. Heo þæt sar ðæra fota of genimð.

## 12

## *Herba artemesia tagantes*

12.1. Wið blædran sar, ond wið þæt man ne mæge gemigan: genim þyssæ wyrte seaw, þe man eac mugwyrt nemneð—seo ys swaþeah oþres cynnes—ond gewyll hy on hatan wætere oððe on wine, ond syle drincan.

12.2. Wið þeona sar: genim þas ylcan wyrte ond gecnuca hy mid smeruwe, ond gewes hy wel mid ecede; gebind syþþan to ðam sare. Ðy þriddan dæge him bið sel.

12.3. Wið sina sare ond wið geswel: genim þa ylcan wyrte *artemesiam;* cnuca hy mid ele wel gewylde; lege þærto. Hyt hælð wundorlice.

mugwort in hand and keep it with him; then he will not feel the hardship of the journey too much. And it also expels demonic possession, and in the house where a person keeps it, it prevents evil medicine, and it also turns away the evil eye.

11.1. For abdominal pain: take this same plant and pound it into a powder and mix it with freshly brewed beer; give to drink. It will soon relieve the soreness in the guts.

11.2. For sore feet: take this same plant and pound it in grease; apply to the feet. It will take away the pain from the feet.

# Chapter 12

# Tansy

12.1. For bladder pain, and if a person cannot urinate: take the juice of this plant, which is also called mugwort—though that is a different kind of plant—and simmer it in hot water or in wine, and give to drink.

12.2. For sore thighs: take this same plant and pound it in grease, and soak it well in vinegar; after this, bind it on the sore area. On the third day the person will be well.

12.3. For sore sinews and for swellings: take this same plant tansy; pound it in well-boiled oil; apply it to them. It will heal them wonderfully.

12.4. Gyf hwa mid fotadle swyþe ond hefelice geswenced sy, þonne genim ðu þysse ylcan wyrte wyrtwalan; syle etan on hunige. Ond eftsona he bið gehæled ond aclænsod swa þæt ðu ne wenst þæt heo mæge swa mycel mægen habban.

12.5. Gyf hwa sy mid feferum gedreht, genime þonne ðysse ylcan wyrte seaw, mid ele, ond smyre hyt. Sona heo þone fefer fram adeþ.

# 13

# Mugwyrt

Ðeos þridde wyrt, þe we *artemesiam leptefilos* ond oðrum naman mucgwyrt nemdon, heo bið cenned abuton dicum ond on ealdum beorgum. Gyf ðu hyre blosðman brytest, he hæfð swæc swylce ellen.

13.1. Wið þæs magan sare: genim þas wyrte ond cnuca hy, ond gewyll hy wel mid amigdales ele þam gemete ðe þu clyþan wyrce. Do þonne on anne clænne cla ond lege þærto. Binnan fif dagum he bið hal. Ond gif þysse wyrte wyrttruma byð ahangen ofer hwylces huses duru, þonne ne mæg ænig man þam huse derian.

13.2. Wið þara sina bifunge: genim þysse ylcan wyrte seaw gemencged mid ele; smyre hy ðonne þærmid. Hy geswicað þære bifunge, ond hyt ealne ðone leahtor genimeð.

12.4. If a person is severely tormented by foot disease, then take the roots of this same plant; give it to the patient to eat in honey. Soon afterward he will be cured and cleansed in such a way that you would never have thought that it might have such great power.

12.5. If a person is suffering from fever, then take the juice of this same plant, along with some oil, and rub it on. It will soon take the fever away.

# Chapter 13

# Roman wormwood

This third plant, which we have called *artemisia leptophyllos* and by another name mugwort, grows around ditches and in old mounds. If you crush its blossoms, it smells like elder tree.

13.1. For stomachache: take this plant and pound it, and simmer it well in almond oil in the same manner as you would make a poultice. Then put it in a clean cloth and apply it to the stomach. Within five days he will be healed. And if the root of this plant is hung over the door of someone's house, then no one can do harm to the house.

13.2. For trembling sinews: take the juice of this same plant mixed with oil; then rub them with it. It quiets the tremor, and it takes away the whole disorder.

2    Witodlice, þas þreo wyrta þe we *artemesias* nemdon, ys sæd þæt Diana hy findan scolde ond heora mægenu, ond læcedom Chironi centauro syllan, se ærest of þyssum wyrtum lacnunge gesette; ond he þas wyrta of naman ðære Dianan, þæt is *artemesias,* genemnede.

# 14

# Doccæ

Ðeos wyrt, þe man *lapatium* ond oðrum naman docce nemneð, bið cenned on sandigum stowum ond on ealdum myxenum.

14.1. Wið cyrnlu þe on gewealde wexeð: genim þas wyrte *lapatium,* ond cnuca hy mid ealdum rysle buton sealte swa þæt ðæs smeruwes sy twam dælum mare þonne þære wyrte. Swyþe wel gemenged, do hyt þonne sinetrundæl ond befeald on caules leafe, ond berec on hatum ahsum, ond þonne hit hat sy, lege ofer þa cyrnlu ond gewrið ðærto. Þys is selest wið cyrnlu.

Indeed, it is said that Diana found these three plants that 2
we call *artemisia* and discovered their powers, and that she
gave this medicine to the centaur Chiron, who first prepared remedies out of these plants; and he named these
plants *artemisia* after the name of that same Diana.

# Chapter 14

# Dock

This plant, which is called *lapathum* and by another name
dock, grows in sandy places and on old dunghills.

14.1. For hard swellings that grow on the genitals: take this
plant dock and pound it in old unsalted fat so that there
are two parts more of fat than of the plant. When it is well
mixed, work it into a ball and enfold it in a cabbage leaf,
and fumigate it on hot ashes, and when it is hot, lay it on
the swellings and bind it to them. This is the best thing for
swellings.

# 15

# Dracentse

Ðeos wyrt, þe man *dracontea* ond oðrum naman dracentse nemneð, ys sæd þæt heo of dracan blode acenned beon sceolde. Heo bið cenned on ufeweardum muntum þær bærwas beoð, swyþost on haligum stowum ond on þam lande þe man Apulia nemneð. Heo on stanigum lande wyxð; heo ys hnesce on æthrine, ond weredre on byrincge, ond on swæce swylce grene cystel, ond se wyrtruma neoðeweard swylce dracan heafod.

15.1. Wið ealra nædrena slite: genim þysse wyrte dracontean wyrttruman; cnuca mid wine, ond wyrm hyt; syle drincan. Eall þæt attor hyt tofereð.

15.2. Wið banbryce: genim þysse ylcan wyrte wyrtruman ond cnuca mid smerwe þam gelice þe ðu clyþan wyrce. Ðonne atyhð hyt of þam lichoman þa tobrocenan ban. Ðas wyrte þu scealt niman on þam monðe þe man Iulium nemneð.

# Chapter 15

# Dragonwort

This plant, which is called *dracontea* and by another name dragonwort, is believed to have been generated in dragon's blood. It grows on high mountains where there are groves of trees, mainly in holy places and in that country that is called Apulia. It grows in stony soil; it is soft to the touch and sweet to the taste and it smells like green chestnuts, and the end of its root looks like a dragon's head.

15.1. For all kinds of snakebite: take the roots of this same plant dragonwort; pound them in wine and warm them; give to drink. It will take away all the poison.

15.2. For a broken bone: take the roots of this same plant and pound them in grease in the same way as you would make a poultice. It will draw out the broken bone from the body. You should gather this plant in the month called July.

# 16

# Hreafnes leac

Ðeos wyrt, ðe man *satyrion* ond oðrum naman hræfnes leac nemneð, heo bið cenned on hean dunum ond on heardum stowum, ond swa some on mædum ond on beganum landum ond on sandigum.

16.1. Wið earfoðlice wundela: genim þysse wyrte wyrtruman þe we *satyrion* nemdon ond eac sume men *priapisci* hatað, ond cnuca tosomne. Hyt þa wunda aclænsað ond ða dolh gelycð.

16.2. Wiþ eagena sar; þæt is, þonne þæt hwa tornige sy: genim þysse ylcan wyrte seaw ond smyre ða eagan þærmid. Butan yldincge hyt of genimð þæt sar.

# 17

# Feldwyrt

Ðeos wyrt, þe man *gentianam* ond oðrum naman feldwyrt nemneþ, heo bið cenned on dunum, ond heo framað to eallum drenceom. Heo bið hnesce on æthrine ond bittere on byrgingce.

# Chapter 16

# Orchid

This plant, which is called *satyrion* and by another name or-chid, grows on high mountains and in places of hard ground, and likewise in meadows and in cultivated places and on sandy soils.

16.1. For painful wounds: take the roots of this same plant, which we have called *satyrion* and some people likewise call *priapiscus,* and pound them together. It cleanses the wounds and is good for scars.

16.2. For eye pain; that is, when a person is bleary-eyed: take the juice of this same plant and apply it to the eyes. It will immediately take the pain away.

# Chapter 17

# Gentian

This plant, which is called *gentiana* and by another name gentian, grows on hills, and it improves all drinks. It is soft to the touch and bitter to the taste.

17.1. Wið nædran slite: genim þysse ylcan wyrte *gentianam* wyrttruman ond gedrige hine; cnuca ðonne to duste anre tremese gewihte; syle drincan on wine þry scenceas. Hit fremað miclum.

# 18

# Slite

Ðeos wyrt, ðe man *orbicularis* ond oþrum naman slite nemneð, heo bið cenned on beganum stowum ond on dunlandum.

18.1. Wiþ þæt ðæt mannes fex fealle: genim þas ylcan wyrte ond do on þa næsþyrlu.

18.2. Wið innoþes styrunga: genim þas ylcan wyrte; wyrc to salfe; lege to ðæs innoðes sare. Eac heo wið heortece well fremað.

18.3. Wið miltan sare: genim þysse ylcan wyrte seaw anne scenc ond fif sticcan fulle ecedes; syle drincan nigon dagas. Þu wundrasð ðære gefremmincge. Genim eac ðære ylcan wyrte wyrtruman ond ahoh abutan þæs mannes swyran swa þæt he hangie forne gean ða miltan. Hrædlice he bið gehæled. Ond swa hwylc man þysse wyrte seaw þigeð, wundorlicre hrædnysse he ongit þæs innoðes liðunge. Þas wyrte man mæg niman on ælcne sæl.

17.1. For snakebite: take the root of this same plant, gentian, and dry it; pound one tremiss-weight of it into a powder; give three cups of it to drink in wine. It helps a great deal.

# Chapter 18

# Sowbread

This plant, which is called *orbicularis* and by another name sowbread, grows in cultivated places and in hilly areas.

18.1. If a person's hair is falling out: take this same plant and put it into the nostrils.

18.2. To move the bowels: take this same plant; work it into a salve; lay it where the abdomen is sore. It also is very good for heartburn.

18.3. For pain in the spleen: take one cup of the juice of this same plant, plus five spoonfuls of vinegar; give to drink for nine days. You will be surprised by its effects. Also take the root of this same plant and hang it around a person's neck so that it hangs in front against the spleen. He will quickly be healed. And whoever drinks the juice of this plant, he will experience relief in his abdomen with wonderful rapidity. This plant can be gathered at any time.

# 19

# Unfortrædde

Ðeos wyrt, ðe man *proserpinacam* ond oðrum naman unfortredde nemneð, heo bið cenned gehwær on beganum stowum ond on beorgum. Ðas wyrte ðu scealt on sumera nimen.

19.1. Wið þæt man blod spiwe: genim þysse wyrte seaw *proserpinace,* ond butan smice gewyl on swiðe godum ond strangum wine; drince þonne fæstende nigon dagas. Binnan þam fæce þu ongytst on þam wundorlic ðingc.

19.2. Wiþ sydan sare: genim þysse ylcan wyrte seaw mid ele, ond smyre gelomlice. Hit genimð þæt sar.

19.3. Wið titta sar wifa þe beoð melce ond toðundene: genim ða ylcan wyrte ond cnuca hy ond mid buteran geliðga; lege ðonne þærto. Heo todrifð wundorlice ða toðundennysse ond þæt sar.

19.4. Wið eagena sare: ær sunnan upgange, oððe hwene ær heo fullice gesigan onginne, ga to ðære ylcan wyrte *proserpinacam,* ond bewrit hy abutan mid anum gyldenan hringe, ond cweð þæt þu hy to eagena læcedome niman wylle. Ond æfter ðrim dagum ga eft þærto ær sunnan upgange ond genim hy ond hoh onbutan þæs mannes swyran. Heo fremað wel.

# Chapter 19

# Knotgrass

This plant, which is called *proserpinaca* and by another name knotgrass, grows everywhere in cultivated places and on hills. This plant should be gathered in summer.

19.1. If a person vomits blood: take the juice of this same plant knotgrass and simmer it in very good and strong wine, without letting it steam; have the patient drink this for nine days after fasting for the night. Within this time you will perceive a wonderful change in him.

19.2. For a pain in the side: take the juice of this same plant, along with some oil, and rub it on frequently. It will take away the pain.

19.3. For a woman's breasts, if they are sore and full of milk and swollen: take this same plant and pound it and soften it with butter; apply it to them. It will expel the swelling and the pain remarkably well.

19.4. For eye pain: before the sun rises, or just before it begins to set fully, go to this same plant knotgrass, and trace a line around it with a golden ring, and say that you want to pick it to make a medicine for the eyes. And after three days go again before sunrise and pick it and hang it around the person's neck. It will help a good deal.

19.5. Wið earena sar: genim þysse ylcan wyrte seaw, gewlæht; drype on þæt eare. Wundorlice hit þæt sar tofereð. Ond eac we sylfe efenlice ond glæwlice onfunden habbað þæt hit fremað; ond eac witodlice utene þæra earena sar gehælð.

19.6. Wið utsihte: genim þysse ylcan wyrte leafa seaw ond wyll on wætere; syle drincan þam gemete þe ðe þince. He bið hal geworden.

20

# Smerowyrt

Ðeos wyrt, þe man *aristolochiam* ond oðrum naman smerowyrt nemneð, heo bið cenned on dunlandum ond on fæstum stowum.

20.1. Wið attres strengðe: genim þas wyrte *aristolochiam* ond cnuca; syle drincan on wine. Heo oferswið ealle strengþe þæs attres.

20.2. Wiþ þa stiþustan feferas: genim ðas sylfan wyrte ond gedrige hy; smoca þonne þærmid. Heo afligð nalæs þone fefer, eac swylce deofulseocnyssa.

20.3. Wið næsðyrla sare: genim þysse ylcan wyrte wyrtruman ond do on þa næsðyrlu. Hrædlice hyt hi afeormað ond to hæle gelædeð. Witodlice ne magon læceas naht mycel hælan butan þisse wyrte.

19.5. For earache: take juice of this same plant, made luke-warm; drip it into the ear. It will take the pain away wonder-fully. And also we ourselves have found clearly and unequiv-ocally that it helps; and in addition, in truth, it heals sores located outside the ears.

19.6. For diarrhea: take juice of the leaves of this same plant and boil it in water; give it to the patient to drink in the quantity you think suitable. The person will be cured.

# Chapter 20

# Birthwort

This plant, which is called *aristolochia* and by another name birthwort, grows on hilly lands and in hard soils.

20.1. Against the power of poison: take this plant birthwort and pound it; give to drink in wine. It will overcome all the strength of the poison.

20.2. For the most violent fevers: take this same plant and dry it; fumigate the person with it. It will drive away not only the fever, but also possession by devils.

20.3. For sore nostrils: take the roots of this same plant and put them in the nostrils. It will quickly cleanse them and lead to healing. In truth, healers cannot cure much without this plant.

20.4. Wið þæt hwa mid cyle gewæht sy: genim þas ylcan wyrte, ond ele ond swinen smero; do tosomne. Þonne hæfð hit ða strengðe hyne to gewyrmenne.

20.5. Wið nædran slite: genim þysse ylcan wyrte wyrttruman tyn penega gewæge ond healfne sester wines; gewes tosomne; syle drincan gelomlice. Þonne tofereð hit þæt attor.

20.6. Gyf hwylc cyld ahwæned sy: þonne genim þu þas ylcan wyrte ond smoca hit mid. Þonne gedest ðu hit ðe glædre.

20.7. Wið þæt wærhbræde hwam on nosa wexe: genim þa ylcan wyrte, ond *cypressum* ond dracentsan ond hunig; cnuca tosomne; lege þærto. Ðonne bið hit sona gebet.

# 21

# Cærse

21.1. Wið þæt mannes fex fealle: genim þære wyrte seaw þe man *nasturcium* ond oðrum naman cærse nemneð; do on þa nosa; þæt fex sceal wexen. Ðeos wyrt ne bið sawen, ac heo of hyre sylfre cenned bið on wyllon ond on brocen. Eac hit awriten ys þæt heo on sumum landon wið wagas weaxen wylle.

21.2. Wið heafodsar, þæt ys, wið scurf ond wið gicðan: genim þysse ylcan wyrte sæd ond gose smeru; cnuca tosomne. Hit þa hwitnesse þæs scurfes of ðam heafde atyhð.

20.4. For a person weakened by severe cold: take this same plant, plus some oil and pig's grease; mix together. Then it will have the strength to warm him.

20.5. For snakebite: take ten pennyweights of the roots of this same plant and half a sester of wine; soak them together; give to drink often. It will then expel the poison.

20.6. If a child is afflicted: then take this same plant and fumigate him with it. In this way you will make him more cheerful.

20.7. If an ulcer should develop on a person's nose: take this same plant, plus cypress and dragonwort and honey; pound them together; apply this to it. It will soon be healed.

# Chapter 21

# Watercress

21.1. If a person's hair is falling out: take the juice of the plant that is called *nasturtium* or by another name watercress; put it in the nose; the hair will grow. This plant is not sown, but it grows naturally in springs and in brooks. Also it is also written that in some places it will grow next to walls.

21.2. For head sores, that is, for a scurvy scalp and itching: take the seeds of this same plant and some goose grease; pound them together. It draws the whiteness of scurf off the head.

21.3. Wið lices sarnysse: genim þas ylcan wyrte *nasturcium* ond polleian; seoð on wætere; syle drincan. Þonne gebetst ðu þæs lichoman sarnysse, ond þæt yfel tofærð.

21.4. Wið swylas: genim þas ylcan wyrte ond cnuca hy mid ele; lege ofer þa swylas. Nim ðonne þære ylcan wyrte leaf ond lege þærto.

21.5. Wið weartan: genim þas ylcan wyrte ond gyst; cnuca to-somne; lege þærto. Hy beoð sona fornumene.

## 22

## Greate wyrt

Ðeos wyrt, þe man *hieribulbum* ond oðrum naman greate wyrt nemneþ, heo biþ cenned abutan heogan ond on fulum stowum.

22.1. Wið liþa sare: genim þysse ylcan wyrte þe we *hieribul-bum* nemdun syx yntsan, ond gætenes smeruwes ðam be ge-licon, ond of *cypresso* þam treowcynne anes pundes gewihte eles ond twegea yntsa; cnuca tosomne, wel gemengced. Hit genimð þæt sar ge þæs innoðes ge þæra liða.

22.2. Gif nebcorn on wifmannes nebbe wexen: genim þysse sylfan wyrte wyrtruman ond gemengc wið ele; þwea syððan þærmid. Hyt afeormað of ealle þa nebcorn.

21.3. For pain in the body: take this same plant watercress, plus pennyroyal; simmer them in water; give to drink. Thus you will alleviate the body's discomfort, and the complaint will go away.

21.4. For swellings: take this same plant and pound it with some oil; lay it on the swellings. Then take leaves of this same plant and lay them on.

21.5. For warts: take this same plant and some yeast; pound them together; apply to the warts. They will soon be taken away.

# Chapter 22

# Autumn crocus

This plant, which is called *hieribulbum* and by another name autumn crocus, grows around hedges and in miry places of stagnant water.

22.1. For sore joints: take six ounces of this same plant that we have called autumn crocus, and the same amount of goat grease, and one pound and two ounces of oil of the cypress tree; pound them together, well mixed. It takes away the pain of both the internal organs and the limbs.

22.2. If spots should grow on a woman's face: take the roots of this same plant and mix them with oil; then wash them with it. It will clear away all the spots.

# 23

# Glofwyrt

Ðeos wyrt, þe man *apollinarem* ond oðrum naman glofwyrt nemneþ, ys sæd þæt Apollo hy ærest findan sceolde ond hy Esculapio þam læce syllan. Þanon he hyre þæne naman on asette.

23.1. Wið handa sare: genim þas ylcan wyrte *apollinarem;* cnuca hy mid ealdum smerwe, butan sealte; do þærto anne scænc ealdes wines, ond þæt sy gehæt butan smice, ond þæs smerwes sy anes pundes gewihte. Cnuca tosomne þam gemete þe ðu clyþan wyrce ond lege to þære handa.

# 24

# Mageþe

24.1. Wið eagena sare: genime man ær sunnan upgange ðas wyrte, þe man *camemelon* ond oðrum naman mageþe nemneð, ond þonne hy man nime cwepe þæt he hy wille wið flean ond wið eagena sare niman. Nyme syððan þæt wos ond smyrige ða eagan ðærmid.

# Chapter 23

# Poison gooseberry

This plant, which is called *apollinaris* and by another name poison gooseberry, is said to have been first discovered by Apollo, who gave it to Esculapius, the healer. He thus gave it this name.

23.1. For sore hands: take this same plant, poison gooseberry; pound it in old grease, without salt; add to it one cup of old wine, which should be heated without causing steam to rise, and there should be one pound of grease. Pound together in the same manner as you would make a poultice, and apply to the hands.

# Chapter 24

# Chamomile

24.1. For eye pain: the person should take this plant, which is called *chamaemelon* and by another name chamomile, before the sun rises, and when picking it the person should say that he is taking it for albugo and for eye pain. Afterward take the juice and anoint the eyes with it.

## 25

# Heortclæfre

Ðeos wyrt, þe man *chamedris* ond oðrum naman heortclæfre nemneð, heo bið cenned on dunum ond on fæstum landum.

25.1. Gyf hwa tobrysed sy: genim þas wyrte þe we *camedris* nemdon; cnuca hy on trywenum fæte; syle drincan on wine. Eac swylce to slite heo gehæleð.

25.2. Wið nædran slite: genim þas ylcan wyrte; cnuca hy swyþe smæl on duste; syle drincan on ealdum wine. Þearle hyt þæt attor todræfð.

25.3. Wið fotadle: genim þas ylcan wyrte; syle drincan on wearmum wine þam gemete þe we hær beforan cwædon. Wundorlice hyt þæt sar geliþegað ond þa hæle gegearwað. Þas wyrte þu scealt niman on þam monðe þe man Augustus nemneð.

# Chapter 25

# Germander

This plant, which is called *chamaedrys* and by another name germander, grows on hills and on areas of hard ground.

25.1. If a person is bruised: take the plant that we have called germander; pound it in a wooden vessel; give to drink in wine. Likewise it heals cuts.

25.2. For snakebite: take this same plant; pound it very fine into a powder; give to drink in old wine. It will expel the poison thoroughly.

25.3. For foot disease: take this same plant; give to drink in warm wine in the way we have just said. It will relieve the pain wonderfully and will restore health. You should gather the plant in the month called August.

## 26

# Wulfes camb

26.1. Wið liferseocnysse: genim þysse wyrte seaw þe man *chameaeleæ* ond oðrum naman wulfes camb nemneð; syle drincan on wine, ond fefergindum mid wearmum wætere. Wundurlice hyt fremað.

26.2. Wið attres drinc: genim þas ylcan wyrte; cnuca hy to duste; syle drincan on wine. Eal þæt attor tofærð.

26.3. Wið wæterseocnysse: genim þas ylcan wyrte, ond hræfnes fot ond heortclæfran ond henep, ealra ðissa wyrta gelice mycel be gewihte. Cnuca hy to smalon duste; syle þycgean on wine: geongum men fif cuceleras fulle, ond gingrum ond untrumrum ond wifum þry cuculeras, litlum cildum ane. Wundurlice he þæt wæter þurh micgðan forlæteð.

# Chapter 26

# Wild teasel

26.1. For liver disease: take juice of the plant that is called *chamelaea* and by another name wild teasel; give to drink in wine, and in warm water for the feverish. It helps wonderfully.

26.2. If someone has drunk poison: take this same plant; pound it to a powder; give to drink in wine. It will expel all the poison.

26.3. For dropsy: take this same plant, plus figwort and germander and ground pine, an equal amount of all these plants by weight. Pound them into thin powder; give to drink in wine: five spoonfuls for young men, three spoonfuls for youngsters, the ill, and women, and one spoonful for little children. It will expel the fluid remarkably well through the urine.

# 27

# Henep

27.1. Wið wundela: genim þas wyrte, þe man *chamepithys* ond oðrum naman henep nemneð; cnuca, ond lege to ðære wunde. Gyf þonne seo wund swyþe deop sy, genim þæt wos ond wring on ða wunda.

27.2. Wiþ innoðes sare: genim þas ylcan wyrte; syle drincan. Heo þæt sar genimð.

# 28

# Hrefnes fot

28.1. Wið innoð to astyrigenne: genim ðas wyrte, ðe Grecas *chamedafne* ond Engle hræfnes fot nemnað; cnuca to smælon duste; syle drincan on wearmum wætere. Hit ðone innoð astyreð.

# Chapter 27

# Ground pine

27.1. For wounds: take this plant, which is called *chamaepitys* and by another name ground pine; pound it, and apply it to the wound. If, however, the wound is very deep, take the juice and wring it into the wound.

27.2. For abdominal pain: take this same plant; give to drink. It will take the pain away.

# Chapter 28

# Figwort

28.1. To move the bowels: take this plant, which the Greeks call *chamaedaphne* and the English call figwort; pound it into a thin powder; give to drink in warm water. It will move the bowels.

# 29

# Lyðwyrt

Ðeos wyrt, þe man *ostriago* ond oðrum naman lyðwyrt nemneð, bið cenned abutan byrgenne ond on beorgum, ond on
wagum þæra husa þe wið duna standað.

29.1. Wiþ ealle ðingc ðe on men to sare acennede beoð:
genim þas wyrte þe we *ostriago* nemdon ond cnuca hy; lege
to ðam sare. Ealle þa þincg, swa we ær cwædon, þe on ðæs
mannes lichoman to laðe acennede beoð heo ðurhhæleð.

2      Gif ðu þas wyrte niman wylle, ðu scealt clæne beon, ond
eac ær sunnan upgange þu hy scealt niman on ðam monðe þe
man Iulius nemneð.

# 30

# Hæwenhydele

30.1. Wið muðes sare: genim þas wyrtc, þc Grecas *brittanice*
ond Engle hæwenhydele nemneð; cnuca hy swa grene, ond
wring þæt wos. Syle supan ond healde swa on his muðe. Ond
þeah man hwylcne dæl þærof swelge, gelice hit fremað.

# Chapter 29

# Madder

This plant, which is called *ostriago* and by another name madder, grows around burial places and on mounds, and on the walls of houses that are built against hills.

29.1. For all things that grow on a person to his discomfort: take the plant that we have called madder and pound it; apply it to the sore. It will thoroughly heal all the conditions that, as we have said before, have grown in a person's body to his annoyance.

If you want to gather this plant, you must be free of defilement, and in addition you must pick the plant before sunrise in the month we call July.

2

# Chapter 30

# Water dock

30.1. For a mouth sore: take this plant, which the Greeks call *britannica* and the English call water dock; pound it when green, and press out the juice. Give it to the person to drink, but have him keep it in his mouth. And even if he should swallow some, it still will help.

30.2. Eft, wið muþes sare: genim þa ylcan wyrte *bryttanicam;* gyf ðu hy grene næbbe, genim hy dryge. Cnuca mid wine on huniges þicnysse; nim ðonne þam sylfan gemete þe we ær cwædon. Heo hæfð þa sylfan gefremmincge.

30.3. Wið toþa sare, ond gyf hy wagegen: genim þas ylcan wyrte; heo of sumre wundurlicre mihte helpeð. Hyre wos ond hyre dust ys to gehealdenne on wintre for ðam þe heo ælcon timan ne atyweð. Hyre wos þu scealt on rammes horne gehealdan; drige eac þæt dust ond geheald. Witodlice eac hyt scearplice fremað to ðam sylfan bryce mid wine onbyrged.

30.4. Wið fæstne innoð to styrigenne: genim þisse ylcan wyrte seaw; syle drincan be þære mihte þe hwa mæge þurh hit self. Butan frecnesse hit afeormað wundurlice ðone in-noð.

30.5. Wið sidan sare þæt Grecas *paralisis* nemnað: genim þas ylcan wyrte, swa grene, mid wyrttrumum; cnuca hy; syle drincan on wine, twegen scenceas oððe ðry. Hyt is gelyfed þæt heo wundurlice fremige.

30.2. Again, for a mouth sore: take this same plant water dock; if you do not have it fresh, take it dried. Pound it in wine until it is the thickness of honey; then take it in the same way as we said before. It has the same effects.

30.3. For toothache, and if the teeth are loose: take this same plant; it will help thanks to some wonderful power in it. Its juice and its powder are to be preserved through the winter because it is not available at all times. You must keep its juice in a ram's horn; also you should dry its powder and preserve it. In truth, it is perfectly effective for the same purpose when taken in wine.

30.4. To move constipated bowels: take the juice of this same plant; give to drink according to the strength that each patient can tolerate. Without posing any danger it will clear out the bowels wonderfully.

30.5. For the pain in the side the Greeks call *paralysis:* take this same plant, fresh, along with its roots; pound it; give to drink in wine, two or three cupfuls. It is believed that it helps wonderfully.

# 31

# Wudulectric

Ðeos wyrt, þe man *lactucan silfaticam* ond oðrum naman wudulectric nemneð, bið cenned on beganum stowum ond on sandigum.

31.1. Wið eagena dymnesse ys sæd þæt se earn, þonne he up fleon wille to þy þæt he þy beorhtor geseon mæge, þæt he wylle mid þam seawe his eagan hreppan ond wætan. Ond he þurh þæt onfehð þa mæstan beorhtnesse.

31.2. Eft, wið eagena dymnysse: genim þysse ylcan wyrte seaw þe we *lactucam silfaticam* nemdon, mid ealdon wine ond mid hunige gemencged (ond þis sy butan smice gesomnud). Þæt bið selust þæt man þysse wyrte seaw, swa we ær cwædon, ond win ond hunig gemengce tosomne ond on anre glæsenre ampullan gelogige. Bruce þonne him þearf sy; of ðam þu healicne læcedom ongitst.

# Chapter 31

# Prickly lettuce

This plant, which is called *lactuca silvatica* and by another name prickly lettuce, grows in cultivated places and in sandy soils.

31.1. For dimness of the eyes: it is said that when the eagle wants to fly on high, in order to see more clearly it will touch and bathe its eyes with this juice. And by that means it gains the greatest clarity of vision.

31.2. Again, for dimness of the eyes: take the juice of this same plant that we have called prickly lettuce, mixed with old wine and with honey (and this should be collected without smoke). It is best, as we said before, if the juice of this plant is mixed together with wine and honey and saved in a glass bottle. It may be used when someone has need of it; from this you will recognize this to be a remarkable remedy.

# 32

# Garclife

32.1. Wið eagena sare: genim þas wyrte, þe man *argimoniam* ond oðrum naman garclife nemneð; cnuca hy, swa grene, þurh hy selfe. Gyf ðu hy þonne grene næbbe, genim hy drige ond dype on wearmum wætere swa þu eaþelicost hy brytan mæge; smyra þonne þærmid. Ofstlice heo ða tale ond þæt sar of þam eagan adrifð.

32.2. Wið innoðes sare: genim þysse ylcan wyrte wyrtruman þe we *argimoniam* nemdon; syle drincan. Hyt fremað wundorlice.

32.3. Wið cancor ond wið wundela: genim þas ylcan wyrte, swa grene; cnuca hy; lege to þam sare. Gecwemlice heo þone leahtor gehælan mæg. Gyf ðonne seo wyrt drigge sy, dype hy on wearmum wætere; hyt ys gelyfed þæt heo to ðam ylcan fremige.

32.4. Wið nædran slite: genim þysse ylcan wyrte, twegra trymesa gewihte, ond twegen scenceas wines; syle drincan. Wundurlice hyt þæt attor tofereð.

32.5. Wið weartan: genim þas ylcan wyrte; cnuca on ecede; lege þærto. Heo genimð þa weartan.

32.6. Wið miltan sare: genim þas ylcan wyrte; syle þicgean on wine. Heo þæt sar fornimð þære miltan.

# Chapter 32

# Agrimony

32.1. For eye pain: take this plant, which is called *argemonia* and by another name agrimony; pound it, fresh, by itself. If you do not have it fresh, take it dry and dip it in warm water so that you can crush it most easily; then anoint the eyes with it. It will quickly take away the defect and the pain from the eyes.

32.2. For abdominal pain: take the roots of this same plant that we call agrimony; give to drink. It helps wonderfully.

32.3. For ulcerous sores and wounds: take this same plant, fresh; pound it; apply it to the sore. It can cure the disorder easily. If however the plant is dry, dip it into warm water; it is believed that it will do good in the same way.

32.4. For snakebite: take two tremiss-weights of this same plant and two cups of wine; give to drink. It will expel the poison wonderfully.

32.5. For warts: take this same plant; pound it in vinegar; apply it. It takes away the warts.

32.6. For pain in the spleen: take this same plant; give to eat with wine. It will take away the pain from the spleen.

32.7. Gif ðu hwylce þingc of ðam lichoman ceorfan wylle ond ðe þonne þince þæt ðu ne mæge: genim þas ylcan wyrte, gecnucude; lege þærto. Heo hyt geopenað ond gehæleð.

32.8. Wið slege isernes oððe stenges: þeos ylce wyrt, gecnucud ond to gelæd. Heo wundurlice gehæleþ.

# 33

# Wudurofe

33.1. Wið sceancena sare oððe fota: genim þysse wyrte seaw, þe man *astularegia* ond oðrum naman wudurofe nemneð, mid amigdales ele; smyre þær þæt sar sy. Hyt bið wundorlice gehæled. Ond gyf hyt geswell sy, cnuca hy ond wel geliðegode; lege þærto.

33.2. Wið lifre sare: genim þysse sylfan wyrte wyrtruman; syle drincan on geswetton wætere. Hit þæt sar wundorlice of genimð.

32.7. If you want to cut anything from the body but you think you cannot do it: take this same plant, pounded; lay it on the place. It will open it and heal it.

32.8. For a blow from an iron blade or a cudgel: this same plant, pounded and applied to the wound. It will heal it wonderfully.

# Chapter 33

# Asphodel

33.1. For pain in the shanks or in the feet: take the juice of this plant, which is called *hastula regia* and by another name asphodel, along with some almond oil; rub it on where the pain is. It will be wonderfully healed. And if it is a case of swelling, pound the plant and make it very soft; lay it on the swelling.

33.2. For pain in the liver: take the roots of this same plant; give it to drink in sweetened water. It will take the pain away wonderfully.

# 34

# Wududocce

34.1. Gyf hwylc stiðnes on lichoman becume: genim þas wyrte, þe man *lapatium* ond oðrum naman wududocce nemneð, ond eald swynen smeru ond ðone cruman of ofenbacenum hlafe; cnuca tosomne þam gemete ðe ðu clyðan wyrce; lege to ðam sare. Hyt gehælð wundorlice.

# 35

# Eorðgealla oððe curmelle

35.1. Wið liferadle: genim þas wyrte, þe Grecas *centauria maior* ond Angle curmelle seo mare nemnað (ond eac sume men eorðgeallan hatað); seoð on wine; syle drincan. Wundorlice heo gestrangað. Ond wið miltan sare do þis sylfe.

35.2. Wið wunda ond wið cancor: genim þas ilcan wyrte; cnuca hy; lege to þam sare. Ne geþafað heo þæt ðæt sar furður wexe.

# Chapter 34

# Common sorrel

34.1. If a hardening should develop on the body: take this plant, which is called *lapathum* and by another name common sorrel, plus old pig's grease and crumbs of bread baked in the oven; pound them together in the same way as you would make a poultice; apply this where there is pain. It will heal it wonderfully.

# Chapter 35

# Yellowwort

35.1. For liver disease: take this plant, which the Greeks call *centaurea maior* and the English greater centaury (and some people also call it yellowwort); simmer in wine; give to drink. It strengthens the liver wonderfully. And do the same for pain in the spleen.

35.2. For wounds and ulcerous sores: take this same plant; pound it; lay it on the sore. It does not permit the sore to spread.

2  Ðeos sylfe wyrt, *centauria,* ys swyþe scearpnumul niwe wunda ond wide to gehælenne swa þæt þa wunda hrædlice togædere gað; ond eac swa some heo gedeþ þæt flæsc togædere geclifað gyf hyt man on þam wætere gesygð þe heo on bið.

# 36

# Curmelle; feferfuge

Ðeos wyrt, þe man *centauriam minorem* ond oðrum naman curmelle seo læssæ nemneð (ond eac sume men *febrifugam* hatað), heo bið cenned on fæstum landum ond on strangum. Eac ys sæd þæt Chyron centaurus findan sceolde þas wyrta þe we ær *centauriam maiorem* ond nu *centauriam minorem* nemdon; ðanun hy eac þone naman healdað *centaurias.*

36.1. Wið nædran slite: genim þysse ylcan wyrte dust, oððe hy sylfe gecnucude; syle drincan on ealdum wine. Hyt fremað swyðlice.

36.2. Wið eagena sare: genim þysse ylcan wyrte seaw; smyra ða eagan þærmid. Hyt gehælð þa þynnysse þære gesihðe. Gemencg eac hunig þærto. Hyt fremað swa some witodlice dimgendum eagum to þy þæt seo beorhtnys agyfen sy.

36.3. Gyf hwa þonne on þas frecnysse befealle: genim þysse ylcan wyrte godne gripan; seoð on wine oððe on ealoð swa

This same plant, greater centaury, is so efficacious in heal- 2
ing fresh wounds and wide ones that the wounds quickly
close; and also it causes the flesh to knit together if one
soaks the wound with the water containing the plant.

# Chapter 36

# Lesser centaury

This plant, which is called *centaurea minor* and by another
name lesser centaury (and which some people also call *febri-
fugia*), grows in compact and hard soils. It is also said that
Chiron the centaur discovered this plant, which we previously called *centaurea maior* and now call *centaurea minor;*
from him these plants have the name centaury.

36.1. For snakebite: take the powder of the same plant, or
take it pounded by itself; give to drink in old wine. It does a
great deal of good.

36.2. For eye pain: take the juice of this same plant: apply it
to the eyes. It heals weakness of vision. Mix it with honey,
too. Likewise, it will certainly help failing eyesight so much
that clarity is restored.

36.3. If someone is subject to that same danger: take a good
handful of this same plant; simmer it in wine or in ale such

þæt þæs wines sy an ambur full; læt standan þry dagas. Nim þonne æghwylce dæge, þonne ðearf sy, healfne sester; mengc mid hunige; drince ðonne fæstende.

36.4. Wið sina togunge: genim þas ylcan wyrte; seoð on wætere to þriddan dæle; syle drincan swa mycel swa he þonne mæge ond þearf sy. He bið gehæled.

36.5. Wið attres onbyrgingce: genim þas ilcan wyrte; cnuca on ecede; syle drincan. Sona hit þæt attor todrefð. Eac þære sylfan wyrte wyrtruman genim tyn penega gewihte; do on wine; syle drincan þry scenceas.

36.6. Wið þæt wyrmas ymb nafolan dergen: do ealswa we her beforan cwædon wið syna togunge; þæt ys ðonne þæt ðu genime þas ylcan wyrte; seoð on wætere to ðriddan dæle. Heo ða wyrmas ut awyrpð.

# 37

# Bete

37.1. Wið ealle wunda ond wið næddran slitas: genim þysse wyrte seaw, þe man *personaciam* ond oðrum naman bete nemneð; syle drincan on ealdon wine. Ealle nædran slitas hyt wundurlice gehæleð.

that there is a jugful of wine; let it stand for three days. Then every day, when needed, take half a sester; mix it with honey; have him drink it while fasting.

36.4. For sinew spasm: take this same plant; simmer it in water, reducing it by two-thirds; give the person as much to drink as he can and as he needs. He will be cured.

36.5. For the consumption of poison: take this same plant; pound it in vinegar; give to drink. It will soon take away the poison. Also take ten pennyweights of the root of this same plant; put this in wine; give three cupfuls to drink.

36.6. If worms should be doing harm around the navel: do as we have just said as regards sinew spasms; that is, that you should take this same plant; simmer it in water, reducing it by two-thirds. It will expel the worms.

# Chapter 37

# Beet

37.1. For all wounds and for snakebite: take the juice of this plant, which is called *personacia* and by another name beet; give to drink in old wine. It will heal the bites of all snakes remarkably well.

37.2. Wið feferas: genim þysse ylcan wyrte leaf; begyrd to þam fefergendan. Sona hyt wundorlice ðone fefer afligeð.

37.3. Wið þæt cancor on wunde wexe: genim þas wyrte; wyll on wætere; beþe þa wunde ðærmid. Syððan genim þa wyrte, ond sapan ond smeru; cnuca mid ecede; do þonne on clað; lege to ðære wunde.

37.4. Wið innoðes sare: genim þysse ylcan wyrte seawes, anne scenc, ond huniges twegen; syle drincan fæstendum.

37.5. Wið wedehundes slite: genim þysse ilcan wyrte wyrtruman; cnuca mid greatan sealte; lege to ðam slite.

37.6. Wið niwe wunda, þa þe þone wætan gewyrceaþ: genim þisse ylcan wyrte wyrttruman ond hægðornes leaf, ægþres efenmycel; cnuca tosomne; lege to ðam wundum.

# 38

# Streowberian wise

Ðeos wyrt, ðe man *fraga* ond oðrum naman streawbergean nemneð, bið cenned on dihglum stowum ond on clænum, ond eac on dunum.

38.1. Wið miltan sare: genim þysse ylcan wyrte seaw þe we

37.2. For fevers: take the leaves of this same plant; bind them on the feverish person. Soon it will wonderfully put the fever to flight.

37.3. If an ulcerous sore should develop on a wound: take this plant; boil it in water; bathe the wound with it. Then take the plant, along with some soap and some grease; pound in vinegar; then put in a cloth; apply to the wound.

37.4. For abdominal pain: take the juice of this same plant, a single cupful, and two cups of honey; give it to the patient to drink while fasting.

37.5. For the bite of a mad dog: take the roots of this same plant; pound, along with some coarse salt; apply to the bite.

37.6. For fresh wounds, those that seep fluid: take the roots of this same plant and some hawthorn leaves, in an equal amount; pound them together; apply to the wounds.

# Chapter 38

# Wild strawberry

This plant, which is called *fraga* and by another name wild strawberry, grows in out-of-the-way places and in cleared land, and also on hills.

38.1. For pain in the spleen: take the juice of this same plant

*fragam* nemdon, ond hunig; syle drincan. Hyt fremað wundurlice.

38.2. Ðysse ylcan wyrte seaw, wið hunig gemengced mid pipere, fremað myclum, gedruncen, wið nyrwyt ond wið innoðes sare.

# 39

# Merscmealuwe

Ðeos wyrt, þe man *hibiscum* ond oðrum naman merscmealwe nemneð, bið cenned on fuhtum stowum ond on feldum.

39.1. Wið fotadle: genim þas wyrte þe we *hibiscum* nemdon; cnuca mid ealdum rysle; lege to ðam sare. Þy þryddan dæge heo hit gehælð. Þysse wyrte onfundelnysse manega ealdras geseðað.

39.2. Wiþ æghwylce gegaderunga þe on þam lichoman acenned beoð: genim þas ylcan wyrte; seoð mid wyllecærsan ond mid linsæde ond mid melwe; lege to þam sare. Hit tofereð ealle þa stiðnyssa.

that we have called wild strawberry, plus honey; give to drink. It will be wonderfully beneficial.

38.2. The juice of this same plant, drunk with honey mixed with pepper, helps greatly for shortness of breath and abdominal pain.

# Chapter 39

# Marshmallow

This plant, which is called *hibiscus* and by another name marshmallow, grows in damp places and in fields.

39.1. For foot disease: take this plant that we have called marshmallow; pound in old fat; apply to the sore. On the third day it will be healed. Many authorities attest to the proven efficacy of this plant.

39.2. For any accumulations of matter that have grown on the body: take this same plant; simmer with fenugreek and with linseed and with flour; apply this to the sore. It will take away all the hard matter.

# 40

## *Ippirus*

40.1. Wið þæt mon on wambe forwexen sy: genim þysse wyrte seaw, þe Grecas *ippirum* ond Itali *æquiseiam* nemnað, on geswettum wine; syle drincan twegen scenceas. Wel ys gelyfed þæt hyt þæt yfel gehæle.

40.2. Gyf hwa blod swiþe hræce: genime ðysse ylcan wyrte seaw; seoðe on strangum wine butan smice; drince þonne fæstende. Sona hyt þæt blod gewrið.

# 41

## Hocleaf

Ðeos wyrt, þe man *malve erratice* ond oðrum naman hocleaf nemneð, byð cenned æghwær on beganum stowum.

41.1. Wið blædran sare: genim þysse wyrte þe we *malvam erraticam* nemdon, mid hyre wyrtruman, anes pundes gewihte; seoð on wætere þearle to healfan dæle, ond ðæs wæteres sy sester ful oððe mare, ond þæt sy binnan þrim dagum

# Chapter 40

# Horsetail

40.1. If a person has a bloated stomach: take the juice of this plant, which the Greeks call *hippirum* and the Italians call *equisaeta*, in sweetened wine; give two cups of it to drink. It is firmly believed that it will heal the complaint.

40.2. If a person should cough up much blood: take the juice of this same plant; simmer in strong wine without letting it steam; have him drink it while fasting. It will soon staunch the blood.

# Chapter 41

# Mallow

This plant, which is called *malva erratica* and by another name mallow, grows everywhere in cultivated places.

41.1. For bladder pain: take one pound-weight of the plant that we call mallow, along with its roots; boil it vigorously in water until it is reduced by one half, and there should be one sester-full of water or more, and this should be simmered

gewylled, swa we ær cwædon, to healfan dæle; syle drincan
fæstendum. Hyt hyne gehæleð.

41.2. Wið sina sare: genim þas ilcan wyrte; cnuca mid ealdun
rysle. Hyt þæra sina sar wundorlice gehæleð.

41.3. Wið sidan sare: genim þas ylcan wyrte; seoð on ele; ond
syððan þu hy gesoden hæbbe ond togædere gedon, genim
þonne þa leaf; cnuca on anum mortere; do ðonne on anne
cla$; lege þærto swa þæt ðu hyt þrim dagum ne unbinde. Þu
þæt sar gebetst.

41.4. Wið niwe wunda: genim þysse ylcan wyrte wyrttruman;
bærn to duste; do on þa wunda.

# 42

# Hundes tunge

Ðeos wyrt, þe Grecas *buglossam* ond Romane *lingua bubula*
nemnað ond eac Engle glofwyrt ond oðrum naman hundes
tunge hatað, heo bið cenned on beganum stowum ond on
sandigum landum.

42.1. Gif hwylcum men sy þæs þriddan dæges fefer oððe þæs
feorðan: genim þone wyrttruman þysse wyrte ðonne heo
hæbbe þry bogas ðæs sædes; seoð þone wyrttruman on
wætere; syle drincan. Þu hyne gelacnast.

down within three days by half, as we have said; give it to the patient to drink while fasting. It will heal him.

41.2. For sore sinews: take this same plant; pound with some old fat. It will cure the pain of the sinews wonderfully.

41.3. For pain in the side: take this same plant; simmer it in oil; and after you have simmered it and mixed it together, then take the leaves; pound them in a mortar; then place them in a cloth; apply them and do not take this off for three days. You will alleviate the pain.

41.4. For fresh wounds: take the roots of this same plant; burn them to ashes; put this on the wounds.

# Chapter 42

# Bugloss

This plant, which the Greeks call *buglossa* and the Romans *lingua bubula* and in addition the English call bugloss and by another name hound's tongue, grows in cultivated places and in sandy soils.

42.1. If a person has the fever that returns every third or every fourth day: take the root of this plant when it has three pods of seeds; then simmer the root in water; give to drink. You will cure him.

2    Seo eac ðe hæfð þæs sædes feower bogas fremað þam ge-lice þe we her beforan cwædon.

3    Ðonne ys oþer wyrt þysse gelic; seo hæfð sume dæle læs-san leaf ðonne docce. Þære wyrte wyrttruma on wætere geðyged wiðræð iceom ond næddrum.

42.2. Wið nyrwyt: genim þas ylcan wyrte, ond hunig ond hlaf þe sy mid smeruwe gebacen, þam gelice þe þu clyðan wyrce. Wundorlice hyt þæt sar toslit.

# 43

# Glædene

43.1. Wið wæterseocnysse: genim þas wyrte, þe man *bulbiscil-litici* ond oðrum naman glædene nemneð, ond gedryge hy syððan eal onbutan. Genim þonne innewearde; seoð on wætere; ðonne hyt wearm sy, gemengc eac þærto hunig ond eced; syle þry scenceas fulle. Swyðe hraðe sceal seo seocnys beon ut atogen þurh migðan.

43.2. Wið liþa sare: genim þas ylcan wyrte swa we ær cwædan innewearde; wyll on ele; smyra þæt sar ðærmid. Sona hyt fre-mað.

43.3. Wið þa adle þe Grecas *paronichias* nemnað: genim þysse ylcan wyrte wyrttruman; cnuca mid ecede ond mid hlafe; lege to þam sare. Wundorlice hyt hy gehæleð.

Also, the plant that has four pods will be of benefit in the 2
same way as we have just said.

Then there is another plant similar to this one; it has 3
somewhat smaller leaves than dock. The root of this plant
when consumed in water will counteract frogs and snakes.

42.2. For shortness of breath: take this same plant, along
with honey and bread that has been baked with grease, in
the same way as if you were making a poultice. It will elimi-
nate the pain wonderfully.

# Chapter 43

# Squill

43.1. For dropsy: take this plant, which is called *bulbus scilliticus* and by another name squill, and then dry it all over. Then
take its inner part; simmer in water; when it is warm, mix in
honey and vinegar; give three cupfuls to drink. The sickness
will be drawn out very quickly through the urine.

43.2. For sore joints: take the inner part of this plant, as we
said before; simmer in oil; rub the sore area with it. It will
soon help.

43.3. For the disease the Greeks call *paronykia:* take the roots
of this same plant; pound them, along with vinegar and
bread; apply to the sore area. It will cure the disease wonderfully.

43.4. Wið þæt man ne mæge wæterseoces mannes þurst gecelan: genim þysse sylfan wyrte leaf; lege under þa tungan. Sona heo þone þurst forbyt.

# 44

## *Umbilicum*

Ðeos wyrt, ðe Grecas *cotiledon* ond Romane *umbilicum veneris* nemnað, byð cenned on hrofum ond on beorgum.

44.1. Wið swylas: genim þas wyrte ond swinen smeru—wifum swaðeah ungesylt—ægþres gelice micel be wihte; cnuca tosomne; lege to þam swylum. Hyt hy tofereð. Þas wyrte þu scealt niman on wintertide.

43.4. If one cannot quench the thirst of a person suffering from dropsy: take a leaf of this same plant; put it under the tongue. It will soon slake the thirst.

# Chapter 44

# Navelwort

This plant, which the Greeks call *cotyledon* and the Romans *umbilicus veneris,* grows on rooftops and on mounds.

44.1. For swellings: take this plant and some pig's grease—unsalted, however, when it is for women—an equal amount of each by weight; pound them together; apply to the swellings. It will take them away. You should gather this plant in wintertime.

# 45

# Attorlaðe

Ðeos wyrt, þe man *gallicrus* ond oðrum naman attorlaðe nemneð, bið cenned on fæstum stowum ond wið wegas.

45.1. Wiþ hundes slite: genim þas wyrte; cnuca mid rysle ond mid heorðbacenum hlafe; lege to ðam slite. Sona hyt bið gehæled. Eac þys sylfe fremað wið heard geswell, ond hyt eal tofereð.

# 46

# Harehune

46.1. Wið geposu ond wið þæt man hefelice hræce: genim ðas wyrte, ðe Grecas *prassion* ond Romane *marubium* nemnað, ond eac Angle harehune hatað; seoð on wætere; syle drincan þam ðe hefelice hræcen. Heo hine gehæleð wundorlice.

46.2. Wið magan sare: genim þysse ylcan wyrte seaw; syle drincan. Hyt þæs magan sar fram adeð. Ond gif him fefer

# Chapter 45

# Cockspur grass

This plant, which is called *gallicrus* and by another name cockspur grass, grows in hard soils and along roads.

45.1. For dog bite: take this plant; pound it together with some fat and with bread baked on the hearth; apply it to the bite. It will soon be healed. This same plant is also good for a hard swelling, and it will take it all away.

# Chapter 46

# Horehound

46.1. For a cold in the head and for severe coughing: take this plant, which the Greeks call *prasion* and the Romans *marrubium,* and which in addition the English call horehound; simmer it in water; give it to drink to those who are coughing violently. It will heal the person wonderfully.

46.2. For stomachache: take the juice of this same plant; give to drink. It will take the stomachache away. And if someone

derige, syle him þas ylcan wyrte wel drincan on wætere. Heo hyne arærð.

46.3. Wið rengwyrmas abutan nafolan: genim þas ylcan wyrte *marubium,* ond wermod ond elehtran, ealra þyssa wyrta gelice fela be gewihte. Seoð on geswetton wætere ond mid wine; twie oððe þriwa lege to þam nafolan. Hit cwelð þa wyrmas.

46.4. Wið liþa sare ond wið geþind: genim þas ylcan wyrte; bærn to ahsan; do to þam sare. Sona hit gehælð.

46.5. Wið attres ðigne: genim þysse ylcan wyrte wos; syle on ealdum wine drincan. Sona þæt attor tofærð.

46.6. Wið sceb ond wið teter: genim þas ylcan wyrte; seoð on wætere; ðweh þone lichoman þærmid þær þæt sar sy. Heo of genimð þone scruf ond þone teter.

46.7. Wið lungenadle: genim þas ylcan wyrte; seoð on hunige; syle þiggean. He bið wundorlice gehæled.

46.8. Wið ealle stiðnessa þæs lichoman: genim þas ylcan wyrte; cnuca mid rysle; lege to þam sare. Heo hælð wundorlice.

is suffering from fever, give him this same plant to drink in water; it will revive him.

46.3. For harmful worms around the navel: take this same plant horehound, plus wormwood and lupin, all these plants in the same amount by weight. Simmer in sweetened water and with wine; apply it to the navel two or three times. It will kill the worms.

46.4. For pain in the joints and for swelling: take this same plant; burn to ashes; put on the sore. It will soon heal it.

46.5. For ingesting poison: take the juice of this same plant; give to drink in old wine. Soon the poison will depart.

46.6. For scabies and impetigo: take this same plant; simmer in water; wash the body with it where there are sores. It will remove the scabies and the impetigo.

46.7. For lung disease: take this same plant; simmer in honey; give to eat. The person will be cured in a wonderful manner.

46.8. For all kinds of hardness of the body: take this same plant; pound it in fat; lay it on the sore. It will cure wonderfully.

# 47

# Foxes fot

47.1. Wið uncuðe springas þe on lichoman acennede beoð: genim þysse wyrte wyrttruman, þe man *xifion* ond oðrum naman foxes fot nemneð, þreora yntsena gewihte, ond smedman six yntsena gewihte, ecedes twegean scenceas, ond foxes smeoruwes ðreora yntsena gewihte; cnuca tosomne on wine; dec þonne anne clað þærof; lege to ðam sare. Þu wundrast þære lacnunge.

47.2. Wið heafodbryce: genim þas ylcan wyrte ufewerde; gedryge hy ond cnuca. Genim þonne be gewihte efenmycel wines; meng tosomne; lege to þam sare. Hyt ðonne þa forbrocenan ban ut atyhð. Eac gif hwæt on þam lichoman dergende byð, hyt wel wið þæt fremað; oððe gif hwa mid his fet ofstepð ættrig ban snacan oððe næddran, ðeos sylfe wyrt is swyþe scearpnumul wið þæt attor.

# 47

# Sword grass

47.1. For strange carbuncles that grow on the body: take three ounces of the root of this plant which is called *xiphion* and by another name sword grass, and six ounces of fine flour, two cups of vinegar, and three ounces of fox's grease; pound them together in wine; daub a cloth in it; lay it on the sore. The healing will amaze you.

47.2. For a fractured skull: take the upper part of this plant; dry it and pound it. Then take the same amount by weight of wine; mix them together; apply to the sore. It will draw out the fractured bone. Also if anything in the body is doing harm, it will be of use for this; or if someone treads with his foot on the poisonous bones of a snake or an adder, this same plant is very efficacious for this poison.

# 48

# Wæterwyrt

48.1. Gyf swylas fæmnum derigen: genim ðas wyrte, þe man *gallitricum* ond oðrum naman wæterwyrt nemneð; cnuca hy syndrige; lege to þam sare. Heo hyt hælð.

48.2. Gif mannes fex fealle: genim þas ylcan wyrte; cnuca on ele; smyra ðonne þæt fex þærmid. Hyt sona bið fæst.

# 49

# Syngrene

Ðeos wyrt, þe man *temolum* ond oðrum naman singrene nemneð, þæs þe Omerus sægð ys wyrta beorhtust ond þæt Mercurius hy findan sceolde. Ðysse wyrte wos ys swyðe fremful, ond hyre wyrttruma ys synewealt ond sweart, eac on ðære mycele þe leaces.

49.1. Wiþ cwiþan sare: genim þas wyrte; cnuca, ond lege þærto. Heo geliþegað þæt sar.

# Chapter 48

# True maidenhair

48.1. If swellings should be hurting virgins: take this plant which is called *callitrichon* or by another name true maiden-hair; pound it by itself; lay it on the sore. It will heal it.

48.2. If a person's hair is falling out: take this same plant; pound in oil; smear the hair with it. Soon the hair will remain in place.

# Chapter 49

# Garlic

This plant, which is called moly and by another name garlic, is, according to what Homer says, the most excellent of plants, and Mercury is said to have discovered it. The juice of this plant is extremely beneficial, and its root is round and black, and it is also the same size as a leek.

49.1. For pain in the womb: take this plant; pound it, and apply it. It will relieve the pain.

## 50

# Sigelhweorfa

Ðeos wyrt, þe Grecas *æliotrophus* ond Romane *vertamnum* nemnað ond eac Angle sigelhweorfa hatað, byþ cenned ge-hwar on beganum stowum ond on clænum ond eac on mæ-dum. Þeos wyrt hæfð mid hyre sume wundorlice godcund-nysse, þæt is þonne þæt hyre blostman hig æfter þare sunnan ryne wændað, swa þæt þa blosman, þonne seo sunne gesihð, hy sylfe beclysað, ond eft þonne heo up gangeð, hy selfe ge-openiað ond tobrædaþ. Ond heo fremað to þyssum læcedo-mum þe we her wiðæftan awryten habbað.

50.1. Wið ealle attru: genim þas sylfan wyrte; cnuca to swiðe smalon duste, oððe hyre wos syle drincan on godum wine. Wundorlice heo þæt attor tofereð.

50.2. Wiþ flewsan: genim ðysse ylcan wyrte leaf; cnuca, ond lege to ðam sare. Hyt is gelyfed þæt heo scearplice gehæle.

# Chapter 50

# Heliotrope

This plant, which the Greeks call *heliotropium* and the Romans *vertumnum* and in addition the English call heliotrope, grows everywhere in cultivated places and in land that has been cleared of overgrowth, and also in meadows. This plant has a certain wonderful divine quality, namely, that its blossoms change according to the course of the sun, so that when the sun sets, they close themselves, and in turn when the sun rises, they open themselves and expand. And it is of use in the remedies that we have written here in the following passages.

50.1. For all poisons: take this same plant; pound it into fine powder, or give its juice to drink in good wine. It will expel the poison in a remarkable fashion.

50.2. For flux: take the leaves of this same plant; pound, and apply to the sore area. It is believed that this will cure it efficaciously.

# 51

# Mædere

Ðeos wyrt, þe man *gryas* ond oðrum naman mædere nemneð, byð cenned fyrmust in Lucania. Heo hæfð hwites mar-
man bleoh, ond heo bið gefrætewud mid feower readum
stælum.

51.1. Wið banece ond wið banbryce: genim þas ylcan wyrte;
cnuca hy; lege to þam bane. Þy þriddan dæge him bið sel,
swylce þær clyþa to gelæd wære.

51.2. Eac þysse wyrte wyrttruma fremað wið ælc sar þe þam
lichoman dereð; þæt ys, ðonne þæt man þone wyrttruman
cnucige ond to ðam sare gelecge, eal þæt sar he gehælð.

# 52

# Hymele

Ðeos wyrt, ðe man *politricum* ond oðrum naman hymele
nemneð, byþ cenned on ealdum husstedum ond eac on fuhtum stowum.

## Chapter 51

# Madder

This plant, which is called *grias* and by another name madder, is native to Lucania. It has the color of white marble, and it is adorned with four red stalks.

51.1. For pain in the bones and for broken bones: take this same plant; pound it; apply it to the bone. On the third day he will be better, as if a poultice had been applied there.

51.2. Also the roots of this plant are beneficial for any pain that afflicts the body; that is, when one pounds the root and lays it on the sore, it will heal all the pain.

## Chapter 52

# Common maidenhair

This plant, which is called *polytrichum* and by another name common maidenhair, grows on ruins and also in moist places.

52.1. Wið innoðes sare: genim þysse wyrte leaf þe we *politri-cum* nemdon; hyre twigu beoð swylce swinen byrst. Cnuca ðonne þa leaf, ond nigon pipercorn ond coliandran sædes nigon corn, eall tosomne; syle drincan on godum wine, ond þys sy ðonne he gange to bæðe. Eac þeos ylce wyrt gedeþ þæt ægþer ge wera ge wifa feax wexeþ.

# 53

# Wuduhrofe

53.1. Wið þæt man on wombe forwexen sy: genim þysse wyrte wyrttruman, ðe Grecas *malochinagria* ond Romane *astularegia* nemnað ond eac Ængle wudurofe hatað; cnuca mid wine; syle drincan. Sona þu ongitst þysses fremfulnysse.

53.2. Wið innoðes flewsan: genim þysse wyrte sæd þe we *astularegia* nemdun gemencged mid stiþum ecede; syle drincan. Hyt gewrið þone innoð.

52.1. For abdominal pain: take the leaves of this plant that we have called common maidenhair; its twigs are like the bristles of a pig. Then pound the leaves, plus nine peppercorns and nine coriander seeds, all together; give it to the patient to drink in good wine, and this should be when the person is going to bathe. Also this same plant causes the hair of either men or women to grow.

# Chapter 53

# Asphodel

53.1. If a person has a bloated stomach: take the roots of this plant, which the Greeks call *moloche agria* and the Romans *hastula regia* and which in addition the English call asphodel; pound in wine; give to drink. You will soon see the benefit of this.

53.2. For intestinal flux: take the seeds of this plant that we have called asphodel, mixed with strong vinegar; give to drink. It will settle the insides.

# 54

# Popig

54.1. Wið eagena sare, þæt ys, þæt we cweðað tornige: genim þysse wyrte wos, ðe Grecas *moetorias* ond Romane *papaver album* nemnað ond Engle hwit popig hatað, oððe þone stelan mid þam wæstme; lege to þam eagan.

54.2. Wið þunwonga sare oððe þæs heafdes: genim þysse sylfan wyrte wos; cnuca mid ecede; lege ofer þone andwlatan. Hyt geliþegað þæt sar.

54.3. Wið slæpleaste: genym þysse ylcan wyrte wos; smyre þone man mid. Sona þu him þone slep onsenst.

# 55

# *Oenantes*

55.1. Gyf hwa gemigan ne mæge: genim þysse wyrte wyrttruman, þe man *oennantes* ond oðrum naman [. . .] nemneð, to duste gecnucude; syle drincan on wine twegean scenceas fulle. Hyt fremað healice.

# Chapter 54

# White poppy

54.1. For sore eyes, that is, for what we call bleary eyes: take the juice of this plant, which the Greeks call *meconium* and the Romans *papaver album,* and which the English call white poppy, or else take the stem along with its fruit; apply to the eyes.

54.2. For pain in the temples or for headache: take the juice of this same plant; pound it with vinegar; apply it to the temples. It will relieve the pain.

54.3. For sleeplessness: take the juice of this same plant; rub it on the person. You will soon send him off to sleep.

# Chapter 55

# Dropwort

55.1. If a person cannot urinate: take the roots of this plant, which is called *oenanthe* and by another name [. . .], pounded into a powder; give to drink in wine, two cupfuls. It is highly beneficial.

55.2. Gyf hwa swyþe hræce: genime þysse ylcan wyrte wyrttruman; þicge þam gemete þe we nu her beforan cwædun.
Hyt geliðigað þone hracan.

# 56

# Halswyrt

56.1. Wið þa wunda þe on þam men beoð acenned: genim
þysse wyrte wyrttruman, ðe man *narcisum* ond oþrum naman halswyrt nemneð, mid ele ond mid meluwe gecnucudne
þam gelice þe þu clyðan wyrce; lege to þare wunde. Hit hælð
wundorlice.

# 57

# Brunewyrt

57.1. Wið miltan sare: genim þysse wyrte wyrttruman, þe
Grecas *splenion* ond Romane *teucrion* nemnað ond eac Engle
brunewyrt hatað; cnuca to swiðe smalan duste; syle drincan
on liþum wine. Healic þingc ðu þærmid ongitst. Eac ys sæd

55.2. If a person is coughing badly: take the roots of this same plant; have him drink it in the manner that we have just specified. It will relieve the cough.

# Chapter 56

# Throatwort

56.1. For sores that have developed on a person: take the roots of this plant, which is called *narcissus* and by another name throatwort, pounded along with oil and flour, just as you would make a poultice; lay it on the sore. This will heal it wonderfully.

# Chapter 57

# Spleenwort

57.1. For pain in the spleen: take the roots of this plant, which the Greeks call *splenion* and the Romans call *teucrion,* and which in addition the English call spleenwort; pound to a very fine powder; give to drink in light wine. You will

þæt heo þus funden wære: þæt is ðonne þæt hyt gelamp hwilon þæt man þearmas mid þære miltan uppan þas wyrte gewearp; þa sona geclyfude seo milte to þysse wyrte ond heo hrædlice þa miltan fornam. For ðy heo eac fram sumum mannum *splenion* geciged ys; þæt ys on ure geðeode milte nemned. For þam þæs—þe man sægð—þa swin þe hyre wyrttruman etað þæt hy beon butan milten gemette.

2    Sume eac sæcgeað þæt heo stelan mid twigum hysopan gelicne hæbbe ond leaf beanum gelice; þanon hy sume men þam sylfum naman nemnað hysopan. Þa wyrte man nimeð þonne heo blewð; swiðust heo ys gehered on þam muntlandum þe man Cilicia ond Pisidia nemneð.

# 58

## *Polion*

Ðeos wyrt, þe man *polion* ond oðrum naman [. . .] nemneð, bið cenned on unsmeþum stowum.

58.1. Wið monoðseoce: genim þysse wyrte seaw þe we *polion* nemdun; gemengc wið eced; smyra þærmid þa ðe þæt yfel þoligen toforan þam þe hyt hym to wylle. Ond þeah þu hyre leaf ond hyre wyrttruman do on anne clænne claþ ond gewriðe onbutan þæs mannes swyran þe þæt yfel ðolað. Hyt deþ onfundelnysse þæs sylfan þinges.

experience something remarkable with it. It is also said that it was discovered in this way: that is, that it once happened that someone threw intestines, along with the spleen, onto this plant; then the spleen straightway adhered to the plant and the plant quickly consumed the spleen. For this reason some people call this plant *splenion;* that is what is called the spleen in our language. Because of this—some people say— pigs that eat the roots of this plant are found to be without a spleen.

Some people also say that this plant has a stalk with twigs   2 like hyssop and leaves like broad beans; for this reason some people call the plant hyssop by this same name. This plant is gathered when in bloom; it is especially praised in the mountainous lands called Cilicia and Pisidia.


# Chapter 58

## Felty germander

This plant, which is called *polion* and by another name [. . .], grows in rough places.

58.1. For a person suffering from lunacy: take the juice of the plant we call *polion;* mix with vinegar; smear it on persons who are afflicted by this evil before it comes upon them. But also put its leaves and its roots in a clean cloth and fasten it around the neck of the person who suffers from that ailment. This will give proof of the same thing.

# 59

# Cneowholen

59.1. Wið þone dropan ond wið þæs magan sare: genim twegen scenceas fulle woses ðysse wyrte, þe man *victoriole* ond oðrum naman cneowholen nemneð; syle drincan fæstendum, wið hunig gemenged. Sona hyt ðone dropan gewæceð.

# 60

# Galluc

Ðeos wyrt, þe man *confirman* ond oðrum naman galluc nemneð, bið cenned on morum ond on feldum ond eac on mædum.

60.1. Wið wifa flewsan: genim þas wyrte *confirmam;* cnuca to swyþe smalon duste; syle drincan on wine. Sona se flewsa ætstandeþ.

60.2. Gyf hwa innan toborsten sy: genime þysse ylcan wyrte wyrttruman; gebræde on hatan axan; þicge þonne on hunige, fæstende. He bið gehæled, ond eac hyt þone magan ealne afeormað.

## Chapter 59

# Butcher's broom

59.1. For rheum and for stomachache: take two cupfuls of the juice of this plant, which is called *victoriola* and by another name butcher's broom; give it to the patient to drink while fasting, mixed with honey. It will quickly alleviate the rheum.

## Chapter 60

# Comfrey

This plant, which is called *confirma* and by another name comfrey, grows on moors and in fields and also in meadows.

60.1. For women's excessive flow of menstrual blood: take this plant comfrey; pound into very fine powder; give to drink in wine. The flow will soon come to a halt.

60.2. If someone has an internal rupture: take the roots of this same plant; roast them in hot ashes; have him partake of them in honey, while fasting. He will be healed, and in addition, this will clean the whole stomach.

60.3. Wið magan sare: genim þas ylcan wyrte ond gemeng wið hunig ond wið eced. Þu ongitst mycele fremfulnysse.

# 61

## *Asterion*

Ðeos wyrt, þe man *asterion* ond oðrum naman [. . .] nemneð, byð cenned betweoh stanum ond on unsmeþum stowum. Ðeos wyrt scineð on nihte swilce steorra on heofone, ond se ðe hy nytende gesihð, he sægð þæt he scinlac geseo. Ond swa afæred, he bið tæled fram hyrdum ond fram swylcum mannum swylce þære wyrte mihta cunnun.

61.1. Wið fylleseocnysse: genim þysse wyrte bergean þe we *asterion* nemdon; syle etan on wanigendum monan. Ond sy þæt ðonne þære sunnan ryne beo on þam tacne þe man Virgo nemneð—þæt bið, on þam monðe þe man Augustus hateð—ond hæbbe ðas sylfan wyrte on his swyran ahangene. He bið gelacnud.

60.3. For stomachache: take this same plant and mix it with honey and with vinegar. You will perceive a great benefit.

# Chapter 61

# Aster

This plant, which is called *asterion* and by another name [. . .], grows in between stones and in rough places. This plant shines at night like the stars in the skies, and someone who sees it without knowledge of it will say that he has seen an apparition. And frightened in this manner, he will be mocked by shepherds and by those who know more about the powers of this plant.

61.1. For epilepsy: take the berries of this plant that we call *asterion;* give to eat when the moon is waning. And this should be when the orbit of the sun is in the sign that is called Virgo—that is, in the month called August—and have the person keep this same plant hung around the neck. He will be cured.

## 62

# Haran hyge

62.1. Wið innoþes fæstnysse: genim ðas wyrte, þe man *leporis pes* ond oðrum naman haran hige nemneð; gedryge hy; cnuca þonne to duste; syle drincan on wine gif he unfeferig sy; gyf he þonne on fefere sy, syle drincan him on wætere. Sona seo fæstnys toslypeð.

## 63

# *Dictamnus*

Ðeos wyrt, þe man *dictamnum* ond oðrum naman [. . .] nem- neð, byþ cenned on ðam iglande þe man Crete hateð ond on þam munte þe man Ida nemneð.

63.1. Gyf hwylc wif hæbbe on hyre innoðe deadboren tud- dur: genim þysse wyrte wos þe we *dictamnum* nemdun; gif heo butan fefere sy, syle drincan on wine; gif hyre þonne fe- fer derige, syle drincan on wearmum wætere. Sona hit þæt tuddur ut asendeþ butan frecnysse.

# Chapter 62

# Hare's-foot clover

62.1. For constipation: take this plant, which is called *pes leporis* and by another name hare's-foot clover; dry it; then pound it into a powder; give it to the patient to drink in wine if he has no fever; if he has a fever, give it to him to drink in water. Soon the constipated condition will be dispersed.

# Chapter 63

# Dittany of Crete

This plant, which is called *dictamnus* and by another name [. . .], grows on the island called Crete and on the mountain called Ida.

63.1. If a woman is carrying a dead fetus in her womb: take the juice of this plant we have called *dictamnus;* if she does not have a fever, give it to the patient to drink in wine; if she is suffering from fever, give to drink in warm water. This will soon discharge the fetus without any danger.

63.2. Eft, wið wunda, som hy syn of iserne, som hi syn of stence oððe fram nædran: genim þysse ylcan wyrte wos; do on þa wunda, ond syle drincan. Sona he byð hal.

63.3. Eft, soðlice, wið næddran slite: genim þysse ylcan wyrte seaw; syle drincan on wine. Sona hyt þæt attor tofereð.

63.4. Gyf hwa attor þicge: genime þysse ylcan wyrte wos; drince on wine. Witodlice swa mycel ys þysse wyrte strengð swa na þæt an þæt heo mid hyre andweardnysse næddran ofslyhð, swa hwær swa hy hyre gehende beoð, ac forþon of hyre stence þonne he mid winde ahafen bið. Swa hwær swa hy beoð, ond hy þone swæc gestincað, hy sculon sweltan.

2    Eac ys sæd be þysse sylfan wyrte, gyf man on huntuþe ran oððe rægean mid flane oððe oðrum wæpne gewæceþ, þæt hy wyllon þas wyrte etan swa hy hraþost to cuman mægen, ond heo sona þa flane ut adeð ond ða wunde gehæleþ.

63.5. Wið niwe wunda: genim þas ylcan wyrte, ond æþelferðincwyrte ond hindehæleþan; cnuca mid buteran; lege to þære wunde. Þu wundrast on eallum þingum ðysse wyrte gefremincge.

63.2. Again, for wounds, whether they are caused by iron or by a pole or by a snake: take the juice of this same plant; put it on the wounds and give it to drink. The person will be well quickly.

63.3. Again, indeed, for snakebite: take the juice of this same plant; give to drink in wine. It will quickly expel the poison.

63.4. If someone should ingest poison: take the juice of this same plant; give it to the patient to drink in wine. In truth, the strength of this plant is so great that it not only kills snakes by its presence, wherever they are in its proximity, but even by its smell when it is carried by the wind. Wherever they are, if they smell its odor, they will die.

It is also said of this same plant that, if during a hunt someone wounds a roebuck or a roe deer with an arrow or another weapon, the animals will eat this plant as soon as they can get it, and it will soon cause the arrow to drop out and will heal the wound.

63.5. For fresh wounds: take this same plant, along with stitchwort and hindheal; pound in butter; apply to the wound. You will be amazed at the beneficial effects of this plant in all regards.

# 64

## *Solago maior*

64.1. Wið næddran slite ond wið *scorpiones* stincg: genim þas wyrte, þe man *solago maior* ond *helioscorpion* nemneð; dryge hy þonne ond cnuca to swyþe smalon duste; syle drincan on wine. Ond genim þa wyrte gecnucude; lege to þære wunde.

# 65

## *Solago minor*

65.1. Wiþ ðæt rængcwyrmas dergen ymb nafolan: genim þas wyrte, þe man *solago minor* ond oþrum naman *æliotropion* nemneð, gedrigede; cnuca to duste; syle drincan on wearmum wætere. Heo þa wyrmas ofslihð.

# 64

# Heliotrope

64.1. For snakebite and scorpion's sting: take this plant, which is called *solago maior* and *helioscorpios;* dry it and then pound it into a very fine powder; give to drink in wine. And take the plant when pounded; apply it to the wound.

# Chapter 65

# Dyer's croton

65.1. If worms should be doing harm around the navel: take this plant, which is called *solago minor* and by another name *heliotropium,* dried; pound it into a powder; give to drink in warm water. It will kill the worms.

# 66

## *Pionia*

Ðeos wyrt, ðe man *peoniam* nemneð, wæs funden fram Peonio, þam ealdre, ond heo þone naman of him hæfð. Heo bið cenned fyrmest in Creta, þa eac se mæra ealdor Homerus on hys bocum amearcode. Heo bið funden swyþost fram hyrdum; ond heo hæfð corn þære mycelnysse þe *maligranati,* ond heo on nihte scineð swa leohtfæt, ond eac hyre corn beoð gelice coccele. Ond heo byð, swa we ær cwædon, oftust fram hyrdum on nihte gemet ond gegaderod.

66.1. Wið monoðseocnysse: gyf man þas wyrte *peoniam* þam monoðseocan ligcgendon ofer alegð, sona he hyne sylfne halne up ahefð. Ond gyf he hig mid him hafað, næfre seo adl him eft genealæceð.

66.2. Wiþ hypebanece: genim þysse ylcan wyrte sumne dæl wyrttruman ond mid linenan claðe gewrið to þam sare. Hyt gehælð.

# Chapter 66

# Peony

This plant, which is called peony, was discovered by Peonio, the chief, and it gets its name from him. It is native to Crete, as Homer, the great authority, writes in his books. It is found mostly by shepherds; and it has seeds the size of *malum granatum,* and it shines at night like a lantern, and also its seeds are like cockles. And as we said before, it is most often found and gathered by shepherds at night.

66.1. For lunacy: if one lays this plant peony over a lunatic when he is lying down, he will quickly raise himself up healed. And if he has the plant with him, the illness will never come near him again.

66.2. For hipbone ache: take a portion of the root of this same plant and tie it onto the sore area with a linen cloth. This will heal it.

# 67

## *Berbena*

Ðeos wyrt, þe man *peristereon* ond oðrum naman *berbenam* nemneð, heo ys culfron swiðe hiwcuð; þanun hy eac sum þeodsciþe *columbinam* hateð.

67.1. Gyf hwa þas wyrte mid him hafað þe we *peristereon* nemdon, ne mæg he fram hundum beon borcen.

67.2. Wið ealle attru: genim þysse sylfan wyrte dust; sile drincan. Ealle attru heo todrifð. Eac mon sægð þæt dryas to heora cræftum hyre brucen.

# 68

## Hymele

68.1. Wið miltan sare: genim þas wyrte, ðe man *bryonia* ond oþrum naman hymele nemneð; syle þycgean gemang mete. Þonne sceal þæt sar liþelice þurh þone micgþan forð gan. Ðeos wyrt is to þam herigindlic þæt hy man wið gewune drenceas gemencgeað.

# Chapter 67

# Gypsywort

This plant, which is called *peristereon* and by another name *verbena,* is much loved by doves; for this reason some people also call it *columbina.*

67.1. If anyone has with him the plant that we have called *peristereon,* he will not be barked at by dogs.

67.2. For all poisons: take the powder of this same plant; give to drink. It will expel all poisons. It is also said that magicians use it for their practices.

# Chapter 68

# Bryony

68.1. For pain in the spleen: take this plant, which is called *bryonia* and by another name bryony; give it to eat mixed in food. Then the pain will gently exit the body through the urine. This plant is so laudable that people mix it with their customary drinks.

# 69

## *Nymfete*

69.1. Wið þæt man on wambe forwexen sy: genim þysse wyrte sæd, þe man *nymfete* ond oðrum naman [. . .] nemneð; cnuca mid wine; syle drincan.

2    Eft, þæt sylfe, be ðam wyrttruman; syle hyne þam seocan þicgean tien dagas.

69.2. Eft, gyf þu þas wyrte sylst þicgean on strangon wine, heo þæs innoðes unryne gewrið.

# 70

## Clæfre

70.1. Wið þæra gomena sare: gyf hwa þysse wyrte wyrttruman, þe man *crision* ond oðrum naman clæfre nemneð, mid him hafað ond on his swyran byrð, næfre him his goman ne deriað.

# Chapter 69

# White water lily

69.1. If a person has a bloated stomach: take the seed of this plant, which is called *nymphaea* and by another name [. . .]; pound in wine; give to drink.

Again, for the same ailment, as regards the roots; give them to the sick person to eat for ten days.

69.2. Again, if you give this plant to someone to drink in strong wine, it will stop diarrhea.

# Chapter 70

# Thistle

70.1. For sore throat: if anyone has with him the roots of this plant, which is called *crision* and by another name thistle, and carries them on his neck, his throat will never trouble him.

# 71

## *Ysatis*

Ðas wyrte Grecas *isatis* ond Romane *alutam* nemnaþ, ond eac Angle hateð *ad serpentis morsum.*

71.1. Wið næddran slite: genim þysse wyrte leaf þe Grecas *issatis* nemnað; cnuca on wætere; lege to þære wunde. Heo fremað ond þæt sar of genimð.

# 72

## *Scordea*

72.1. Wið nædran slite: genim þas wyrte þe man *scordean* ond oðrum naman [. . .] nemneð; seoð on wine; syle drincan. Cnuca þonne þa wyrte ond lege to þære wunde.

72.2. Wið sina sare: genim þas sylfan wyrte; cnuca hy, ond gewyld mid ðam ele ðe sy of lawertreowe gewrungan. Hyt þæt sar of animð.

72.3. Wið þam fefore þe dæghwamlice oþþe ðy þriddan on man becymð: genim þas ylcan wyrte ond gewrið hy onbutan þæs mannes lichoman. Heo of animð þone dæghwamlican ond þæs þriddan dæges fefor.

# Chapter 71

# Woad

The Greeks call this plant *isatis* and the Romans call it *aluta*, and the English also call it *ad serpentis morsum*.

71.1. For snakebite: take the leaves of this same plant the Greeks call *isatis;* pound in water; apply to the wound. It will help and will take the pain away.

# Chapter 72

# Water germander

72.1. For snakebite: take this plant that is called *scordion* and by another name [. . .]; simmer in wine; give it to drink. Then pound the plant and lay it on the wound.

72.2. For sore sinews: take this same plant; pound it, and boil it with oil pressed from a laurel tree. It will take the pain away.

72.3. For the fever that comes on a person either daily or every third day: take this same plant and fasten it around the person's body. It will take away daily fever and third-day fever.

# 73

# Feltwyrt

Ðeos wyrt, þe man *verbascum* ond oðrum naman feltwyrt nemneð, bið cenned on sandigum stowum ond on myxenum. Þas wyrte ys sæd þæt Mercurius sceolde Ulixe þam ealdormen syllan þa he com to Circean, ond he na syþþan ænige hyre yfelan weorc ondred.

73.1. Gyf hwa mid hym þysse wyrte ane tealgre byrð, ne bið he breged mid ænigum ogan, ne him wildeor ne dereþ, ne ænig yfel geancyme.

73.2. Wiþ fotadle: genim þas ylcan wyrte *verbascum,* gecnucude; lege to þam sare. Binnan feawum tidum heo gehælp þæt sar to ðam scearplice þæt he eac gan dyrre ond mæge. Eac ure ealdras cwædon ond sædun þæt ðeos gesetednys healicost fremade.

# Chapter 73

# Great mullein

This plant, which is called *verbascum* and by another name great mullein, grows in sandy soils and on dunghills. Mercury, it is said, is supposed to have given this plant to Ulysses, the chief, when he approached Circe, and never after did he fear any of her evil deeds.

73.1. If anyone carries with him one twig of this plant, he will not be frightened by any terror, nor will any wild beast scare him, nor will any evil approach him.

73.2. For foot disease: take this same plant great mullein, well pounded; apply it to the sore. Within a few hours it will heal the pain so efficaciously that he will even dare to walk and will be able to. Also, ancient authorities have declared and said that this preparation has been of the highest value.

# 74

## *Heraclea*

74.1. Se þe wylle ofer langne weg feran, hæbbe mid him on þam wege þas wyrte, þe man *heraclean* ond oðrum naman [. . .] nemneþ. Þonne ne ondrædeþ he hym ænigne sceaþan, ac heo hy aflygeþ.

# 75

## Cyleþenie

75.1. Wiþ eagena dymnysse, ond sarnysse, ond ofertogennysse: genim þysse wyrte seaw, þe man *celidoniam* ond oðrum naman þam gelice cyleþenie nemneð; gecnucud of þam wyrttruman mid ealdum wine ond hunige ond pipore, ond sy þæt wel tosomne gepunud; smyre þonne þa eagan innan. Eac we onfundun þæt sume men of ðære meolce þysse ylcan wyrte heora eagan smyredon, ond him þy sel wæs.

75.2. Eft, wið dymgendum eagum: genim þysse ylcan wyrte wos, oþþe ða blostman, gewrungene ond gemengced mid hunige; gemengc þonne liþelice weallende axan þærto, ond

# Chapter 74

# Ironwort

74.1. If a person wants to travel a long way, on the journey he should keep with him this plant, which is called *heraclea* and by another name [. . .]. Then he will not fear any robbers, but it will put them to flight.

# Chapter 75

# Greater celandine

75.1. For dimness and soreness and albugo of the eyes: take the juice of the roots of this plant, which is called *chelidonia* and by another name that resembles it, celandine; pound it with old wine and honey and pepper, and let it be well pounded together; then rub it into the eyes. Moreover, we have found that some people have applied the milk of this same plant to their eyes, and by this means they got better.

75.2. Again, for dimness of the eyes: take the juice of this same plant, or else the blossoms, pressed and mixed with honey; then mix in some gently boiling ashes, and simmer

seoð þær tosomne on ærenum fæte. Ðys is synderlic læcedom wið eagena dymnysse. Eac is gewis þæt sume men, swa we ær cwædon, þæs woses synderlice brucað.

75.3. Wið cyrnlu: genim þas ylcan wyrte; cnuca mid rysle; lege to þam cyrnlun, swa þæt hi ærest syn mid wætere gebeþode.

75.4. Wiþ heafodece: genim þas sylfan wyrte; cnuca mid ecede; smyre þone andwlatan ond þæt heafod.

75.5. Wið þæt man gebærned sy: genim þas ylcan wyrte; cnuca mid gætena smerwe, ond lege þærto.

# 76

## *Solsequia*

76.1. Wið geswel: genim þas wyrte, þe man *solate* ond oðrum naman solosece nemneð, gecnucude ond mid ele gemengcede; lege þærto. Hyt fremað.

76.2. Wiþ earena sare: genim ðysse ylcan wyrte wos; gemengc mid þam ele of cypro ond gewyrm hyt, ond swa wlæc dryþe on þæt eare.

76.3. Wið toðece: syle etan ðysse sylfan wyrte croppas.

76.4. Wið blodryne of nosum: genim þysse ylcan wyrte wos, ond dype anne linenne cla, ond forsete þa næsðyrlu þærmid. Sona þæt blod ætstent.

them together in a brass vessel. This is an exceptionally good recipe for dimness of the eyes. Moreover, it is certain that some people, as we said before, use the juice by itself.

75.3. For hard swellings: take this same plant; pound with some fat; lay it on the swellings, which should first have been bathed with water.

75.4. For headache: take this same plant; pound in vinegar; apply to the face and to the head.

75.5. If a person is burned: take this same plant; pound with goat's grease, and apply it to the burn.

# Chapter 76

# Black nightshade

76.1. For swellings: take this plant, which is called *solata* and by another name black nightshade, pounded and mixed with oil; apply it to the swelling. It will be of benefit.

76.2. For earache: take the juice of this same plant; mix with oil of henna and warm it, and when lukewarm drip it into the ear.

76.3. For toothache: give berries of this same plant to eat.

76.4. For nosebleed: take the juice of this same plant, and dip a linen cloth in it, and block the nostrils with it. Soon the blood will stop.

# 77

# Grundeswylige

Þeos wyrt, ðe man *senecio* ond oðrum naman grundeswylige nemneð, byþ cenned on hrofum ond onbutan wagum.

77.1. Wið wunda, þeah hy syn swyþe ealde: genim þas wyrte þe we *senecio* nemdun; cnuca mid ealdum rysle; lege to þam wundum. Hyt hæleþ sona.

77.2. Gyf hwa mid iserne geslegen sy: genim þas ylcan wyrte on ærnemergen oððe to middan dæge; cnuca hy, swa we ær cwædon, mid ealdum rysle; lege to þære wunde. Sona heo þa wunde geopenað ond afeormað.

77.3. Wið fotadle: genim þas ilcan wyrte; cnuca mid rysle; lege to þam fotum. Hyt geliþegað þæt sar; eac hit fremað mycelum wið þæra sina sare.

77.4. Wiþ lendena sare: genim þas ylcan wyrte; cnuca mid sealte þam gemete ðe þu clyþan wyrce; lege to ðam lendenum. Þam gelice hyt fremað eac wið þæra fota sare.

# Chapter 77

# Groundsel

This plant, which is called *senecio* and by another name groundsel, grows on rooftops and along walls.

77.1. For wounds, even if they are very old: take this plant we have called groundsel; pound with some old fat; apply it to the wounds. It will soon heal them.

77.2. If a person has been struck with an iron blade: take this same plant in the early morning or at midday; pound it, as we said before, with old fat; apply it to the wound. Soon it will open the wound and cleanse it.

77.3. For foot disease: take this same plant; pound with some fat; lay it on the feet. It will relieve the pain; it also is of much benefit for sore sinews.

77.4. For sore loins: take this same plant; pound it with salt as you would for a poultice; apply it to the loins. In the same manner, it also is good for sore feet.

# 78

# Fern

78.1. Wið wunda: genim þysse wyrte wyrttruman, þe man *filicem* ond oðrum naman fearn nemneþ, gecnucudne; lege to þære wunde; ond æþelferþincgwyrte twegea trymessa gewæge syle drincan on wine.

78.2. Wið þæt geong man healyde sy: genim þas ylcan wyrte þær heo on becenan treowes wyrttruman gewexen sy; cnuca mid rysle ond gedec anne cla ð þærmid, ond gewri ð to ðam sare, swa þæt he þa hwyle upweard sy gewend. Þy fiftan dæge he bi ð gehæled.

# 79

# Cwice

79.1. Wi ð miltan sare: genym þysse wyrte leaf, þe man *gramen* ond oðrum naman cwice nemne ð, ond geseo ð hy. Smyre þonne anne cla ð þærmid; lege to þære miltan. Þu ongytest fremfulnysse þærof.

# Chapter 78

# Fern

78.1. For wounds: take the roots of this plant, which is called *filix* and by another name fern, pounded; apply them to the wound; and give two tremiss-weights of stitchwort to drink in wine.

78.2. If a young man is ruptured: take the same plant from where it grows on the roots of a beech tree; pound with some fat and soak a cloth in it, and tie the cloth on the sore, always keeping the cloth turned upward. By the fifth day he will be healed.

# Chapter 79

# Couch grass

79.1. For pain in the spleen: take the leaves of this plant, which is called *gramen* and by another name couch grass, and simmer them. Then apply them to a cloth; lay it on the spleen. You will perceive benefit from this.

# 80

## Glædene

80.1. Wið blædran sare, ond wið þæt man gemigan ne mæge: genim þysse wyrte wyrttruman utewearde, ðe man *gladiolum* ond oþrum naman glædene nemneþ; drige hyne þonne ond cnuca, ond gemengc ðærto twegean scenceas wines ond þry wæteres; syle drincan.

80.2. Wið miltan sare: genim þas ilcan wyrte *gladiolum* þonne heo geong sy; drige hy ond cnuca to swyþe smalan duste; syle þicgean on liþum wine. Hyt is gelyfed þæt hit wundorlice þa miltan gehæleþ.

80.3. Wiþ innoþes sare ond þæra breosta: genim þysse sylfan wyrte bergean, gecnucude ond on gætenre meolce, oððe gyt selre on wine, gewlehte; syle drincan. Þæt sar geswiceþ.

# 81

## Boðen

Ðeos wyrt, þe man *rosmarim* ond oðrum naman boþen nemneþ, byþ cenned on sandigum landum ond on wyrtbeddum.

# Chapter 80

# Iris

80.1. For bladder pain, and if a person cannot urinate: take the outer part of the roots of this plant, which is called *gladiolus* and by another name iris; then dry it and pound it, and mix with it two cups of wine and three of water; give to drink.

80.2. For pain in the spleen: take this same plant iris when it is young; dry it and pound it into a very fine powder; give to drink in light wine. It is believed that it heals the spleen wonderfully.

80.3. For abdominal and chest pain: take the berries of this same plant, pounded and made lukewarm in goat's milk or, even better, in wine; give to drink. The pain will cease.

# Chapter 81

# Rosemary

This plant, which is called *ros marinum* and by another name rosemary, grows in sandy soils and in garden plots.

81.1. Wið toþece: genim þysse wyrte wyrtwalan þe we *rosmarim* nemdun; syle etan. Butan yldincge he genimð þæra toða sar. Ond healde þæt wos on his muþe; sona hyt gehælþ þa teð.

81.2. Wið adligende: genim þas wyrte *rosmarinum;* cnuca mid ele; smyre ðone adligendan. Wundorlice þu hine gehælest.

81.3. Wið gicþan: genim þas ylcan wyrte; gecnuca hy, ond gemengc hyre wos wið eald win ond wið wearm wæter; syle drincan þry dagas.

81.4. Wið liferseocnysse ond þæs innoðes: genim þysse sylfan wyrte sumne gripan; scearfla on wæter, ond gemencg þærto nardis twa handfulla ond rudan sumne stelan; seoð tosomne on wætere; syle drincan. He bið hal.

81.5. Wið niwe wunda: genim þas ylcan wyrte þe we *rosmarim* nemdun; cnuca mid rysle; lege to þam wundum.

# 82

# Feldmoru

Þeos wyrt, þe man *pastinace silvatice* ond oðrum naman feldmoru nemneþ, bið cenned on sandigum stowum ond on beorgum.

81.1. For toothache: take the root of this plant we call rosemary; give to eat. Without delay it will take away the pain. And have the person keep its juice in his mouth; it will soon heal the teeth.

81.2. For the sickly: take this plant rosemary; pound with oil; smear it over the person who is ill. You will heal him wonderfully.

81.3. For itching: take this same plant; pound it, and mix its juice with old wine and with warm water; give to drink for three days.

81.4. For liver and abdominal disease: take a handful of this same plant; shred it into water, and mix with it two handfuls of spikenard and some stalks of rue; simmer together in water; give to drink. He will be well.

81.5. For fresh wounds: take this same plant that we have called rosemary; pound with some fat; apply it to the wounds.

# Chapter 82

# Wild carrot

This plant, which is called *pastinaca silvatica* or by another name wild carrot, grows in sandy soils and on hills.

82.1. Wið þæt wifmen earfuðlice cennen: genim þas wyrte
þe we *pastinacam silvaticam* nemdun; seoð on wætere; syle
þonne þæt se man hyne þærmid beðige. He bið gehæled.

82.2. Wið wifa afeormungæ: genim þas ylcan wyrte *pastinacam;* seoð on wætere, ond þonne heo gesoden beo, mengc hy
wel ond syle drincan. Hy beoð afeormude.

# 83

# Dolhrune

Ðeos wyrt, þe man *perdicalis* ond oðrum naman dolhrune
nemneð, byþ cænned wið wegas, ond wið weallas, ond on
beorgum.

83.1. Wið fotadle ond wið cancor: genim þas wyrte þe we *perdicalis* nemdun; seoð on wætere; beþe þonne þa fet ond þa
cneowa. Cnuca syððan þa wyrte mid rysle; do on ænne cla,
ond lege to þam fotum ond to þam cneowum. Þu hy wel gehælst.

82.1. If women have difficulty in giving birth: take this plant that we have called wild carrot; simmer in water; give it so she can bathe herself with it. She will be healed.

82.2. For women's cleansing: take this plant wild carrot; simmer in water, and when it is cooked, mix it well and give to drink. They will be cleansed.

# Chapter 83

# Pellitory-of-the-wall

This plant, which is called *perdicalis* and by another name pellitory-of the-wall, grows along roads, against walls, and on hills.

83.1. For foot disease and for ulcerous sores: take this plant that we have called pellitory-of-the-wall; simmer in water; then bathe the feet and the knees. After this pound the plant with some fat; put it on a cloth; and lay this on the feet and on the knees. You will heal them well.

# 84

# Cedelc

84.1. Wið þæs innoðes heardnysse: genim þas wyrte, þe man *mercurialis* ond oðrum naman cedelc nemneð, on wætere gegnidene; syle þam ðolegendum. Sona heo ða heardnysse ut atyhð ond ðone magan afeormað. Þam gelice þæt sæd fremað.

84.2. Wið eagena sar ond geswel: genim ðysse sylfan wyrte leaf, gecnucude on ealdum wine; lege to þam sare.

84.3. Gif wæter on earan swiðe gesigen sy: genim þysse ylcan wyrte seaw, wlæc; drype on þæt eare. Sona hyt toflyhð.

# 85

# Eforfearn

Ðeos wyrt, þe man *radiolum* ond oðrum naman eforfearn nemneð, ys gelic fearne, ond heo byð cenned on stanigum stowum ond on ealdum husstedum. Ond heo hæfð on æghwylcum leafe twa endebyrdnyssa fægerra pricena, ond þa scinað swa gold.

# Chapter 84

# Dog's mercury

84.1. For hardening of the abdomen: take this plant, which is called *mercurialis* and by another name dog's mercury, crushed in water; give it to the sufferer. It will soon draw out the hardness and will purge the stomach. The seeds are of similar benefit.

84.2. For pain and swelling of the eyes: take the leaves of this same plant, pounded in old wine; apply to the sore area.

84.3. If water has penetrated into the ears: take juice of this same plant, lukewarm; drip into the ear. Soon the water will disappear.

# Chapter 85

# Polypody

This plant, which is called *radiolum* and by another name polypody, is like a fern, and it grows in stony places and among ruins. And on each leaf it has two rows of beautiful spots, and they shine like gold.

85.1. Wið heafodece: genim þas wyrte þe we *radiolum* nemdun, swiðe clæne afeormude; seoð on ecede þearle; smyre þonne þæt heafud þærmid. Hyt geliðegað þæt sar.

# 86

# Wuducerville

86.1. Wið blædran sare oððe geswelle: genim þysse wyrte wyrttruman, þe man *sparagiagrestis* ond oðrum naman wuducerfilla nemneð; seoð on wætere to feorðan dæle; drince ðonne fæstende seofan dagas. Ond he manegum dagum bæþes bruce, ond na on caldum wætere cume, ne he cealdne wætan ne þicge. Wundorlice he hæle ongyt.

86.2. Wið toðece: genim þysse sylfan wyrte seaw þe we *sparagi* nemdun; syle supan, ond healde hyt swa on hys muðe.

86.3. Wið æddrena sare: genim þysse ylcan wyrte wyrtwalan, gecnucude on wine; syle drincan. Hyt fremað.

86.4. Gyf hwylc yfeldæde man þurh ænigne æfþancan oþerne begaleþ: genim þysse sylfan wyrte wyrttruman, gedrigide; syle þicgean mid wyllewætere, ond besprengc hyne mid þam wætere. He bið unbunden.

85.1. For headache: take this plant that we have called polypody, cleaned very well; simmer thoroughly in vinegar; anoint the head with it. It will relieve the pain.

# Chapter 86

# Asparagus

86.1. For pain or swelling of the bladder: take the roots of this plant, which is called *asparagus agrestis* and by another name asparagus; simmer in water until reduced to a fourth part; have the person then drink it for seven days while fasting. He should bathe for many days, but he is not to get into cold water nor to drink cold water. He will feel healthy in a wonderful way.

86.2. For toothache: take the juice of this same plant that we have called asparagus; give him this to sip, and have him hold it in his mouth.

86.3. For sore veins: take the roots of this same plant, pounded in wine; give to drink. It will be of benefit.

86.4. If any wicked person enchants another person out of malice: take roots of this same plant, dried; give to eat with spring water, and sprinkle the person with the water. He will be unbound.

# 87

## Savine

87.1. Wið þa cynelican adle, þe man *aurignem* nemneð—þæt ys, on ure geþeode, þæra syna getoh ond fota geswel: genim þas wyrte þe man *sabinam* ond oðrum naman wel þam gelice *savinam* hateþ; syle drincan mid hunige. Heo tofereþ þæt sar. Þæt sylfe heo deþ mid wine gecnucud.

87.2. Wiþ heafodece: genim þas ylcan wyrte *sabinam,* eornlice gecnucude mid ecede ond mid ele gemencgede; smyra þonne þæt heafud ond þa þunwonga. Healice hyt fremað.

87.3. Wiþ deadspringas: genim þas wyrte *sabinam,* mid hunige gecnucude; smyre þonne þæt sar.

# 88

## Hundes heafod

88.1. Wið eagena sar ond geswel: genim þysse wyrte wyrtwalan þe man *canis caput* ond on ure geþeode hundes heafod hatað; seoð on wætere, ond syþþan mid þam wætere þa eagan gebeþa. Hrædlice hyt þæt sar geliþigað.

# Chapter 87

# Savine

87.1. For the king's disease, which is called *aurigo*—that is, in our language, spasm of the sinews and foot swelling: take the plant that is called *sabina* and by another similar name *savina;* give to drink with honey. It will drive away the pain. It does the same thing when pounded in wine.

87.2. For headache: take this same plant savine, well pounded in vinegar and mixed with oil; then smear it on the head and the temples. It will be of particular benefit.

87.3. For carbuncles: take the plant savine, pounded in honey; smear the sore with it.

# Chapter 88

# Snapdragon

88.1. For soreness and swelling of the eyes: take the roots of the plant that is called *canis caput* and in our language snapdragon; simmer in water, and after this bathe the eyes with this water. It will quickly relieve the pain.

# 89

# Bremel

89.1. Wið earena sar: genim þas wyrte, þe man *erusti* ond oþrum naman bremel nemneð, swa mearwe; gecnuca. Nim þonne þæt wos, gewleht; drype on þæt eare. Hyt þæt ge-wanað ond gewislice gehæleþ.

89.2. Wið wifes flewsan: genim þysse ylcan wyrte croppas, swa mearwe, ond þæra syn þriwa seofeone; seoð on wætere to þriddan dæle; syle drincan fæstende þry dagas, swa þæt ðu þeah æghwylce dæg þone drenc niwie.

89.3. Wið heortece: genim þysse ylcan wyrte leaf, gecnucude þurh hy sylfe; lege ofer þone wynstran tit. Þæt sar tofærð.

89.4. Wið niwe wunda: genim þysse sylfan wyrte blostman; lege to ðam wundum. Butan ælcre yldincge ond frecenysse hy þa wunda gehælað.

89.5. Wiþ liþa sare: genim þysse ylcan wyrte sumne dæl; seoð on wine to þriddan dæle, ond of þam wine syn þonne þa lyþu gebeðede. Ealle þæra liða untrumnysse hyt geliðigaþ.

89.6. Wið næddran slite: genim þysse ilcan wyrte leaf þe we *erusti* nemdun, swa niwe, gecnucude; lege to ðam sare.

# Chapter 89

# Bramble

89.1. For earache: take this plant, which is called *erusti* and by another name bramble, when it is tender; pound it. Then take the juice, lukewarm; drip it into the ear. It will lessen the pain and will surely provide healing.

89.2. For a woman's excessive flow of menstrual blood: take the sprigs of this same plant, when they are very soft, and there should be three times seven of them; simmer in water, reducing it to a third; give to drink for three days while she is fasting, making sure, however, that you make a fresh drink every day.

89.3. For heart pain: take the leaves of this same plant, pounded by themselves; lay this on the left nipple. The pain will disappear.

89.4. For fresh wounds: take the flowers of this same plant; lay them on the wounds. Without any delay or danger they will heal the wounds.

89.5. For pain in the joints: take a portion of this same plant; simmer in wine reducing it to one-third, and let the joints be bathed with this wine. It will relieve all the infirmity in the joints.

89.6. For snakebite: take leaves of this same plant we call bramble, tender ones, pounded; apply them to the bite.

90

# Gearwe

90.1. Ðas wyrte, þe man *millefolium* ond on ure geþeode gearwe nemneþ, ys sæd þæt Achilles, se ealdorman, hy findan scolde, ond he mid þysse sylfan wyrte gehælde þa þe mid iserne geslegene ond gewundude wæran. Eac heo of sumum mannum for þy genemned ys *achylleos*. Mid þære wyrte ys sæd þæt he eac sumne man gehælan sceolde þam wæs Thelephon nama.

90.2. Wið toðece: genim þysse wyrte wyrtwalan ðe we *millefolium* nemdun; syle etan fæstendum.

90.3. Wiþ wunda þe mid iserne syn geworhte: genim þas ylcan wyrte, mid rysle gecnucude; lege to þam wundum. Heo þa wunda afeormaþ ond gehæleð.

90.4. Wiþ geswell: genim þas ylcan wyrte *myllefolium,* mid buteran gecnucude; lege to þam geswelle.

90.5. Wið þæt hwylc man earfoðlice gemigan mæge: genim þysse ylcan wyrte wos mid ecede; syle drincan. Wundurlice heo hæleþ.

90.6. Gif wund on men acolod sy: genim þonne ða sylfan wyrte *millefolium* ond gnid swyþe smale ond mengc wið buteran; lege ðonne on ða wunda. Heo cwicaþ sona ond wearmað.

# Chapter 90

# Yarrow

90.1. It is said that Achilles, the chief, discovered this plant, which is called *millefolium* and in our language yarrow, and with this same plant he healed those who had been struck and wounded by iron weapons. And for this reason this plant is called *achilleos* by some people. It is said that with that plant Achilles also cured a man whose name was Telephos.

90.2. For toothache: take the roots of this plant that we have called yarrow; give to the patient to eat while fasting.

90.3. For wounds that have been inflicted by an iron weapon: take this same plant, pounded in fat; apply it to the wounds. It will cleanse and heal the wounds.

90.4. For swellings: take this same plant yarrow, pounded with some butter; apply it to the swelling.

90.5. If a person has difficulty urinating: take the juice of this same plant, along with vinegar; give to drink. It will cure it wonderfully.

90.6. If a person's wound has grown cold: take this same plant yarrow and crush very fine and mix with butter; then apply it to the wound. It will quickly recover feeling and will warm up.

90.7. Gif men þæt heafod berste oððe uncuð swyle on ge-
sitte: nime þysse ylcan wyrte wyrtwalan; binde on þone
swyran. Ðonne cymeð hym þæt to godre freme.

90.8. Eft, wið þam ylcan: genim þas ylcan wyrte; wyrc to
duste; do on ða wunde. Þonne byþ heo sona hatigende.

90.9. Gyf hwylcum men ædran aheardode syn oððe his mete
gemyltan nelle: nym þysse ylcan wyrte seaw; mengc ðonne
win ond wæter ond hunig ond þæt seaw eall tosomne; syle
hyt him ðonne wearm drincan. Ðonne byþ him sona bet.

90.10. Eft, wið þæra ðearma ece ond wið ealles þæs innoðes:
nim þas ylcan wyrte; dryg hy þonne ond gegnid to duste,
swyþe smale. Do ðonne þæs dustes fif cuculeras fulle ond
ðreo full godes wines; syle hym ðonne drincan þæt. Ðonne
deah hyt him wið swa hwylcum earfoðum swa him on innan
bið.

90.11. Gyf ðonne æfter ðam men sy sogoþa getenge oððe
hwylc innan gundbryne, genim ðonne þysse wyrte wyrtwalan ond gecnuca swyþe wel. Do ðonne on swyþe god beor;
syle hyt him þonne wlacu supan. Ðonne wene ic þæt hyt him
wel fremie ge wið sogoðan ge wið æghwylcum incundum
earfoðnyssum.

90.12. Wið heafodece: genim þas ylcan wyrte; wyrc clyþan
þærof; lege ðonne on þæt heafod. Ðonne genimð hyt sona
þæt sar onweg.

90.13. Wiþ þam næddercynne ðe man *spalangius* hateð:
genim þysse ylcan wyrte twigo ond þa leaf; seoð on wine.

90.7. If a swelling on someone's head has burst or a strange swelling develops there: take the roots of this same plant; fasten them around the neck. This will be to his considerable benefit.

90.8. Again, for the same: take this same plant; work it into a powder; put it on the wound. It will soon grow warm.

90.9. If a person's veins are hardened or he has difficulty digesting food: take the juice of this same plant; mix wine and water and honey and the juice all together; then give him this to drink while warm. Then he will soon be better.

90.10. Again, for sore intestines and for pain anywhere in the abdomen: take this same plant; then dry it and crush it into a powder, very fine. Then take five spoonfuls of the powder and three of good wine; give him this to drink. This will serve him well whatever discomforts he is experiencing internally.

90.11. If, however, after taking this the person is afflicted by heartburn or by some purulent inflammation, then take the roots of this plant and pound them very well. Then put them in some very good beer; give him this to drink, lukewarm. I believe that this will truly benefit him, both for heartburn and for any internal discomforts.

90.12. For headache: take this same plant; make a poultice with it; then lay it on the head. It will soon take the pain away.

90.13. For the type of venomous creature called *spalangius:* take twigs and leaves of this same plant; simmer in wine.

Gnid þonne swipe smale ond lege on wunde, gyf heo tosomne hleapan wolde. Ond ðonne æfter þam genim ða wyrte
ond hunig; menge tosomne; smyre þa wunde ðærmid. Þonne
hatað heo sona.

90.14. Wið næddran slite: gyf hwylc man hyne begyrdeþ mid
þysse wyrte ond hy on wege mid him bereþ, he bið gescylded
fram æghwylcum næddercynne.

90.15. Wið wedehundes slite: genim ðas ylcan wyrte; gnid,
ond hwæten corn; lege on þa wunde. Ðonne halað heo sona.

90.16. Eft, wið nædran slite: gyf seo wund forþunden sy,
genim þysse sylfan wyrte telgran; seoð on wæter, gnid þonne
swyþe smale; gesodene, lege þonne on ða wunde. Ðonne
þæt dolh open sy, genim þa ylcan wyrte unsodene; gnid
swyþe smale; mengc wið hunig; lacna þonne þa wunde þærmid. Ðonne byð heo sona hal.

# 91

# Rude

91.1. Gif blod of nosum flowc: gcnim ðas wyrtc, þc man
*rutam* ond þam gelice oðrum naman rudan nemneþ; do gelomlice on þa næsðyrlu. Wundorlice heo þæt blod of ðam
næsðyrlun gewrið.

Then crush them very fine and lay this on the wound if it is closing up too quickly. Then after this take the plant and some honey; mix them together; smear the wound with this. It will soon grow warmer.

90.14. For snakebite: if a person is girded with this plant and he keeps it with him on his way, he will be shielded from every kind of serpent.

90.15. For the bite of a mad dog: take this same plant; crush it, along with grains of wheat; apply it to the wound. It will soon heal.

90.16. Again, for snakebite: if the wound is badly swollen, take twigs of this same plant; simmer in water, then crush them very fine; then apply them, boiled, to the wound. If, however, the wound is open, then take this same plant uncooked; pound very fine; mix with honey; treat the wound with this. It will soon be well.

# Chapter 91

# Rue

91.1. If blood should flow from the nose: take this plant, which is called *ruta* and similarly by another name, rue; put it often in the nostrils. It will staunch the nosebleed in a wonderful way.

91.2. Wið toðundennysse: genim þas ylcan wyrte *rutam;* syle hy dælmelum swa grene etan, oððe on drince þicgean.

91.3. Wið þæs magan sare: genim þysse ylcan wyrte sæd, ond swefel ond eced; syle þicgean fæstendum.

91.4. Wið eagena sare ond geswel: genim þas ylcan wyrte rutan, wel gecnucude; lege to ðam sare. Eac se wyrttruma, gecnucud ond ðærmid gesmyred, þæt sar hyt wel gebet.

91.5. Wið þa adle ðe man *litargum* hateð; þæt ys on ure ge-þeode ofergytulnys cweden: genim þas ylcan wyrte *rutam,* mid ecede gewesede; begeot þonne ðæne andwlatan ðær-mid.

91.6. Wiþ eagena dymnysse: genim ðysse sylfan wyrte leaf; syle etan fæstendum, ond syle hy drincan on wine.

91.7. Wið heafodece: genim ðas ylcan wyrte; syle drincan on wine. Cnuca eft þas sylfan wyrte ond wring þæt wos on eced; smyre ðonne þæt heafod þærmid. Eac þeos wyrt fremað wið deadspringas.

91.2. For swellings: take this same plant rue; give it to the patient to eat, fresh, a little at a time, or to consume in a beverage.

91.3. For stomachache: take the seeds of this same plant, plus sulfur and vinegar; give to the patient to consume while fasting.

91.4. For pain and swelling of the eyes: take this same plant rue, pounded well; lay it on the sore area. Also, the root, pounded and smeared around the area, will cure the pain well.

91.5. For the illness called *lethargus,* which is forgetfulness in our language: take this same plant rue, soaked in vinegar; moisten the forehead with it.

91.6. For dimness of the eyes: take the leaves of this same plant; give them to the patient to eat while fasting, and give them to be drunk in wine.

91.7. For headache: take this same plant; give to drink in wine. Afterward pound this same plant and press out the juice into vinegar; then smear the head with it. This plant is also good for carbuncles.

## 92

# Minte

92.1. Wið earena sare: genim þysse wyrte wos, þe man *mentastrum* ond oðrum naman [. . .] hateþ, mid strangon wine gemengced; do on þæt eare. Þeah ðær beon wyrmas on acennede, hi þurh ðis sceolon beon acwealde.

92.2. Wið hreoflan: genim þysse ylcan wyrte leaf; syle etan. Gewislice he bið gehæled.

## 93

# Wælwyrt, oððe ellenwyrt

93.1. Wið þæt stanas on blædran wexen: genim þas wyrte, þe man *ebulum* ond oðrum naman ellenwyrte nemneþ (ond eac sume men wealwyrt hatað); gecnuca hy þonne swa mearwe mid hyre leafum; syle drincan on wine. Heo ut anydeþ ða untrumnysse.

93.2. Wið næddran slite: genim þas ylcan wyrte þe we *ebulum* nemdun; ond ær þam ðe þu hy forceorfe, heald hy on þinre handa ond cweð þriwa nigon siþan: *Omnes malas bestias canto.*

# Chapter 92

# Horsemint

92.1. For earache: take the juice of this plant, which is called *mentastrum* and by another name [. . .], mixed with strong wine; put into the ear. Even if worms are growing there, they will be killed by this.

92.2. For a severe skin disease: take the leaves of this same plant; give to eat. The disease will surely be cured.

# Chapter 93

# Dwarf elder, or danewort

93.1. If stones should grow in the bladder: take this plant, which is called *ebulus* and by another name dwarf elder (and moreover some people call it danewort); pound it with its leaves when tender; give to drink in wine. It will expel the infirmity.

93.2. For snakebite: take this same plant that we have called dwarf elder; but before you cut it up, hold it in your hand and say this three times nine times: *Omnes malas bestias canto.*

Þæt ys þonne on ure geþeode: "Besing ond ofercum ealle yfele wilddeor." Forceorf hy ðonne mid swyþe scearpon sexe on þry dælas; ond þa hwile þe þu ðis do, þenc be þam men þe þu ðærmid þencst to gelacnienne; ond þonne þu þanon wende, ne beseoh þu þe na. Nim ðonne þa wyrte ond cnuca hy; lege to þam slite. Sona he bið hal.

93.3. Wiþ wæterseocnysse: genim þysse ylcan wyrte wyrtwalan, gecnucude; wring þonne þærof swa þæt þu hæbbe þærof feower scenceas, ond wines healfne sester; syle drincan æne on dæg. Hyt fremað myclum þam wæterseocan. Eac hyt bynnan healfon geare ealne þone wætan ut atyhþ.

# 94

# Dweorgedweosle

Ðeos wyrt, þe man *pollegium* ond oþrum naman dweorgedwosle nemneþ, hæfð mid hyre manega læcedomas, þeah hy fela manna ne cunne. Þonne ys þeos wyrt twegea cynna, þæt is, wer ond wif. Se wer hafaþ hwite blostman ond þæt wif hafaþ reade oþþe brune; æghwæþer ys nytlic ond wundorlic. Ond hi on him habbaþ wundorlice mihte: mid þam mæstan bleo hy blowaþ ðonne nealice oþre wyrta scrincaþ ond weorniað.

94.1. Wiþ ðæs innoþes sare: genim þas ylcan wyrte *pollegium,* ond cymen; cnuca tosomne mid wætere ond lege to þam nafolan. Sona he bið gehæled.

That is in our language: "I enchant and overcome all evil beasts." Then cut it up into three parts with a very sharp knife; and while you do this, think of the person you want to cure by this means; and when you turn from there, do not look around yourself. Then take the plant and pound it; lay it on the bite. He will soon be well.

93.3. For dropsy: take the roots of this same plant, pounded; wring out enough of it so that you have four cups, plus a half sester of wine; give one cup a day to drink. It does much good for a person suffering from dropsy. Also, within half a year it will draw out all the fluid.

# Chapter 94

# Pennyroyal

This plant, which is called *pulegium* and by another name pennyroyal, has many healing powers within it, although not many people know them. This plant is of two kinds, that is, masculine and feminine. The masculine variety has white flowers and the feminine has red or purple; each is useful and remarkable. And they have a wonderful virtue in them: they bloom with the brightest colors when nearly all other plants are withering and fading.

94.1. For abdominal pain: take this same plant pennyroyal, plus cumin; pound them together with water and apply this to the navel. The person will soon be healed.

94.2. Eft, wið þæs magan sare: genim þas sylfan wyrte *pollegium;* cnuca hy ond mid wætere gewæs; syle drincan on ecede. Hyt þone wlættan þæs magan wel geliþigaþ.

94.3. Wið gicþan þæra gesceapa: genim þas ylcan wyrte; seoð on weallendon wætere; let þonne colian swa oðþæt hyt man drincan mæge, ond hyt þonne drince. Hyt geliþegaþ þone gicþan.

94.4. Eft, wið þæs innoðes sare: þeos sylfe wyrt fremaþ wel geetan, ond to þam nafolan gewriþen swa þæt heo fram þam nafolan feallan ne mæge. Sona heo þæt sar tofereþ.

94.5. Wið þam fefore þe þy ðryddan dæge on man becymeþ: genim þysse ylcan wyrte twigu; befeald on wulle; ster hyne þærmid toforan þam timan þe se fefor hym to wylle. Ond gyf hwa hys heafod mid þysse wyrte onbutan bewindeþ, heo þæt sar þæs heafodes geliðigaþ.

94.6. Gyf deadboren cyld sy on wifes innoðe: genim þysse ylcan wyrte þry cyþas, ond þa syn niwe swa hy swyþost stincen; cnuca on ealdon wine; syle drincan.

94.7. Gyf hwa on scipe wlættan þolige: genime þas ylcan wyrte polleian, ond wermod; cnucie tosomne, mid ele ond mid ecede; smyrige hyne þærmid gelomlice.

94.8. Wið blædran sare, ond wið þæt stanas þaron wexen: genim þas ylcan wyrte polleian, wel gecnucude, ond twegen scenceas wines; gemencg tosomne; syle drincan. Sona seo blædder to selran gehwyrfeð, ond binnan feawum dagum heo þa untrumnysse gehæleþ ond þa stanas þe þæron weaxeþ ut anydeð.

94.2. Again, for stomachache: take this same plant penny-royal; pound it and soak it in water; give to the patient to drink in vinegar. It will relieve nausea of the stomach.

94.3. For itching genitals: take this same plant; cook it in boiling water; then let it cool to the point where one can drink it, and then drink it. It will relieve the itching.

94.4. Again, for abdominal pain: this same plant is of benefit when consumed, and when fastened to the navel so that it cannot fall off. It will soon take the pain away.

94.5. For the fever that comes on a person every third day: take twigs of this same plant; wrap them in wool; fumigate the patient with this before the time when the fever is due to strike. And if a person winds this plant around his head, it will relieve the headache.

94.6. If a stillborn child should remain inside a woman: take three sprigs of this same plant, and they should be fresh so that they smell most strongly; pound in old wine; give to drink.

94.7. If a person suffers nausea on shipboard: take this same plant pennyroyal, plus wormwood; pound them together, along with oil and vinegar; rub him with this frequently.

94.8. For bladder pain, and if stones should grow there: take this same plant pennyroyal, pounded well, plus two cups of wine; mix them together; give to drink. The bladder will soon get better, and within a few days this will heal the infirmity and will expel the stones that have grown inside.

94.9. Gyf hwa onbutan his heortan oþþe on his breoston sar þolie, þonne ete he þas ylcan wyrte *polleium* ond drince hy fæstende.

94.10. Gyf hwylcum men hramma derige: genim þas ylcan wyrte ond twegen scenceas ecede; drince fæstende.

94.11. Wið þæs magan toþundennysse ond þæra innoþa: genim þas ylcan wyrte *pollegium,* gecnucude ond on wætere oððe on wine gewyllede, oþþe þurh hy sylfe; syle þicgean. Sona byþ seo untrumnys forlæten.

94.12. Wið miltan sare: genim þas ylcan wyrte *polleium;* seoð on ecede; syle drincan swa wearm.

94.13. Wiþ lendena ece ond wið þara þeona sare: genim þas ylcan wyrte *polleium* ond pipor, ægþres gelice mycel be gewihte; cnuca tosomne. Ond þonne þu on bæþe sy, smyre þærmid þær hyt swyþost derige.

# 95

# Nepte

Ðas wyrte man *nepitamon* ond oþrum naman nepte nemneþ, ond eac Grecas hy *mente orinon* hataþ.

95.1. Wiþ nædran slite: genim þas wyrte ðe we *nepitamon* nemdon; cnuca mid wine; wring þonne þæt wos ond syle

94.9. If a person should suffer pain around the heart or in the chest, then he should eat this same plant pennyroyal and drink it while fasting.

94.10. If a person suffers from a spasm: take this same plant, plus two cups of vinegar; he should drink it while fasting.

94.11. For swelling of the stomach and the abdomen: take this same plant pennyroyal, pounded and simmered in water or in wine, or by itself; give to the patient to consume. The disease will soon go into remission.

94.12. For pain in the spleen: take this same plant pennyroyal; simmer in vinegar; give it to drink, warm.

94.13. For pain in the loins and for sore thighs: take this same plant pennyroyal, plus pepper, the same amount of each by weight; pound them together. And when you are bathing, smear it on where it hurts the most.

# Chapter 95

# Catmint

This plant is called *nepeta* and by another name catmint, and the Greeks also call it *menta orinon*.

95.1. For snakebite: take this plant that we have called catmint; pound in wine; then wring out the juice and give to the

drincan on wine. Genim eac þa leaf þysse sylfan wyrte, gec-
nocode; lege to þære wunde.

# 96

# Cammoc

Ðas wyrte man *peucedanum,* ond oðrum naman cammoc
nemneþ. Ðeos wyrt þe we *peucedanum* nemdun mæg nædran
mid hyre swæce aflian.

96.1. Wið nædran slite: genim þas ylcan wyrte *peucedanum,*
ond betonican ond heortes smeoruw (oððe þæt mearh) ond
eced; do tosomne; lege þonne to þære wunde. He bið gehæled.

96.2. Wið þa adle þe Grecas *frenesis* nemnað, þæt is on ure
geþeode gewitlest þæs modes, þæt byþ ðonne þæt heafod
aweallen byþ: genim þonne þas ylcan wyrte *peucedanum;*
cnuca on ecede; begeot þonne þæt heafod þærmid. Hyt fremaþ healice.

patient to drink in wine. Also take the leaves of this same plant, pounded; lay them on the wound.

# Chapter 96

# Hog's fennel

This plant is called *peucedanum* and by another name hog's fennel. This plant that we have called hog's fennel can put snakes to flight with its smell.

96.1. For snakebite: take this same plant hog's fennel, plus betony and deer's grease (or marrow) and vinegar; mix together; then apply to the wound. He will be healed.

96.2. For the disease that Greeks call *frenesis,* which is in our language witlessness, that comes about when the head is hot with disease: take this same plant hog's fennel; pound in vinegar; sprinkle the head with this. It will be of premier benefit.

# 97

# Sperewyrt

97.1. Wið blædran sare: genim þas wyrte þe man *hinnula campana,* ond oþrum naman sperewyrte, nemneþ, ond merces sæd ond eorðnaflan ond finules wyrtwalan; cnuca tosomne; syle þonne wlæc drincan. Scearplice hyt fremað.

97.2. Wið toþa 'sare ond wagunge: genim þas ylcan wyrte; syle etan fæstendum. Heo þa teþ getrymeð.

97.3. Wið þæt ymb þæne nafolan syn rengwyrmas: genim þas ylcan wyrte hinnulan; cnuca on wine; lege to ðam innoðe.

# 98

# Ribbe

Ðas wyrte, þe man *cynoglossam* ond oðrum naman ribbe nemneþ, ond hy eac sume men *linguam canis* hateþ.

# Chapter 97

# Elecampane

97.1. For bladder pain: take this plant, which is called *inula Campana* and by another name elecampane, plus wild celery seeds and asparagus and fennel roots; pound them together; then give to the patient to drink lukewarm. It will help efficaciously.

97.2. For toothache and loose teeth: take this same plant; give to the patient to eat while fasting. It will firm up the teeth.

97.3. If there are harmful worms around the navel: take this same plant elecampane; pound in wine; apply it to the abdomen.

# Chapter 98

# Ribwort

This plant is called *cynoglossum* and by another name ribwort, and in addition some people call it *lingua canis*.

98.1. Wið nædran slite: þeos wyrt þe we *cynoglossam* nemdun wel fremað, gecnucud ond on wine geþiged.

98.2. Wið þam fefore ðe þy feorþan dæge on man becymeþ: genim þas ylcan wyrte *cynoglossam,* ða þe feower leaf hæbbe; cnuca hy; syle drincan on wætere. Heo alyseþ þone man.

98.3. Wiþ ðæra earena unnytlicnysse, ond wið þæt man wel gehyran ne mæge: genim þas ylcan wyrte *cynoglossam,* gecnucude ond on ele gewlæhte; drype on þæt eare. Wundorlice hyt hæleþ.

# 99

# Sundcorn

Ðeos wyrt, ðe man *saxifragam* ond oþrum naman sundcorn nemneð, byþ cenned on dunum ond on stænihtum stowum.

99.1. Wið þæt stanas on blædran wexen: genim þas wyrte þe we *saxifragam* nemdun; cnuca on wine; syle drincan þam þoligendan ond ðam fefergendan on wearmum wætere. Swa andweard heo ys, þæs þe is sæd of ðam þe his afandedon, þæt heo þy ylcan dæge þa stanas forbrycð ond hy ut atyhð ond þone man to hys hæle gelædeþ.

98.1. For snakebite: this plant that we call ribwort is of good benefit, pounded and taken in wine.

98.2. For the fever that comes on a person every fourth day: take this same plant ribwort, the one that has four leaves; pound it; give it to drink in water. It will relieve the person.

98.3. For uselessness of the ears, and if a person cannot hear well: take this same plant ribwort, pounded and warmed in oil; drip into the ear. It will cure it wonderfully.

# Chapter 99

# Saxifrage

This plant, which is called *saxifraga* and by another name saxifrage, grows on hills and stony places.

99.1. For stones growing in the bladder: take the plant that we have called saxifrage; pound it in wine, give it to the person who is suffering, or who is feverish, to drink in warm water. It is so effective, as is said of it by those who have tried it, that that same day, it breaks up the stones and expels them and restores health to the person.

# 100

# Eorðyfig

100.1. Wið þæt stanas on blæddran wexen: genim þysse wyrte, þe man *hederan nigran* ond oþrum naman eorðifig nemneþ, seofon berian oððe endlufon, on wætere gegnidene; syle drincan. Wundorlice þa stanas on þære blædran gegaderað ond hy tobricð ond þurh migþan ut atyhð.

100.2. Wið heafodsar: genim þas ylcan wyrte *hederam,* ond rosan wos on wine gewesed; smyre þonne þa ðunwonga ond þone andwlatan. Þæt sar geliðigaþ.

100.3. Wið miltan sare, genim þysse ilcan wyrte croppas: ærest þry, æt oþrum sæle fif, æt þam þryddan sæle seofone, æt þam feorþan cyrre nigon, æt þam fiftan cyrre endlufon, æt þam sixtan cyrre þreotyne, ond æt þam seofoðan cyrre fiftyne, ond æt þam ehteoþan cyrre seofontyne, ond æt þam nigoþan cyrre nigontyne, æt þam teoþan sæle an ond twentig. Syle drincan dæghwamlice on wine; gyf he þonne on fefore sy, syle drincan on wearmum wætere. Mycelon he byþ gebet ond gestrangod.

100.4. Wið þæra wyrma slite þe man *spalangiones* nemneð: genim þysse sylfan wyrte seaw þæs wyrtwalan þe we *hederam* nemdun; syle drincan.

100.5. Eft, wið þæra wunda lacnunge: genim þas ylcan wyrte; seoð on wine; lege to þam wundum.

# Chapter 100

# Ground ivy

100.1. For stones growing in the bladder: take from this plant, which is called *hedera nigra* and by another name ground ivy, seven or eleven berries, crushed in water; give to drink. It will collect the stones in the bladder remarkably well and will break them up and expel them through the urine.

100.2. For headache: take this same plant ivy, plus rose juice soaked in wine; then rub this on the temples and on the face. The pain will be alleviated.

100.3. For pain in the spleen, take sprigs of this same plant: first three, the second time five, the third time seven, the fourth time nine, the fifth time eleven, the sixth time thirteen, the seventh time fifteen, the eighth time seventeen, the ninth time nineteen, the tenth time twenty-one. Give it to the patient to drink daily in wine; if the patient is feverish, give it to him to drink in warm water. He will be much improved and fortified.

100.4. For the bite of those venomous creatures that are called *spalangiones:* take the juice of the roots of this same plant that we have called ivy; give to drink.

100.5. Again, for the healing of those wounds: take this same plant; simmer in wine; apply to the wounds.

100.6. Wiþ þæt næsþyrlu yfele stincen: genim þysse sylfan wyrte seaw, wel ahlytred; geot on þa næsþyrlu.

100.7. Wið þæra earena unnytlicnysse ond wið þæt man ne mæge well gehyran: genim þysse ylcan wyrte seaw—swyþe clæne—mid wine; drype on þa earan. He bið gelacnud.

100.8. Wiþ þæt heafod ne ace for sunnan hætan: genim þysse sylfan wyrte leaf, swyþe hnesce; cnuca on ecede; smyre þonne þone andwlatan þærmid. Eac hyt fremaþ ongean ælc sar þe þam heafode dereþ.

# 101

# Organe

101.1. Wið þæs heafodes sare: genim þysse wyrte seaw, þe man *serpillum* ond oþrum naman organe nemneþ, ond ele ond gebærned sealt, to swyþe smalan duste gebryt; gemengc ealle tosomne; smyre þæt heafod þærmid. Hyt byþ hal.

101.2. Eft, wið heafodece: genim þas ylcan wyrte *serpillum*, gesodene; cnuca on ecede; smyre þærmid þa ðunwonga ond þone andwlatan.

101.3. Gif hwa forbærned sy: genim þas ylcan wyrte *serpillum*, ond æscþrote anne wrid, ond anre yntsan gewihte geswyrfes of seolfre, ond rosan þreora yntsena gewihte; gepuna þonne

100.6. If the nostrils emit a bad smell: take the juice of this same plant, well clarified; pour it into the nostrils.

100.7. For uselessness of the ears and if a person cannot hear well: take the juice of this same plant—very clean—in wine; drip it into the ears. The patient will be healed.

100.8. So that the head will not ache because of the sun's heat: take leaves of this same plant, very fresh; pound in vinegar; then smear the face with them. This will also be of benefit for every pain that afflicts the head.

# Chapter 101

# Wild thyme

101.1. For headache: take the juice of this plant, which is called *serpyllum* and by another name wild thyme, plus oil and burned salt, crushed into a very fine powder; mix them all together; smear the head with it. It will be well.

101.2. Again, for headache: take this same plant wild thyme, cooked; pound in vinegar; smear the temples and the forehead.

101.3. If a person is burned: take this same plant wild thyme, and one stalk of vervain, and one ounce of silver filings and three ounces of roses; pound all together in a mortar. Add

eall tosomne on anum mortere. Do þonne ðærto wex, ond healfes pundes gewihte beran smeruwes ond heortenes; seoð ealle tosomne, feorma hyt, ond lege to þam bærnette.

## 102

# Wermod

Ðeos wyrt, þe man *absinthium* ond oþrum naman wermod nemneð, byþ cenned on beganum stowum ond on dunum ond on stænihtum stowum.

102.1. Wið þæt man læla ond oðre sar of lichaman gedo: genim þas wyrte *absinthium;* seoð on wætere; do þonne on anne clað; lege to þam sare. Gyf þonne se lichoma mearu sy, seoð on hunige; lege to þam sare.

102.2. Wið þæt rengwyrmas ymbe þone nafolan derigen: genim þas ylcan wyrte *absinthium,* ond harehunan, ond elechtran, ealra gelice mycel; seoð on geswettum wætere oþþe on wine; lege tuwa oððe þriwa to þam nafolan. Hyt cwelþ þa wyrmas.

wax to it, and half a pound of bear and deer grease; simmer all these things together, filter it, and apply it to the burn.

# Chapter 102

# Wormwood

This plant, which is called *absinthium* and by another name wormwood, grows in cultivated places, on hills, and in rocky places.

102.1. To remove bruises and other sores from the body: take this plant wormwood; simmer in water; then put it in a cloth; put it on the sore. If the flesh is tender, simmer in honey; put it on the sore.

102.2. If intestinal worms are doing harm around the navel: take this same plant wormwood, and horehound, and lupins, all in the same quantity; simmer in sweetened water or in wine; lay it two or three times on the navel. It will kill the worms.

# 103

# Salvie

103.1. Wið gicþan þæra gesceapa: genim þas wyrte ðe man salvian nemneð; seoð on wætere, ond mid þam wætere smyre þa gesceapu.

103.2. Eft, wið gicþan þæs setles: genim þas ylcan wyrte salfian; seoð on wætere, ond mid ðam wætere beþa þæt setl. Hyt geliðigað ðone gicþan healice.

# 104

# Celendre

104.1. Wið þæt rengwyrmas ymb ðone nafolan wexen: genim þas wyrte, þe man *coliandrum* ond oðrum naman þam gelice cellendre nemneð; seoð on ele to þryddan dæle; do to þam sare ond eac to ðam heafode.

104.2. Wið þæt wif hrædlice cennan mæge: genim þysse ylcan coliandran sæd, endlufon corn oððe þreottyne; cnyte mid anum ðræde on anum clænan linenan claþe. Nime ðonne an man þe sy mægðhades man, cnapa oþþe mægden,

# Chapter 103

# Sage

103.1. For itching genitals: take this plant that is called sage; simmer it in water; and smear the genitals with this water.

103.2. Again, for itching anus: take this same plant sage; simmer it in water; bathe the anus with this water. It relieves the itching remarkably.

# Chapter 104

# Coriander

104.1. If harmful worms should develop around the navel: take this plant, which is called *coriandrum* and similarly by another name, coriander; simmer it down in oil by two-thirds; put it on the sore area and also on the head.

104.2. So that a woman may give birth quickly: take the seeds of this same plant coriander, eleven or thirteen grains; bind them with a thread to a clean linen cloth. Then take a person who is a virgin, either a boy or a girl, and have him

ond healde æt þam wynstran þeo neah þam gewealde. Ond
sona swa eall seo geeacnung gedon beo, do sona þone læcedom aweg þy læs þæs innoðes dæl þæræfter filige.


# 105

## *Porclaca*


105.1. Wið swiðlicne flewsan þæs sædes: fremað wel þeos
wyrt, þe man *porclaca,* ond oðrum naman [. . .] nemneþ,
ægþer ge þurh hy sylfe geþiged ge eac mid oþrum drenceon.


# 106

## Cearfille


106.1. Wið þæs magan sare: genim þysse wyrte, þe man
*cerefolium* ond oþrum naman þam gelice cerfille nemneþ, ðry
croppas swa grene, ond dweorgedwoslan; cnuca on anum
trywenan mortere, ond anne cuculere fulne ameredes huniges ond grene popig; wyll tosomne; syle ðicgean. Hyt þone
magan hrædlice gestrangaþ.

keep it on the left thigh near the genitals. And as soon as the delivery is completed, quickly take the remedy away so that part of the internal organs does not follow.

# Chapter 105

# Purslane

105.1. For excessive flow of semen: the plant that is called *porcilaca,* and by another name [. . .], helps effectively, either consumed by itself or along with other preparations.

# Chapter 106

# Chervil

106.1. For stomachache: take three fresh sprigs from this plant, which is called *chaerephyllum* and similarly by another name, chervil, plus pennyroyal; pound them in a wooden mortar, adding one spoonful of refined honey and fresh poppies; boil together; give it to the patient to consume. It will quickly fortify the stomach.

# 107

# Brocminte

107.1. Wið þære blæddran sare, ond wið þæt man gemigan ne mæge: genim þysse wyrte wos, þe man *sisimbrium* ond oþrum naman brocminte nemneþ; syle þam þoligendan on wearmum wætere ðicgean gyf he feforgende sy; gyf he þonne ne sy, syle him on wine drincan. Ðu hine gelacnust wundorlice.

# 108

# *Olisatra*

108.1. Eft, wið þære blædran sare ond wið þæt man gemigan ne mæge: genim þas wyrte, þe man *olisatrum* ond oþrum naman [. . .] nemneþ; cnuca on gewylledan wine; syle drincan. Heo ðone migðan mihtelice gebet.

# Chapter 107

# Brook mint

107.1. For bladder pain, and if a person cannot urinate: take the juice of this plant, which is called *sisymbrium* and by another name brook mint; give it to the patient to drink in warm water if he is feverish; if he is not, give it to him to drink in wine. You will cure him wondrously.

# Chapter 108

# Alexanders

108.1. Again, for bladder pain and if a person cannot urinate: take this plant, which is called *holus atrum* and by another name [. . .]; pound in boiled wine; give to drink. It will greatly improve urination.

# 109

# Lilie

Ðas wyrte, þe man *crinion* ond oðrum naman *lilium* nemneþ.

109.1. Wið nædran slite: genim þas wyrte þe we *lilium* nemdun, ond *bulbum* þa wyrte, ða man eac oþrum naman halswyrt hateþ; cnuca tosomne; syle drincan. Nim þonne *bulbum* þa wyrte, gecnucude; lege to ðam slite. He byð gehæled.

109.2. Wið geswel: genim lilian leaf, gecnucude; lege to þam geswelle. Scearplice hyt hæleþ ond þæt geswel geliðigaþ.

# 110

# *Lacterida*

Ðeos wyrt, þe man *titymallos calatites* ond oþrum naman *lacteridam* nemneþ, biþ cenned on wætum stowum ond on ofrum.

110.1. Wið þæra innoða sare: genim þysse wyrte wrid *titymalli;* cnuca on wine swa þæt þæs wines syn twegen

# Chapter 109

# Lily

This plant is called *crinon* and by another name *lilium*.

109.1. For snakebite: take this plant that we have called lily, plus the plant *bulbum,* which is also called by another name, throatwort; pound together; give to drink. Then take the plant throatwort, pounded; apply it to the bite. It will be cured.

109.2. For swellings: take lily leaves, pounded; apply them to the swelling. It will cure it efficaciously and will reduce the swelling.

# Chapter 110

# Caper spurge

This plant, which is called *tithymallos calatites* and by another name *lacterida,* grows in damp places and on shores.

110.1. For abdominal pain: take a shoot of this plant *tithymallos;* pound it in wine so that there are two cupfuls of wine;

scenceas; do þonne of þære wyrte þæs woses þærto twegen cuculeras fulle; drince ðonne fæstende. He byþ gehæled.

110.2. Wið weartan: genim þysse ylcan wyrte meolc ond clufþungan wos; do to þære weartan. Þy þriddan dæge hyt þa weartan gehæleþ.

110.3. Wið hreoflan: genim ðysse sylfan wyrte croppas, mid tyrwan gesodene; smyre þærmid.


# III

# Wuduþistel


Ðeos wyrt, ðe man *carduum sylfaticum* ond oðrum naman wuduþistel nemneð, bið cænned on mædum ond wið wegas.

111.1. Wið þæs magan sare: genim þysse wyrte þe we *carduum silfaticum* nemdum ðone crop ufeweardne, swa mearune ond swa grenne; syle þicgean on geswetton ecede. Hyt geliðigað þa sarnysse.

111.2. Wiþ þæt ðu nane yfele geancymas ðe ne ondræde: genim þas ylcan wyrte *carduum silfaticum* on ærnemergen, þonne seo sunne ærest up gange—ond þæt sy þonne se mona sy in Capricornu—ond heald hy mid þe. Swa lange swa ðu hy mid þe byrst, nanwiht yfeles þe ongean cymeð.

then add to it two spoonsful of the juice of this plant; then drink while fasting. He will be cured.

110.2. For warts: take the milk of this same plant, plus juice of celery-leaved crowfoot; put it on the wart. By the third day it will heal the wart.

110.3. For a severe skin disease: take sprigs of this same plant, simmered in resin; smear this on.

# Chapter 111

# Sow thistle

This plant, which is called *carduus silvaticus* and by another name sow thistle, grows in meadows and along paths.

111.1. For stomachache: take the upper part of the flower head of this plant that we have called sow thistle, when it is soft and green; give it to the patient to eat in sweetened vinegar. It will alleviate the soreness.

111.2. So that you do not fear any evil encounters: take this same plant sow thistle in the early morning, when the sun first rises—and this should be done when the moon is in Capricorn—and keep it with you. As long as you carry it with you, no evil will come upon you.

## 112

## *Lupinum montanum*

Ðeos wyrt, þe man *lupinum montanum* ond oþrum naman [. . .] nemneþ, byþ cenned wið hegas ond on sandigum stowum.

112.1. Wið þæt wyrmas ymb ðone nafolan derigen: genym þas wyrte *lupinum montanum,* gecnucude; syle drincan on ecede, anne scenc fulne. Butan yldingce heo ða wyrmas ut awyrpeð.

112.2. Gyf þonne cildun þæt sylfe derige: genim ðas ylcan wyrte *lupinum* ond wermod; cnuca tosomne; lege to ðam nafolan.

## 113

## Gyðcorn

Þeos wyrt, þe man *lactyridem* ond oþrum naman giðcorn nemneð, byð cenned on beganum stowum ond on sandigum.

113.1. Wið þæs innoþes heardnysse: genim þysse wyrte sæd— þæt syndon, ða corn—wel afeormude; syle drincan on wearmum wætere. Sona hyt þone innoð astyreþ.

# Chapter 112

# Lupin

This plant, which is called *lupinus montanus* and by another name [. . .], grows along hedges and in sandy places.

112.1. If worms are doing harm around the navel: take this plant *lupinus montanus,* pounded; give to drink in vinegar, one cupful. It will expel the worms without any delay.

112.2. If the same problem is affecting children: take this same plant *lupinus montanus,* plus wormwood; pound together; apply to the navel.

# Chapter 113

# Spurge laurel

This plant, which is called *latirida* and by another name spurge laurel, grows in cultivated places and in sandy soils.

113.1. For hardening of the abdomen: take the seeds of this plant—that is, the grains—well cleansed; give to drink in warm water. It will soon move the bowels.

# 114

## *Lactuca*

Ðeos wyrt, þe man *lactucam leporinam* ond oþrum naman þam gelice *lactucam* nemneþ, bið cenned on beganum stowum ond on sandigum. Be ðysse wyrte ys sæd þæt se hara, ðonne he on sumura for swiðlicre hætan geteorud byþ, mid þysse wyrte hyne sylfne gelacnað. For þy heo ys *lactuca leporina* genemned.

114.1. Wið feforgende: genim þas wyrte *lactucam leporinam;* lege him nytendum under his pyle. He byþ gehæled.

# 115

## Hwerhwette

Ðeos wyrt, þe man *cucumerem silvaticum* ond oþrum naman hwerhwette nemneþ, byþ cenned neah sæ ond on hatum stowum.

115.1. Wið þæra sina sare ond wið fotadle: genim wyrtwalan þysse wyrte þe we *cucumerem silfaticum* nemdun; seoð on ele to þriddan dæle; smyre þærmid.

## Chapter 114

## Lettuce

This plant, which is called *lactuca leporina* and similarly by another name, *lactuca,* grows in cultivated places and sandy soils. About this plant it is said that the hare, when it is tired in summer on account of the extreme heat, restores itself with this same plant. For this reason, it is called *lactuca leporina.*

114.1. For a person who is feverish: take this plant *lactuca leporina;* lay it under his pillow without his knowing it. He will be cured.

## Chapter 115

## Squirting cucumber

This plant, which is called *cucumis silvaticus* and by another name squirting cucumber, grows near the sea and in warm places.

115.1. For sore joints and for foot disease: take the roots of this same plant that we have called squirting cucumber; simmer down in oil by two-thirds; smear it on.

115.2. Gif cild misboren sy: genim ðysse ylcan wyrte wyrttruman to þriddan dæle gesodenne; þweah ðonne þæt cild þærmid. Ond gyf hwa þysse wyrte wæstm fæstende þygeð, hyt
him becymð to frecnysse; for ði gehwa hine forhæbbe þæt
he hi na fæstende ete.

# 116

# Henep

Ðeos wyrt, þe man *cannave silfatica* ond oþrum naman henep
nemneþ, byþ cenned on wiþerrædum stowum ond wið
wegas ond hegas.

116.1. Wið þæra breosta sare: genim þas wyrte *cannavem
silvaticam,* gecnucude mid rysle; lege to þam breostan. Heo
tofereþ þæt geswel; ond gyf þær hwylc gegaderung biþ, heo
þa afeormaþ.

116.2. Wið cile bærnettes: genim þysse ylcan wyrte wæstm,
mid netelan sæde gecnucudne ond mid ecede gewesed; lege
to þam sare.

115.2. If a child is born prematurely: take the roots of this same plant simmered down by two-thirds; then wash the child with it. Also, if a person eats the fruit of this plant while fasting, it will endanger him; therefore, everyone should refrain from eating it while fasting.

# Chapter 116

# Hemp

This plant, which is called *cannabis silvatica* and by another name hemp, grows on rough ground and along paths and hedges.

116.1. For sore breasts: take this plant hemp, pounded in fat; lay it on the breasts. It will diminish the swelling; and if there is some kind of abscess there, it will cleanse it.

116.2. For frostbite: take the fruit of this same plant, pounded with nettle seeds and soaked in vinegar; apply to the sore.

# 117

# Rude

Ðeos wyrt, þe man *rutam montanam* ond oþrum naman þam gelice rudan nemneþ, byþ cenned on dunum ond on unbeganum stowum.

117.1. Wið eagena dymnysse ond wið yfele dolh: genim þysse wyrte leaf þe we *rutam montanam* nemdon, on ealdum wine gesodene; do þonne on an glæsen fæt; smyre syþþan þærmid.

117.2. Wiþ ðæra breosta sare: genim þas ylcan wyrte *rutam silvaticam;* cnuca on trywenan fæte; nim þonne swa mycel swa ðu mid ðrim fingron gegripan mæge; do on an fæt ond þærto anne scenc wines ond twegen wæteres; syle drincan; gereste hyne þonne sume hwile. Sona he byð hal.

117.3. Wið lifersare: genim þysse ylcan wyrte anne gripan ond oþerne healfne sester wæteres ond ealswa mycel huniges; wyll tosomne; syle drincan þry dagas, ma gyf him þearf sy. Þu hine miht gehælan.

117.4. Wið þæt man gemigan ne mæge: genim þysse ylcan wyrte *rute silvatice* nigon stelan ond wæteres ðry scenceas; cnuca tosomne ond ecedes healfne sester; wyll eal tosomne; syle drincan singallice nigon dagas. He byð gehæled.

117.5. Wið þære nædran slite þe man *scorpius* hateþ: genim þysse ylcan wyrte sæd *rute silvatice;* cnuca on wine; syle drincan. Hyt geliðigaþ þæt sar.

# Chapter 117

# Wild rue

This plant, which is called *ruta montana* and similarly by another name, wild rue, grows on hills and in uncultivated places.

117.1. For dimness of the eyes and for a severe ulcer: take the leaves of this plant that we have called wild rue, cooked in old wine; then put it in a glass vessel; after this smear it on.

117.2. For a sore chest: take this same plant wild rue; pound in a wooden vessel; then take as much as you can pick up with three fingers; put it into a vessel and add one cup of wine and two of water; give to drink; let the person then rest for some while. He will soon be well.

117.3. For liver pain: take one handful of this same plant and one and a half sesters of water and the same amount of honey; boil these together; give to drink for three days, more if the person needs it. You can cure him.

117.4. If a person cannot urinate: take nine stalks of this same plant, wild rue, and three cups of water; pound them together with half a sester of vinegar; simmer them all together; give this to drink continually for nine days. He will be healed.

117.5. For the bite of the venomous creature called scorpion: take the seeds of this same plant wild rue; pound in wine; give to drink. It will relieve the pain.

## 118

# Seofenleafe

Ðeos wyrt, þe man *eptafilon* ond oðrum naman *septifolium* nemneð (ond eac sume men seofonleafe hatað), byþ cenned on beganum stowum ond on sandigum landum.

118.1. Wið fotadle: genim þas wyrte *septifolium,* gecnucude ond wið croh gemengcgede; smyre ðonne þa fet mid þam wose. Þy ðryddan dæge hyt þæt sar genimeþ.

## 119

# Mistel

119.1. Wið heafodece: genim þas wyrte, þe man *ocimum* ond oðrum naman mistel nemneþ; cnuca mid rosan wose oððe wyrtriwes oððe mid ecede; lege to ðam andwlatan.

119.2. Eft, wið eagena sare ond geswel: cnuca ðas sylfan wyrte on godum wine; smyre þa eagan þærmid. Þu hy gehælst.

# Chapter 118

# Tormentil

This plant, which is called *heptaphyllum* and by another name *septifolium* (and in addition some people call it tormentil), grows in cultivated places and in sandy soils.

118.1. For foot disease: take this plant tormentil, pounded and mixed with saffron; then smear the feet with the juice. By the third day this will take the pain away.

# Chapter 119

# Basil

119.1. For headache: take this plant, which is called *ocimum* and by another name basil; pound it with rose juice or myrtle juice or with vinegar; apply it to the forehead.

119.2. Again, for pain and swelling of the eyes: pound this same plant in good wine; spread it around the eyes. You will cure them.

119.3. Wið ædrena sare: do þæt sylfe; syle drincan on rinde þæs æples þe man *malum granatum* nemneþ.

# 120

# Merce

120.1. Wið eagena sare ond wið geswel: nim ðas wyrte, þe man *appium* ond oðrum naman merce nemneþ, wel ge-cnucude mid hlafe; lege to þam eagon.

# 121

# Yfig

Ðeos wyrt, þe man *hedera crysocantes* ond oðrum naman ifig nemneþ, is gecweden *crysocantes* for ðy þe heo byrð corn golde gelice.

121.1. Wið wæterseocnysse: genim þysse wyrte twentig corna; gnid on anne sester wines, ond of þam wine syle drin-can þry scenceas seofon dagas. Seo untrumnys ðurh þone migðan byð aidlud.

119.3. For kidney pain: do the same thing; give to drink in the skin of the kind of apple that is called *malum granatum*.

# Chapter 120

# Wild celery

120.1. For pain and swelling of the eyes: take this plant, which is called *apium* and by another name wild celery, pounded well along with bread; apply it to the eyes.

# Chapter 121

# Ivy

This plant, which is called *hedera crisocantes* and by another name ivy, is named *crisocantes* because it bears seeds that are like grains of gold.

121.1. For dropsy: take twenty grains of this plant; crush them into one sester of wine, and give the patient three cups of this wine to drink for seven days. The infirmity will be drained through the urine.

## 122

# Minte

122.1. Wið teter ond pypylgende lic: genim ðysse wyrte seaw, þe man *mentam* ond þam gelice oþrum naman mintan nemneð; do þonne þærto swefel ond eced; cnuca eal tosomne; smyre mid anre feþere. Sona þæt sar geliðigað.

122.2. Gyf yfele dolh oððe wunda on heafde syn: genim þas ylcan wyrte *mentam,* gecnucude; lege to þam wundum. Heo hy gehæleþ.

## 123

# Dile

123.1. Wið gicðan ond wið sar þæra gesceapa: genim þas wyrte, ðe man *anetum* ond oþrum naman dyle nemneþ; bærn to duste. Nim þonne þæt dust ond hunig; mengc tosomne. Beþa ærest þæt sar mid wætere, þweah syþþan mid wearmum wyrtrywenum wose; lege þonne ða lacnunge þærto.

123.2. Gyf þonne wifmen hwæt swylces derige: do hyre man fram hyre byrþþinene þone sylfan læcedom þære wyrte þe we nu her beforan cwædon.

# Chapter 122

# Mint

122.1. For impetigo and a pimply body: take the juice of this plant, which is called *menta* and similarly by another name, mint; add to it sulfur and vinegar; pound all this together; apply it with a feather. The pain will soon be relieved.

122.2. If there are bad open sores or wounds on the head: take this same plant mint, pounded; apply it to the wounds. It will heal them.

# Chapter 123

# Dill

123.1. For itching and pain in the genitals: take this plant, which is called *anethum* and by another name dill; burn it to ashes. Then take the ashes and some honey; mix together. First bathe the sore area with water; after that wash it with warm myrtle oil; then apply the preparation to it.

123.2. If anything is troubling a woman: have her midwife give her the same medication, made from the plant we have just named.

123.3. Wið heafodece: genim þysse ylcan wyrte blostman; seoð mid ele; smyre ða þunwonga, ond gewrið þæt heafod.

# 124

# Organe

124.1. Ðeos wyrt, þe man *origanum* ond oðrum naman þam gelice organan nemneþ, is hattre gecynde ond swyðlicre, ond heo gebræceo ut atyhð, ond heo ælc yfel blod ond ðone dropan gewyldeþ, ond heo wyþ nyrwet ond liferseocum wel fremað.

124.2. Wið gebræceo: genim þas ylcan wyrte organan; syle etan. Þu wundrast hyre fremfulnysse.

# 125

# Sinfulle

125.1. Wið ealle gegaderunga þæs yfelan wætan of þam lichoman: genim þas wyrte þe man *sempervivum* ond oðrum

123.3. For headache: take the flowers of this same plant; simmer in oil, smear it on the temples, and tie around the head.

# Chapter 124

# Marjoram

124.1. This plant, which is called *origanum,* and similarly by another name, marjoram, is rather hot by nature and rather strong, and it suppresses coughs, and it also subdues bad blood and rheum, and it is helpful for shortness of breath and liver disease.

124.2. For coughs: take this same plant marjoram; give it to eat. You will be astonished at its usefulness.

# Chapter 125

# Houseleek

125.1. For all accumulations of corrupted matter in the body: take this plant, which is called *sempervivum* and by another

naman sinfulle nemneþ, ond rysle ond hlaf ond coliandran; cnuca eal tosomne þam gelice þe ðu clyþan wyrce; lege to þam sare.

# 126

## Finol

126.1. Wið gebræceo ond wyð nyrwyt: genim þysse wyrte wyrttruman, þe man *fenuculum* ond oðrum naman finul nemneþ; cnuca on wine; drince fæstende nigon dagas.

126.2. Wið blædran sare: genim þysse ylcan wyrte þe we *fenuculum* nemdun, anne gripan swa grene, ond merces wyrttruman grenne ond eorðnafolan wyrtruman grene; do on anne niwne croccan ond wæteres anne sester fulne; wyl tosomne to feorðan dæle; drince þonne fæstende seofon dagas oþþe ma. Ond he bæþes bruce, na swaþeah coles, ne he colne wætan þicge. Butan yldincge þære blæddran sar byð geliðigod.

name houseleek, and fat and bread and coriander; pound all together in the same way as you would make a poultice; apply to the sore area.

# Chapter 126

# Fennel

126.1. For coughs and shortness of breath: take the roots of this plant, which is called *feniculum* and by another name fennel; pound in wine; have the patient drink it for nine days while fasting.

126.2. For bladder pain: take this same plant that we have called fennel, one handful when fresh, plus fresh roots of wild celery and fresh roots of asparagus; put into a new earthenware pot along with one sester of water; boil together, reducing by three fourths; have the person then drink this for seven days or more while fasting. And he should take a bath: not, however, a cold one, nor should he drink cold liquids. The bladder pain will be relieved without delay.

## 127

# Liðwyrt

Ðeos wyrt, þe man *erifion* ond oþrum naman liðwyrt nemneþ, byþ cenned fyrmest in Gallia—þæt is, on Franclande—
on þam munte þe man Soractis hateþ. Heo hæfð merces gelicnysse, ond heo hafað blostman readne swylce cærse, ond
heo hafaþ seofon wyrttruman ond swa fela stelena, ond
heo hy sylfe tobrædeð on unbeganum stowum ond na on
wætum. Heo byþ ælcon timan blowende, ond heo hafað sæd
swylce beana.

127.1. Wiþ lungenadle: genim þas wyrte *erifion,* gecnucude
þam gelice þe þu clyþan wyrce; lege to þam sare. Heo hit gehæleþ. Nim þonne þæt wos þisse sylfan wyrte; syle drincan.
Þu wundrast þæs mægenes þysse wyrte.

## 128

# Halswyrt

128.1. Wið wifes flewsan: genim þas wyrte, þe man *sinfitum
album* ond oþrum naman halswyrt nemneþ; gedrige hy ond
cnuca to swiþe smalan duste; syle drincan on wine. Sona heo
þa flewsan gewrið.

# Chapter 127

# Rue

This plant, which is called *eriphion* and by another name rue, is native to Gaul—that is, to the land of the Franks—on the mountain called Soractis. It looks like wild celery, and it has red flowers like watercress, and it has seven roots and just as many stems, and it propagates itself in uncultivated places but not in wetlands. It blooms in all seasons, and it has seeds like broad beans.

127.1. For lung disease: take this same plant rue, pounded in the same way as you would make a poultice; lay it on the sore area. It will heal it. Then take the juice of this same plant; give to drink. You will be amazed at the power of this plant.

# Chapter 128

# Comfrey

128.1. For women's excessive flow of menstrual blood: take this plant, which is called *symphytum album* and by another name comfrey; dry it and pound it into a very fine powder; give to drink in wine. It will soon staunch the flow.

# 129

# Petersilie

Ðas wyrte man *triannem* ond oþrum naman *petroselinum* nemneþ, ond eac hy sume men þam gelice petersilie hateþ.

129.1. Wið næddran slite: genim of ðysse wyrte *petroselini* swyþe smæl dust, anes scyllincges gewihte; syle drincan on wine. Nim ðonne þa wyrte, gecnucude; lege to þære wunde.

129.2. Wiþ ðæra sina sare: genim þas ylcan wyrte *petroselinum,* gepunude; lege to þam sare. Heo geliþigað þæt sar þæra sina.

# 130

# Cawel

130.1. Wið ealle geswel: genim þysse wyrte croppas þe man *brassicam silvaticam* ond oðrum naman caul nemneþ; cnuca mid ealdon rysle; gemencg ðonne swylce ðu clyðan wyrce; do on anne þicne linenne clað; lege to þam sare.

130.2. Wið sidan sare: genim þas ylcan wyrte *brassicam silvaticam;* lege to þam sare, swa gemencged swa we her beforan cwædon.

# Chapter 129

# Parsley

This plant is called *triannis* and by another name *petroselinum,* and in addition, similarly, some people call it parsley.

129.1. For snakebite: take the very fine powder of this same plant parsley, one shilling-weight; give to drink in wine. Then take the plant itself, pounded; apply to the wound.

129.2. For sore sinews: take this same plant parsley, pounded; apply to the sore area. It will relieve the pain in the sinews.

# Chapter 130

# Cabbage

130.1. For all swellings: take the shoots of this plant called *brassica silvatica* and by another name cabbage; pound it in old fat; mix as if you were making a poultice; put on a thick linen cloth; apply to the sore spot.

130.2. For pain in the side: take this same plant cabbage; lay it on the sore area, mixed as we have just said.

130.3. Wið fotadle: genim þas sylfan wyrte *brassicam,* on þa ylcan wisan þe we ær cwædon; ond swa se læcedom yldra byþ swa he scearpnumulra ond halwendra byþ.

# 131

# Nædderwyrt

131.1. Ðeos wyrt, þe man *basilisca* ond oðrum naman nædder-wyrt nemneþ, byð cenned on ðam stowum þær seo nædre byþ þe man þam ylcan naman nemneð *basiliscus.* Witodlice nys heora cyn an, ac hi sindon þreora cynna. An ys *olocry-seis*—þæt is on ure geðeode gecweden þæt heo eall golde scine. Ðonne is oðer cyn *stillatus*—þæt is on ure geþeode dropfah; seo ys swylce heo gyldenum heafde sy. Þæt ðridde cyn ys *sanguineus*—þæt is, blodread; eac swilce heo gylden on heafde sy. Ealle ðas cyn þeos wyrt basilisca hæfð. Þonne gyf hwa þas wyrte mid him hafað, þonne ne mæg him nan ðyssa næddercynna derian.

2 Seo forme næddre, *olocryssus,* is genemned *eriseos:* seo swa hwæt swa heo gesihð, heo toblæwð ond anæleþ. Ðonne seo oþer, *stillatus,* is soðlice gecweden *crysocefalus asterites:* þeos swa hwæt swa heo gesyhð hyt forscrincð ond gewiteþ. Þonne is seo ðridde genemned *hematites* ond *crysocefalus:* swa hwæt swa ðeos gesyhð oþþe hrepeð, hyt toflewð swa ðæt

130.3. For foot disease: take this same plant cabbage, prepared in the same way as we have just said; and the older the preparation is, the more efficacious and the more salutary it will be.

# Chapter 131

# Sweet basil

131.1. This plant, which is called *basilica* and by another name sweet basil, grows in those places where the snake lives that is called by the same name *basiliscus*. Indeed, there is not only one type of them, but there are three kinds. One is *olocryseis*—that is, as is said in our language, that it shines like gold. Then the second kind is *stillatus*—that is, spotted in our language; it looks as if it had a gilded head. The third kind is *sanguineus*—that is, bloodred; this one too looks as if it were gilded on its head. The plant *basilica* comprises all these kinds. If anyone has this plant with him, then none of these kinds of snakes can harm him.

The first snake, *olocryssus,* is named *eriseos:* whatever it sees, it blows on and sets on fire. The second, *stillatus,* is indeed called *chrysocefalus asterites:* whatever it sees shrinks up and dies. The third is called *hematites* and *chrisocefalus:* whatever it sees or touches melts away, so that nothing is left but

þær nanwiht belifeþ buton þa ban. Þonne hæfð þeos wyrt *basilisca* ealle heora strengða; gyf hwylc man þas wyrte mid him hafað, wið eall næddercyn he biþ trum.

3    Þeos wyrt ys rudan gelic, ond heo hæfð meolc reade swylce celidonie, ond heo hæfð wolcenreade blostman. Ond se þe hy niman wylle, he hyne sylfne clænsie ond hy bewrite mid golde, ond mid seolfre, ond mid heortes horne, ond mid ylpenbane, ond mid bares tuxe, ond mid fearres horne; ond mid hunige geswette wæstmas þær onbutan gelecge.

# 132

# *Mandragora*

Ðeos wyrt þe man *mandragoram* nemneþ ys mycel ond mære on gesihþe, ond heo ys fremful. Ða þu scealt þyssum gemete niman. Þonne þu to hyre cymst, þonne ongist þu hy be þam þe heo on nihte scineð ealswa leohtfæt. Þonne ðu hyre heafod ærest geseo, þonne bewrit þu hy wel hraþe mid iserne þy læs heo þe ætfleo. Hyre mægen ys swa mycel ond swa mære þæt heo unclænne man, þonne he to hyre cymeþ, wel hraþe forfleon wyle. For ðy þu hy bewrit, swa we ær cwædon, mid iserne, ond swa þu scealt onbutan hy delfan, swa ðu hyre mid þam iserne na æthrine, ac þu geornlice scealt mid ylpenbanenon stæfe ða eorðan delfan. Ond þonne þu hyre handa ond hyre fet geseo, þonne gewrið þu hy; nim

the bones. The plant *basilica* has all the powers of these snakes; if anyone has this plant with him, he will be able to resist all kinds of snakes.

This plant looks like rue, and it has red, milky juice like celandine, and it has purple flowers. And whoever wants to pick it, he should cleanse himself and should scratch a line around it with gold, with silver, with deer antler, with ivory, with a boar's tusk, and with the horn of an ox; and he should strew fruit sweetened with honey round about the place.

# Chapter 132

# Mandrake

This plant called *mandragoras* is large and glorious in appearance, and it is beneficial. You must pick it in this manner. When you approach the plant, you will recognize it because it shines in the night like a lantern. As soon as you see its head, very quickly scratch a line around it with an iron tool, lest it flee from you. Its power is so great and so glorious that it will quickly flee if any unclean person approaches it. Because of this you must score a line around it with an iron tool, as we have just said, and then you must dig around it, being careful not to touch it with the iron, but you should diligently dig the earth with an ivory staff. When you see its hands and feet, then tie them up; then take the other end of

þonne þæne oþerne ende ond gewrið to anes hundes swyran, swa þæt se hund hungrig sy; wurp him syþþan mete toforan, swa þæt he hyne aræcan ne mæge buton he mid him þa wyrte up abrede.

2    Be þysse wyrte ys sæd þæt heo swa mycele mihte hæbbe þæt swa hwylc þincg swa hy up atyhð, þæt hyt sona scyle þam sylfan gemete beon beswycen. For þy, sona swa þu geseo þæt heo up abroden sy ond þu hyre geweald hæbbe, genim hy sona on hand swa, andwealc hi, ond gewring þæt wos of hyre leafon on ane glæsene ampullan. Ond þonne ðe neod becume þæt þu hwylcon men þærmid helpan scyle, þonne help þu him ðyssum gemete.

132.1. Wið heafodece ond wið þæt man slapan ne mæge: genim þæt wos; smyre þone andwlatan. Ond seo wyrt swa some þam sylfan gemete þone heafodece geliðigaþ; ond eac þu wundrast hu hrædlice se slæp becymeþ.

132.2. Wið þæra earena sare: genim þysse ylcan wyrte wos, gemænged mid ele þe sy of nardo; geot on ða earan. Þu wundrast hu hrædlice he byþ gehæled.

132.3. Wið fotadle, þeah ðe heo hefegust sy: genim of þære swyþran handa þysse wyrte ond of þære wynstran, of ægþerre handa þreora penega gewihte; wyrc to duste; syle drincan on wine seofon dagas. He byþ gehæled; na þæt an þæt þæt geswel geset, ac eac þæra sina togunge to hæle gelædeþ ond þa sar; butu wundurlice gehæleþ.

132.4. Wið gewitleaste; þæt is, wið deofulseocnysse: genim of þam lichoman þysse ylcan wyrte *mandragore* þreora penega gewihte; syle drincan on wearmum wætere swa he eaðelicost mæge. Sona he byþ gehæled.

the rope and fasten it around the neck of a dog, seeing to it that the dog is hungry; after that, throw some food before the dog, so that he cannot reach it unless he pulls the plant up with him.

About this plant it is said that it has such great powers that whatever pulls it up will be deceived in that same way. For this reason, as soon as you see that it has been pulled up and you have it in your power, take it quickly into your hand, twist it, and wring the juice of its leaves into a glass bottle. And when you have a need to help a person with it, then help him in this way.

132.1. For headache and sleeplessness: take the juice; smear it on the forehead. And in this same manner, the plant like-wise relieves headache; and moreover, you will be amazed at how quickly sleep will come.

132.2. For earache: take the juice of this same plant, mixed with oil of spikenard; drip it into the ears. You will be amazed at how quickly he will be cured.

132.3. For foot disease, even if it is very severe: take from the right and from the left hand of this plant, three penny-weights from each hand; work it into a powder; give to drink in wine for seven days. The patient will be cured; not only will the swelling go down, but this will also ameliorate the sinew spasms and the pain; both will be wondrously cured.

132.4. For insanity, that is, for possession by devils: take three pennyweights from the body of this same plant *man-dragoras;* give it to the patient to drink in warm water when he can most easily do so. He will soon be cured.

132.5. Eft, wið sina togunge: genim of ðam lichoman þysse wyrte anre yntsan gewihte; cnuca to swyþe smalan duste; gemencg mid ele; smyre þonne þa þe ðas foresprecenan untrumnysse habbað.

132.6. Gyf hwa hwylce hefige yfelnysse on his hofe geseo: genime þas wyrte *mandragoram* onmiddan þam huse, swa mycel swa he þonne hæbbe. Ealle yfelu heo ut anydeð.

# 133

# Læcewyrt

133.1. Ðeos wyrt, ðe man *lichanis stefanice* ond oðrum naman læcewyrt nemneþ, hafað lange leaf ond geþufe ond hæwene; ond hyre stela byð mid geþufum bogum, ond heo hafað on ufeweardum þam stelan geoluwe blostman. Þysse wyrte sæd, on wine geseald, fremað wel ongean eal næddercyn ond wið *scorpiones* stincg, to ðam swyþe þæs ðe sume men secgeað þæt gyf hy man ofer þa *scorpiones* gelegð, þæt heo him unmihtignesse ond untrumnysse on gebrincge.

132.5. Again, for sinew spasm: take the weight of one ounce from the body of this plant; pound into a very fine powder; mix with oil; smear it on those who have the infirmity just mentioned.

132.6. If one should see any grievous evil in his home: have him take the plant *mandragoras* within the house, however much he has of it. It will expel all evils.

# Chapter 133

# Rose campion

133.1. This plant, which is called *lychnis stephanomatica* and by another name rose campion, has leaves that are long and luxuriant and of a blue-green hue; and its stem has bushy shoots, and it has yellow flowers on the upper part of the stem. The seeds of this plant, when given in wine, are of much benefit against all snakes and against a scorpion's sting, so much so that some people say that if a person puts it on a scorpion, it will cause it to be weak and incapacitated.

# 134

## Action

Ðeos wyrt, ðe man *action* ond oðrum naman [. . .] nemneð, hafað gelice leaf cyrfættan, ac hy beoð maran ond heardran; ond heo hafað wið þone wyrttruman greatne stelan ond twegea fæðma lange; ond heo hafað on ufeweardon þam stelan sæd ðistele gelic, ac hyt byð smælre ond read on bleo.

134.1. Wið þæt man blod ond worsm gemang hræce: genim þysse wyrte feower penega gewiht sædes, ond cyrnlu of pintrywenum hnutum; cnuca tosomne þam gelice þe þu anne æppel wyrce; syle þicgean þam untruman. Hyt hyne gehæleð.

134.2. Wið þæra liða sare: genim þas ylcan wyrte, gecnucude ond to clyþan geworhte; lege to ðam sare; heo hyt geliðigað. Eac þam sylfan gemete heo ealde wunda gehæleþ.

# Chapter 134

# Burdock

This plant, which is called *arcion* and by another name [. . .], has leaves like a gourd, but they are larger and tougher; and toward its root it has a large stalk two cubits long; and on the upper part of the stem it has seeds like the thistle, but they are smaller and red in color.

134.1. If a person coughs up blood and foul matter together: take four pennyweights of the seeds of this plant, plus kernels from pine-tree cones; pound them together in the same way that you would make a dumpling; give this to the sick person to eat. It will heal him.

134.2. For pain in the joints: take this same plant, pounded and made into a poultice; lay it on the sore; it will relieve the pain. It also heals old wounds in the same way.

## 135

# Suþernewuda

Ðeos wyrt, þe man *abrotanum* ond oðrum naman suð-ernewuda nemneþ, ys twegea cynna. Þonne is þæt oðer cyn greaton bogum ond swyþe smælon leafon, swylce heo ma fexede gesewen sy, ond heo hafað blostman ond sæd swyþe gehwæde; ond heo is godes swæces ond myceles ond biterre on byrgynge.

135.1. Wyð nyrwyt, ond wið banece, ond wið þæt man ear-foðlice gemigan mæge: þysse wyrte sæd wel fremað, ge-cnucud ond on wætere geðiged.

135.2. Wið sidan sare: genim ðas ylcan wyrte ond betonican; cnuca tosomne; syle drincan.

135.3. Wið attru ond wið nædrena slite: genim ðas ylcan wyrte *abrotanum;* syle drincan on wine. Heo helpeð wel. Cnuca hy eac mid ele ond smyre ðone lichoman þærmid. Eac heo wið þone colan fefor wel fremað; eac þæt sæd þysse wyrte stranglice afligeþ, gindstred oððe onæled.

135.4. Wið þæra nædrena slite þe man *spalangiones* ond *scor-piones* nemneð: þeos sylfe wyrt wel fremað.

135.5. Wið eagena sare: genim þas ylcan wyrte *abrotanum,* gesodene mid ðære wyrte þe man *melacidoniam* ond oðrum naman *codoniam* hateþ, ond ðonne mid hlafe gecnucude þam gelice þe þu clyþan wyrce; lege to þam sare. Hyt byð ge-liðigod.

# Chapter 135

# Southernwood

This plant, which is called *habrotonum* and by another name southernwood, is of two kinds. The second kind has thick boughs and very small leaves that look like hair, and it has very meager flowers and seeds; but this kind has a good and intense smell and is bitter to the taste.

135.1. For shortness of breath, for pain in the thigh, and for difficulty urinating: the seeds of this plant are of much benefit, pounded and drunk in water.

135.2. For pain in the side: take this same plant, plus betony; pound together; give to drink.

135.3. For poison and for snakebite: take this same plant southernwood; give to drink in wine. It helps very much. Also, pound it in oil and smear the body with it. It also helps with feverish chills; moreover, the seeds of this plant, whether scattered about or set alight, will drive out all evils.

135.4. For bites of those venomous creatures that are called *spalangiones* and of scorpions: this same plant is effective.

135.5. For eye pain: take this same plant southernwood, simmered along with the plant called *malum cydoneum* and by another name *cotoneum,* and then pounded along with bread in the same manner as you would make a poultice; lay this on the sore area. The pain will be relieved.

2  Þeos wyrt is, swa we her beforan cwædon, twegea cynna:
oðer ys wif, oðer wer, ond hy habbað on eallum þingon gelice
mihte ongean þa ðincg ðe we her beforan sædon.

# 136

# Laber

Ðeos wyrt, þe man *sion* ond oðrum naman laber nemneþ,
byð cenned on wætum stowum.

136.1. Wið þæt stanas on blædran wexen: genim ðas wyrte;
syle etan oððe gesodene oððe hræwe. Heo þa stanas þurh
migþan ut atyhð.

136.2. Eac ðeos sylfe wyrt wel fremað wið utsiht ond wið þæs
innoþes astyrunge.

As we have stated above, this plant is of two kinds: one is female, the other male; they have the same powers in all regards against the diseases that we have just mentioned.

2

# Chapter 136

# Water parsnip

This plant, which is called *sium* and by another name water parsnip, grows in watery places.

136.1. If stones should grow in the bladder: take this plant; give to eat either cooked or uncooked. It will draw out the stones through the urine.

136.2. This same plant also helps for diarrhea and for movement of the bowels.

# 137

# Sigilhweorfa

Ðeos wyrt, þe man *eliotropus* ond oðrum naman sigilhweorfa nemneð, byþ cenned on fættum landum ond on beganum. Ond heo hafað leaf, neah swylce mistel, þa beoð ruge ond brade; ond heo hafað sæd sinewealt ond þæt byð þreora cynna bleos.

137.1. Wið ealra næddercynna slitas ond wið *scorpiones:* genim þysse wyrte wyrttruman *eliotropos;* syle drincan on wine, ond gecnucude, lege to þære wunde. Heo fremað mycelon.

137.2. Wyð þæt wyrmas ymb ðone nafolan on þam innoðe derigen: genim ðas ylcan wyrte, ond ysopan ond *nytrum* ond cærsan; cnuca tosomne ealle; syle drincan on wætere. Heo acwelleþ ða wyrmas.

137.3. Wið weartan: genim þas ylcan wyrte ond sealt; cnuca tosomne; lege to þam weartum. Heo hy fornimeþ. Þanon heo eac *verrucaria* genemned is.

# Chapter 137

# Heliotrope

This plant, which is called *heliotropium* and by another name heliotrope, grows in rich soils and in cultivated places. And it has leaves, much like basil, that are hairy and broad; and it has round seeds that are of three different colors.

137.1. For the bites of all kinds of snakes and for scorpions: take the roots of this plant heliotrope; give to drink in wine, and lay it on the wound, pounded. It will help a great deal.

137.2. If worms around the navel are harming the intestines: take this same plant, plus hyssop and niter and watercress; pound them all together; give to drink in water. It will kill the worms.

137.3. For warts: take this same plant and some salt; pound together; apply to the warts. It will take them away. For this reason it is also called *verrucaria*.

# 138

## *Spreritis*

Ðeos wyrt, ðe man *spreritis* ond oðrum naman [. . .] nemneþ, hæfð gehwæde leaf ond geðufe, ond heo of anum wyrt-truman manega bogas asendeþ, ond þa beoð neah ðære eorðan alede. Ond heo hafað geoluwe blostman, ond gyf þu hy betweonan þinum fingrum gebrytest, þonne hafað heo swæc swylce myrre.

138.1. Wið þone colan fefor: genim þas wyrte *spreritis;* seoð on ele. Ond to ðam timan ðe se fefor to þam men ge-nealæcean wylle, smyre hyne þærmid.

138.2. Wyð wedehundes slite: genim þas ylcan wyrte; cnuca to duste; nim ðonne anne cuculere fulne; syle drincan on wearmum wætere. He byð hal.

138.3. Wyþ miltan sare: genim þysse sylfan wyrte anne godne gripan ond anne sester fulne meolce; wyll tosomne. Syle drincan, healf on mergen healf on æfen, þa hwyle þe him þearf sy. Seo milte byð gelacnud.

# Chapter 138

# Field marigold

This plant, which is called *spieritis* and by another name [. . .], has small and bushy leaves, and from one root it sends out many branches, and they extend close to the ground. And it has yellow flowers, and if you crush them between your fingers they smell like myrrh.

138.1. For a feverish chill: take this plant *spieritis;* simmer it in oil. And toward the time when the fever begins its onset, smear the person with it.

138.2. For the bite of a mad dog: take this same plant; pound into a powder; then take a spoonful; give to drink in warm water. The patient will be well.

138.3. For pain in the spleen: take a good handful of this plant and one sester of milk; simmer them together. Give to drink, half in the morning, half in the evening, as long as the patient needs it. The spleen will be cured.

# 139

## *Aizos minor*

Ðeos wyrt, þe man *ayzos minor* ond oðrum naman [. . .] nemneþ, byð cenned on wagum, ond on stænigum stowum, ond
on dunum, ond on ealdum byrgenum. Ond heo of anum
wyrttruman manega gehwæde bogas asendeþ, ond þa beoð
fulle of gehwædum leafum ond langum ond scearpum ond
fættum ond wel wosigum. Ond þysse wyrte wyrttruma ys
unnytlic.

139.1. Wið oman, ond wið eagena sare, ond wið fotadle:
genim þas wyrte, butan wyrttruman; cnoca mid smedman
ðam gelice þe þu cliðan wyrce; lege to þisum untrumnyssum.
Hit hy geliðigað.

139.2. Wyð heafodece: genim þysse ylcan wyrte wos, ond
rosan wos; mængc tosomne; smyre þæt heafod þærmid. Þæt
sar byð geliðigud.

139.3. Wyð þæra wyrma slite þe man *spalangiones* hateþ:
genim þas ylcan wyrte *aizos,* on wine gecnucude; syle drincan. Hyt fremað nytlice.

139.4. Wið utsiht ond wið innoðes flewsan, ond wyð wyrmas
þe on ðam innoþe deriað, þeos sylfe wyrt wel fremað.

139.5. Eft, wyð gewhylce untrumnysse þæra eagena: genim
þysse ylcan wyrte wos; smyra ðonne þa eagan þærmid.
Nytlice hyt fremað.

# Chapter 139

# Stonecrop

This plant, which is called *aizoon minus* and by another name [. . .], grows along paths, and in stony places, and on hills, and on old burial mounds. And from a single root it sends out many small shoots that are full of leaves that are small but are long and pointed and full-bodied and full of juice. And the root of this plant is of no use.

139.1. For erysipelas and eye pain and foot disease: take this plant, except for the roots; pound it with fine flour in the same way as you would make a poultice; apply it for these infirmities. It will relieve them.

139.2. For headache: take the juice of this same plant, plus juice of roses; mix them together; smear the head with this. The pain will be relieved.

139.3. For the bite of those venomous creatures called *spalangiones:* take this same plant *aizoon,* pounded in wine; give to drink. It will help in a useful way.

139.4. For diarrhea and intestinal flux, and for worms that do harm in the intestines, this same plant does well.

139.5. Again, for every infirmity affecting the eyes: take the juice of this same plant; rub it around the eyes. It will help in a useful way.

# 140

# Tunsingwyrt

140.1. Ðeos wyrt, þe man *elleborum album* ond oðrum naman tunsincgwyrt nemneð (ond eac sume men wedeberge hatað), byð cenned on dunum, ond heo hafað leaf leace gelice. Þysse wyrte wyrttruman man sceal niman onbutan midne sumur, ond eac swa some þa wyrt ealle, for ðy heo is to læcedomum wel gecweme. Þæt is to lufigenne on ðysse wyrte þæt heo hafað gehwædne wyrttruman ond na swa rihtne þæt he be sumum dæle gebyged ne sy; he byþ breaþ ond tidre þonne he gedriged byð, ond þonne he tobrocen byþ he rycþ eal swylce he smic of him asende, ond he byð hwonlice bitterre on byrgincge. Þonne beoð þa maran wyrttruman lange ond hearde ond swyþe bittere on byrgincge, ond hy habbaþ to ðam swyþlice mihte ond frecenfulle, þæt hy foroft hrædlice þone man forþilmiaþ. Ðonne sceal man þysne wyrttruman, swa we ær cwædon, gedrigean ond þa langnysse toceorfan on pysena gelicnysse.

2    Mycel læcedom is to gehwylcum þingum þæt man ðonne þysses wyrttruman genime, tyn penega gewihte; swaðeah ne mæg man æfre for his strengðe hyne syllan þicgean onsundrum, ac mid sumum oðrum mete gemencgedne be þære swylcnysse þe seo untrumnys þonne byð: þæt is, gyf seo untrumnes swa stið beo, syle þicgean on beore oððe on blacan briwe.

# Chapter 140

# White hellebore

140.1. This plant, which is called *elleborus albus* and by another name white hellebore (and also some people call it false hellebore), grows on hills, and it has leaves like a leek. The roots of this plant should be picked about midsummer, and likewise the entire plant, because it is well suited to medications. What is attractive about his plant is that it has a small root, and never so straight that it is not bent in some part; it is brittle and fragile when it is dried, and when it is broken off it emits fumes as if it were sending out smoke, and it is somewhat bitter to the taste. The larger roots, though, are long and hard and very bitter to the taste, and in that regard they have a strong and dangerous power, that very often they quickly choke a person. Thus, as we said before, one must dry this root and cut up its length into pieces like peas.

If one takes ten pennyweights of this root, this will make 2 for a strong medicine for many purposes; because of its strength, however, it should never be given to be consumed by itself, but rather mixed with some other food in the amount suited to the infirmity; that is, if the illness is very serious, give it to drink in beer or in a light soup.

140.2. Gyf he þonne on utsihte sy: syle þicgean on pysena wose oððe mid þære wyrte ðe man *oriza* hateþ, mid smedeman. Þa ealle swaþeah sceolon beon ærost on liðon beore gesodene ond geliðigode.

140.3. Ðeos wyrt soðlice ealle ealde ond hefige ond unlacnigendlice adlu tofereþ, swa þæt he byþ gelacnud þeah he ær his hæle on tolætenesse wære.

# 141

## *Buoptalmon*

141.1. Ðeos wyrt, þe man *buoptalmon* ond oðrum naman [. . .] nemneþ, hafað hnescne stelan ond leaf gelice finule, ond heo hafað geoluwe blostman eal swylce eage, þanon heo eac þone naman onfeng. Heo byþ cenned fyrmest wið Meoniam ða ceastre. Þysse wyrte leaf, gecnucude ond to clyþan geworhte, tolysað gehwylce yfele springas ond heardnyssa.

141.2. Wyþ æwyrdlan þæs lichoman þe cymeþ of togotennysse þæs eallan: genim þysse wyrte wos; syle drincan. Heo agyfð þæt gecyndelice hiw, ond he byð gehywlæht swylce he of swiðe haton bæþe geode.

140.2. If a person is suffering from diarrhea: give it to him to consume in a pea soup or with that plant that is called *oriza,* along with some flour. All these, however, should be first cooked and diluted in light beer.

140.3. Indeed, this plant will cure all long-lasting or grievous or incurable diseases, so that a person will be cured even though he had been in despair about his health.

# Chapter 141

# Corn marigold

141.1. This plant, which is called *buphthalmon* and by another name [. . .], has tender stems and leaves like fennel, and it has a yellow flower that looks just like an eye, from which it got its name. It is native to the city of Meonia. The leaves of this plant, pounded and made into a poultice, will remove all bad boils and hardenings.

141.2. For damage to the body that comes from effusion of bile: take the juice of this plant; give to drink. It will restore the natural complexion, and the person will be colored as if he were coming from a very hot bath.

# 142

# Gorst

Ðeos wyrt, ðe man *tribulus* ond oðrum naman gorst nemneþ, is twegea cynna; oþer byð cenned on wyrtunum, oðer ut on felda.

142.1. Wið mycelne hætan þæs lichaman: genim þas wyrte *tribulum,* gecnucude; lege þærto.

142.2. Wyð þæs muþes ond þæra gomena fulnysse ond forrotudnysse: genim þas wyrte *tribulum,* gesodene; cnuca mid hunige. Heo hæleþ ðone muð ond þa goman.

142.3. Wiþ þæt stanas on blædran wexen: genim þysse ylcan wyrte sæd, swa grene, gecnucud; syle drincan. Wel hyt fremað.

142.4. Wyþ næddran slite: genim þysse ylcan wyrte sæd, swa grene, gecnucud, fif penega gewihte; sile drincan. Eac swylce nim þa wyrte mid hyre sæde, gecnucude; lege to þære wunde. Heo alyseþ hyne of þære fræcenysse.

142.5. Þisse sylfan wyrte sæd eac swylce, on wine gedruncen, is halwende ongean attres drync.

142.6. Wiþ flean: genim þas ylcan wyrte, mid hyre sæde gesodene; sprengc into þam huse. Heo cwelð þa flean.

# Chapter 142

# Land caltrop

This plant, which is called *tribulus* and by another name land caltrop, is of two kinds; one grows in gardens, the other one out in the fields.

142.1. For a high bodily temperature: take this plant land caltrop, pounded; apply it.

142.2. For foulness and putrefaction of the mouth and throat: take this plant land caltrop, boiled; pound it in honey. It will heal the mouth and the throat.

142.3. If stones should grow in the bladder: take the seeds of this same plant, fresh, pounded; give to drink. It will be of much help.

142.4. For snakebite: take seeds of this same plant, fresh, pounded, five pennyweights; give to drink. Moreover, take the plant with its seeds, pounded; apply to the wound. It will deliver the person from danger.

142.5. The seeds of this same plant, drunk in wine, are also salutary against a poisonous drink.

142.6. For fleas: take this same plant, boiled with its seeds; sprinkle it about the house. It will kill the fleas.

# 143

## *Coniza*

143.1. Ðeos wyrt, þe man *conize* ond oðrum naman [. . .] nemneþ, ys twegea cynna, þeah þe oðer sy mare, oþer læsse.
Þonne hafað seo læsse smæle leaf ond gehwæde ond swyþe
gecwemne swæc; ond seo oðer hafað maran leaf ond fætte,
ond hefigne swæc. Ond þyssa wyrta wyrttruman syndon un-
nytlice, ac þysse wyrte stela mid þam leafum, gindstred ond
onæled, nædran afligeþ. Ond eac heo gecnucud ond to
clyþan geworht þæra nædrena slite gehæleþ; ond heo gnættas, ond micgeas, ond flean acwelleþ, ond heo eac swylce
ealle wunda gelacnað. Ond heo earfoðlicnysse þæs migþan
astyreþ, ond heo þa cynelican adle gehæleþ, ond heo on
ecede geseald fylleseocum helpeþ.

143.2. Þeos wyrt *conize,* on wætere gesoden ond sittendum
wife under geled, heo ðone cwiþan afeormaþ.

143.3. Gyf wif cennan ne mæge: nime þysse ylcan wyrte wos,
mid wulle; do on þa gecyndelican. Sona heo þa cennincge
gefremeþ.

143.4. Wyþ ða colan feforas: genim þas ylcan wyrte; seoð on
ele; nim þonne þone ele; smyre þone lichaman. Ða feforas
beoð fram anydde.

143.5. Wiþ heafodece: þyssa wyrta genim ða læssan; wyrc to
clyþan; lege to ðam sare. Heo hyt geliðigaþ.

# Chapter 143

# Fleabane

143.1. This plant, which is called *conyza* and by another name [. . .], has two types, though one is larger and the other smaller. The smaller has small, slight leaves and a very agreeable smell, and the other has larger, succulent leaves and a strong smell. And the roots of this plant are of no use, but the stem of this plant with its leaves, if set alight and strewn about, puts snakes to flight. Also, pounded and made into a poultice, it heals snakebites; and it kills gnats, mosquitoes, and fleas, and it likewise heals all wounds. It stimulates the urine if there is difficulty urinating, and it heals the king's disease, and, drunk in vinegar, it helps those who suffer from epilepsy.

143.2. This plant *conyza,* boiled in water and set under a seated woman, cleanses the womb.

143.3. If a woman cannot bring forth a child: take the juice of this same plant, along with some wool; put it into the genitals. It will soon bring about the birth.

143.4. For feverish chills: take this same plant; simmer in oil; then take the oil; smear the body with it. The fevers will be driven away.

143.5. For headache: take the smaller of these plants; make it into a poultice; apply it where there is pain. It will relieve it.

# 144

# Foxes glofa

144.1. Wið oman: genim þysse wyrte leaf þe man *trycnos manicos* ond oðrum naman foxes glofa nemneþ; wyrc to clyþan; lege to þam sare. Hyt geliðigaþ.

144.2. Wiþ pypelgende lic, þæt Grecas *erpinam* nemnað: genim þas ylcan wyrte ðe we *trycnos manicos* nemdun, ond smedeman; wyrc to clyþan; lege to þam sare. Hyt byþ gehæled.

144.3. Wið heafdes sare, ond wið þæs magan hætan, ond wið cyrnlu: genim þas ylcan wyrte, mid ele gecnucude; smyre þa sar. Hy toslupað.

144.4. Wiþ ðæra earena sare: genim þisse sylfan wyrte seaw, mid rosan seawe; drype on þæt eare.

# Chapter 144

# Thorn apple

144.1. For erysipelas: take the leaf of this plant, which is called *strychnon manicon* and by another name thorn apple; work it into a poultice; lay it on the sore. This will soothe it.

144.2. For a pimply body, which the Greeks call *herpes:* take this same plant that we have called thorn apple, plus fine flour; work it into a poultice; lay it on the sore. It will be healed.

144.3. For headache, for excessive heat in the stomach, and for hard swellings: take this same plant, pounded in oil; smear it on the sore areas. The pains will dissipate.

144.4. For earache: take the juice of this same plant, along with juice of roses; drip it into the ear.

# 145

## *Glycyrida*

145.1. Wið þone drigean fefor: genim þas wyrte ðe man *glycyridam* ond oðrum naman [. . .] nemneþ; wyl on wearmum wætere; syle drincan. Hyt fremaþ nytlice.

145.2. Eac swylce þeos sylfe wyrte ðæra breosta sar, ond þære lifre, ond þære blædran, ond þæra ædrena mid gesodenan wine gehæleþ. Eac heo þyrstendon þone þurst geliþigað.

145.3. Wið leahtras ðæs muþes: þysse ylcan wyrte wyrttruma, geeten oððe gedruncen, wel fremað ond þa leahtras gehæleþ. Eac heo wunda gehæleþ, ðærmid gewesede; ond se wyrttruma swa same þæt sylfe gegearwað, ac na swaþeah swa scearplice.

# Chapter 145

# Licorice

145.1. For a dry fever: take this plant, which is called *glycyr-rhiza* and by another name [. . .]; simmer it; give to drink in warm water. It will help effectively.

145.2. Likewise this same plant, when taken with boiled wine, will heal pain in the chest, and in the liver, and in the bladder, and in the kidneys. And it also relieves thirst among those who are thirsty.

145.3. For disorders affecting the mouth: the root of this same plant, when eaten or drunk, is beneficial and will cure the disorders. It also heals wounds, when they are saturated with it; and the root likewise has the same effect, although not so decisively.

# 146

## *Strutius*

146.1. Wið þæt man gemigan ne mæge: genim ðysse wyrte wyrttruman ðe man *strutium* ond oþrum naman [. . .] nemneþ; syle ðicgean. He þone migðan astyreð.

146.2. Wið liferseocnysse ond wið nyrwyt ond wið swiðlicne hracan: genim þysse wyrte, to duste gecnucudre, anne cuculere fulne; syle drincan on liþan beore. Hyt framað; ond eac hyt þone innoð wið þæs eallan togotenysse gegladað ond þæt yfel forð gelædeþ.

146.3. Wiþ þæt stanas on blædræn wexen: genim ðas sylfan wyrte *strutium,* ond lubastican wyrttruman ond ðære wyrte ðe man *capparis* hateð; cnuca tosomne; syle drincan on liðon beore. Hyt tolyseþ ða blædran ond ða stanas forð gelædeþ; ond eac þære miltan sar hyt tolyseþ.

146.4. Wið hreoflan: genim þas ylcan wyrte, ond meluw ond eced; cnuca togædre; lege to þam hreoflan. He bið gelacnud.

146.5. Eft, ðeos sylfe wyrt, mid berenum meluwe on wine gesoden, ealle yfele heardnyssa ond gegaderunga heo tofereþ.

# Chapter 146

# Soapwort

146.1. If a person cannot urinate: take the root of this plant, which is called *struthium* and by another name [. . .]; give to eat. It will stimulate the urine.

146.2. For liver disease and shortness of breath, and for a bad cough: take one spoonful of this plant, pounded into a powder; give it to the patient to drink in light beer. It will be of benefit; and it will also relieve the intestines from an overabundance of bile and will draw out what is harmful.

146.3. If stones should grow in the bladder: take this same plant *struthium*, plus the roots of lovage and of the plant called *capparis;* pound together; give to drink in light beer. It will relieve the bladder and will draw out the stones; and it will also relieve pain in the spleen.

146.4. For a severe skin disease: take this same plant, plus flour and vinegar; pound together; apply to the diseased areas. The disease will be cured.

146.5. Again, this same plant, boiled in wine with barley meal, will expel all harmful hardenings and abscesses.

# 147

## *Aizon*

147.1. Ðeos wyrt, ðe is *aizon* ond oðrum naman [. . .] gecweden, seo is swylce heo symle cwycu sy. Ond heo hafað elne langne stelan on fincres greatnysse, ond heo is wel wosig, ond heo hafað fætte leaf in fingeres lenge. Heo bið cenned on dunum, ond heo eac byþ hwilon on wealle geseted. Ðeos wyrt, mid meoluwe gecnucud, gehæleþ mænigfealde untrumnyssa ðæs lichoman; þæt is, berstende lic ond forrotudnysse þæs lices ond eagena sarnysse ond hætan ond forbærnednysse. Ealle þas þing heo gehæleþ.

147.2. Wið heafodece: genim þysse ylcan wyrte wos, *aizon*, mid rosan wose gemenged; begeot þæt heafod þærmid. Hyt geliðigaþ þæt sar.

147.3. Wið þære nædran slite þe man *spalangionem* nemneþ: genim þas ylcan wyrte *aizon;* syle drincan on haton wine.

147.4. Eft, do þæt sylfe wið utsiht, ond wið wyrmas on innoðe, ond wið swiðlicne cyle. Hyt fremað.

# Chapter 147

# Orpine

147.1. This plant, which is called *aizoon* and by another name [. . .], gives the appearance of being always alive. And it has a stem an ell long and as thick as a finger; it is full of juice, and it has succulent leaves that are the length of a finger. It grows on hills, and also sometimes it is affixed to a wall. Pounded along with flour, this plant will heal various infirmities of the body; that is, a ruptured body and putrefaction of the flesh and eye pain and inflammations and burns. It heals all these conditions.

147.2. For headache: take the juice of this same plant *aizoon*, mixed with juice of roses; moisten the person's head with it. It will soothe the pain.

147.3. For bite of the venomous creature that is called *spalangio:* take this same plant *aizoon;* give to drink in hot wine.

147.4. Again, do the same for diarrhea, and for intestinal worms, and for a severe chill. It will be of benefit.

# 148

# Ellen

148.1. Wið wæterseocnysse: genim þas wyrte þe man *samsu-chon* ond oðrum naman ellen hateþ; syle drincan, gewyllede. Heo gehnæceþ ða anginnu þam wæterseocum; eac swylce heo fremaþ wið þa unmihticnysse þæs migðan ond wið þæra innoða astyrunga.

148.2. Wið springas ond wið toborsten lic: genim þysse ylcan wyrte leaf *samsuchon*, gedrigede ond gecnucude ond mid hunige gemencgede; lege to þam sare. Hyt sceal berstan ond halian.

148.3. Wið *scorpiones* stincg: genim þas ylcan wyrte, ond sealt ond eced; cnuca tosomne ond to plastre gewyrc; lege to ðam stinge. He bið gehæled.

148.4. Wið micele hætan ond wið geswel ðæra eagena: genim ðas sylfan wyrte, mid meluwe gemencgede ond to cliðan geworhte; lege to þam eagon. Hy bið geliþigad.

# Chapter 148

# Sweet marjoram

148.1. For dropsy: take this plant, which is called *sampsuchum* and by another name sweet marjoram; give to drink, boiled. It checks the onset of dropsy; it is also of use for difficulty in urinating and for movement of the bowels.

148.2. For carbuncles and for a ruptured body: take the leaves of this same plant, dried and pounded and mixed with honey; apply this to the sore. It will burst and heal.

148.3. For a scorpion's sting: take this same plant, plus salt and vinegar; pound together and make into a plaster; lay it on the sting. It will be healed.

148.4. For severe inflammation and swelling of the eyes: take this same plant, mixed with flour and made into a poultice; lay it on the eyes. They will be relieved.

# 149

## *Stecas*

Þeos wyrt, ðe man *stecas* ond oþrum naman [. . .] nemneþ, hæfð sæd mycel, ond þæt ys smæl ond gehwæde; ond heo sylf ys boþene gelic, buton þæt heo hafað sumon dæle maran leaf ond stiðeran.

149.1. Genim þas wyrte, gesodene; syle drincan. Heo þæra breosta sar gehæleþ. Eac hyt is gewunelic þæt hy man to manegum godum drenceon gemencge.

# 150

## *Thyaspis*

Ðeos wryrte, ðe man *thyaspis* ond oþrum naman [. . .] nemneþ, hafaþ smæle leaf on fingres lencge ond todælede ond nyþer wið þa eorþan ahyldende. Ond heo hafað ðynne stelan ond langne, ond heo hafað on ufeweardum hæwene blostman, ond þæt sæd byþ cenned gind ealne þone stelan. Eal ðeos wyrt is strangre gecynde ond bitterre; ðysse wyrte wos, wel gewrungen ond an scenc ful gedruncen, ealle þa

# Chapter 149

# French lavender

This plant, which is called *stoechas* and by another name [. . .], has many seeds, ones that are fine and small; and the plant itself looks like rosemary, except that it has somewhat larger and stiffer leaves.

149.1. Take this plant, boiled; give to drink. It will heal chest pain. It is also customary for it to be mixed into many good drinks.

# Chapter 150

# Shepherd's purse

This plant, which is called *thlaspis* and by another name [. . .], has small, divided leaves of a finger's length that incline downward to the earth. And it has a long and thin stem, and it has whitish flowers on the upper part, and its seeds are generated along the entire stem. This whole plant has a strong and bitter nature; a cupful of the juice of this plant, when it is well pressed out and one cupful is drunk, expels all

biternysse ðe of þam geallan cymeþ, heo ðurh ða gemæneli-
can neode ond ðurh spiwðan ut anydeþ.

150.1. Ðeos sylfe wyrt ealle þa yfelan gegaderunge þæs innoþes heo fornimeþ, ond eac swylce heo wifa monoðlican
astyreð.

# 151

# *Omnimorbia*

Þeos wyrt, þe man *polios* ond oþrum naman *omnimorbia*
nemneþ, ond eac sume men [. . .] hataþ, byþ cenned on
dunum. Ond heo of anum wyrttruman manega telgran
asendeþ, ond heo on uferwerdum hafaþ sæd swylce croppas;
ond heo is hefegon swæce ond hwon weredre on byrgincge.

151.1. Wið nædran slite: genim þysse wyrte wos *polios,* on
wætere gesoden; syle drincan. Hyt gehæleþ ðone slite.

151.2. Wið wæterseocnysse: do þæt sylfe. Hyt þone innoð
alyseþ.

151.3. Wið miltan sare: genim þas ylcan wyrte *polios;* seoð on
ecede; syle drincan. Nytlice heo þone miltseocan gehæleþ.
Ðeos sylfe wyrt, on huse gestred oþþe onæled, nædran
afligeþ, ond eac swylce heo niwe wunda fornimeþ.

the bitterness that comes from the bile by natural evacuation of the bowels or by vomiting.

150.1. This same plant takes away all accumulations of harmful matter in the abdomen, and it also stimulates women's menses.

# Chapter 151

# Wood sage

This plant, which is called *polium* and by another name *omnimorbia,* and also some people call it [. . .], grows on hills. And it sends out many shoots from a single root, and on its upper part it has seeds like berries; and it has a strong smell and is somewhat sweet to the taste.

151.1. For snakebite: take the juice of this plant *polium,* boiled in water; give to drink. It will heal the bite.

151.2. For dropsy: do the same. It will relieve the abdomen.

151.3. For pain in the spleen: take this same plant *polium;* simmer in vinegar; give to drink. It will effectively heal the sufferer. This same plant, if strewn or burned in the house, will put snakes to flight, and in like manner it will also heal fresh wounds.

## 152

# *Hypericon*

152.1. Þeos wyrt þe man *hypericon* ond oþrum naman *corion* nemneþ, for gelicnysse cymenes, heo hafaþ leaf rudan gelice, ond of anum stelan manega telgran weaxaþ, ond þa reade; ond heo hafaþ blostman swylce banwyrt; ond heo hafað berian synewealte ond hwon lange on beres mycelnysse on þam ys sæd, ond þæt sweart ond on swæce swylce tyrwe; ond heo bið cenned on beganum stowum. Ðeos wyrt, gecnucud ond gedruncen, þone migþan astyreþ, ond heo þa monoðlican wundorlice deþ gyf hy man ðam gecyndelican lime under gelegeþ.

152.2. Wið þone fefor þe þy feorðan dæge on man becymeþ: genim þas ylcan wyrte, gecnucude; syle drincan on wine.

152.3. Wiþ ðæra sceancena geswel ond ece: genim þysse ylcan wyrte sæd; syle drincan on wine. Binnan feowertigan dagon he byð gehæled.

# Chapter 152

# Saint John's wort

152.1. This plant is called *hypericum* and by another name *corion,* for it looks like cumin. It has leaves like rue, and many twigs grow from one stalk, and they are red; and it has flowers that resemble wallflower; and it has round, somewhat long berries, the size of barley, in which is the seed, and the seed is black and smells like resin; and the plant grows in cultivated places. When pounded and drunk, this plant stimulates the urine, and it also brings on the menses wonderfully if it is put under the genital organs.

152.2. For the fever that comes on a person every fourth day: take this same plant, pounded; give to drink in wine.

152.3. For swelling and pain of the legs: take the seeds of this same plant; give to drink in wine. Within forty days he will be healed.

# 153

## *Acantaleuca*

Ðeos wyrt, þe man *acantaleuce* ond oðrum naman [. . .] nemneþ, byð cenned on stænigum stowum ond on dunum. Ond heo hafaþ leaf swylce wulfes camb, ac hi beoþ mearwran ond hwittran ond eac geþufran; ond heo hafað twegea elne lancne stelan on fingres greatnysse, oððe sumon dæle maran.

153.1. Wið þæt man blode hræce, ond wið þæs magan sare: genim ðas ylcan wyrte *acantaleuce;* cnuca to duste; syle drincan on wætere anne cuculere fulne. Hyt fremað wel.

153.2. Wið þæs migðan astyrunge: genim þas ylcan wyrte swa wosige, gecnucude; syle drincan. Heo ðone migðan forð gelædeþ.

153.3. Wið yfele læla: genim þas ylcan wyrte; wyrc to clyþan; lege to þam sare. Heo hyt afyrreð. Þysse sylfan wyrte syde þæra toþa sar geliðigað, gyf hyne man swa wearmne on þam muþe gehealdeþ.

153.4. Wið hramman: genim þysse ylcan wyrte sæd, gecnucud; syle drincan on wætere. Hyt helpeþ. Se sylfa drenc eac swylce ongean næddrena slite wel fremað. Eac swylce gyf mon þas wyrte on mannes swyran ahehð, heo næddran aflygeþ.

# Chapter 153

# Globe thistle

This plant, which is called *acantha leuca* and by another name [. . .], grows in stony places and on hills. And it has leaves like wild teasel, but they are softer and whiter and also thornier; and it has a stem more than two ells long and as thick as a finger, or somewhat bigger.

153.1. If a person should cough up blood, and for stomachache: take this same plant *acantha leuca;* pound into a powder; give a spoonful of it to drink in water. It will be of much benefit.

153.2. To stimulate the urine: take this same plant when it is juicy, pounded; give to drink. It will bring the urine forth.

153.3. For bad bruises: take this same plant; work it into a poultice; lay it on the sore. It will remove it. A decoction of this same plant will ease toothache, if it is held warm in the mouth.

153.4. For cramps: take the seeds of this same plant, pounded; give to drink in water. It will help. This same drink will likewise be of much benefit for snakebite. Likewise if the plant is worn about the neck, it will put snakes to flight.

# 154

# Beowyrt

Ðeos wyrt, þe man *acanton* ond oþrum naman beowyrt nemneð, byþ cenned on wynsumon stowum ond on wætum ond
eac swylce on stænigum.

154.1. Wið þæs innoþes astyrunge ond þæs migþan: genim
þysse ylcan wyrte wyrttruman, gedrigedne ond to duste gecnucudne; syle drincan on wearmum wætere.

154.2. Wiþ lungenadle, ond wið gehwylce yfelu þe on þam innoðe dereþ: ðeos sylfe wyrt wel fremað, geþiged þam gelice
þe we her beforan cwædon.

# 155

# Cymen

155.1. Wyð þæs magan sare: genim þysse wyrte sæd, þe man
*quimminon* ond oþrum naman cymen nemneþ, on ele gesodene ond mid syfeðon gemencged ond swa togædere gewylled; wyrc þonne to clyþan; lege to ðam innoþe.

# Chapter 154

# Scotch thistle

This plant, which is called *acantion* and by another name Scotch thistle, grows in pleasant places and on damp ground and likewise in stony places.

154.1. To stimulate the bowels and the urine: take the roots of this same plant, dried and pounded into a powder; give to drink in warm water.

154.2. For lung disease, and for any disease that is doing harm in a person's insides: this same plant is of much benefit, if eaten or drunk as we have said before.

# Chapter 155

# Cumin

155.1. For stomachache: take the seeds of this plant, which is called *cuminum* and by another name cumin, boiled in oil and mixed with bran and then simmered together; work it into a poultice; lay it on the abdomen.

155.2. Wyþ nyrwyt: genim þas ylcan wyrte *quiminon,* ond wæter ond eced; meng tosomne; syle drincan. Hyt fremað nytlice. Ond eac, on wine geþiged, heo næddran slite wel gehæleþ.

155.3. Wiþ ðæra innoþa toðundennysse ond hætan: genim þas ylcan wyrte, mid winberian gecnucude oððe mid beanenon meoluwe; wyrc to clyþan. Heo gehæleþ ða toðundennysse.

155.4. Eac swylce blodryne of næsþyrlon heo gewrið mid ecede gemængedum.

# 156

# Wulfes tæsl

Ðeos wyrt, þe man *camelleon alba* ond oþrum naman wulfes tæsl nemneþ, hafað leaf wiþerræde ond þyrnyhte, ond heo hafaþ on middan sumne sinewealtne crop ond þyrnyhtne, ond se biþ brunon blostmun behæfd; ond he hafað hwit sæd ond hwitne wyrtruman ond swyðe gestencne.

156.1. Wiþ þæt wyrmas on þam innoþe ymb þone nafolan dergen: genim ðisse ylcan wyrte wyrttruman seaw oððe dust; syle drincan on wine oððe on wætere þe ær wære organe oððe dweorgedwosle on gewylled. Hyt rume þa wyrmas forð gelædeþ.

155.2. For shortness of breath: take this same plant cumin, plus water and vinegar; mix together; give to drink. It will serve usefully. And also, if drunk in wine, it will heal snake-bite well.

155.3. For swelling and inflammation of the abdomen: take this same plant, pounded with grapes or with broad-bean flour; work it into a poultice. It will heal the swelling.

155.4. Likewise it stops nosebleed when mixed with vinegar.

# Chapter 156

# Carline thistle

This plant, which is called *chamaeleon albus* and by another name carline thistle, has leaves that are rough and thorny, and in its middle it has a round and thorny flower head, and this is surrounded by brown flowers; and it has white seeds and white roots and a strong smell.

156.1. If intestinal worms are doing harm around the navel: take the juice of the roots of this same plant, or its powder; give to drink in wine or in water in which marjoram or pennyroyal was previously boiled. It will expel the worms to a great extent.

156.2. Þysse sylfan wyrte wyrtruman, fif penega gewihte on wine geþiged, þa wæterseocan gedrigeþ. Ðas sylfan strengþe heo hafaþ, gewylled ond gedruncen, wið þæs migþan ear-foðlicnyssa.

# 157

## Scolymbos

157.1. Ðeos wyrt, þe man *scolimbos* ond oþrum naman [. . .] nemneþ, on wine gewylled ond gedruncen, heo þone fulan stenc ðæra oxna ond ealles þæs lichaman afyrreþ.

157.2. Eac swylce ðeos sylfe wyrt ðone fulstincendan migþan forð gelædeþ, ond eac halwendne mete mannum gegearwaþ.

# 158

## Iris Yllyrica

Ðeos wyrt, þe man *iris Illyricam* ond oðrum naman [. . .] nemneþ, is gecweden *iris Illyrica* of ðære misenlicnysse hyre

156.2. Five pennyweights of the roots of this same plant, drunk in wine, will relieve a person suffering from dropsy. When boiled and drunk, it has the same ability to relieve a person's difficulty in urinating.

# Chapter 157

# Golden thistle

157.1. This plant, which is called *scolymos* and by another name [. . .], when simmered in wine and drunk, will remove a foul smell from the armpits or any other part of the body.

157.2. Likewise this same plant will draw out foul-smelling urine, and it will also provide healing food for people.

# Chapter 158

# Iris

This plant, which is called *iris Illyrica* and by another name [. . .], is named *iris Illyrica* for the diversity of its blossoms,

blostmena, for þy þe ys geðuht þæt heo þone heofonlican bogan mid hyre bleo geefenlæce se is on Leden *iris* gecweden, ond heo on Illyrico þam lande swiðost ond strengost wexeþ. Ond heo hafað leaf glædenan gelice, þa Grecas *xifian* hataþ, ond heo hafað trumne wyrtruman ond swyþe gestencne. Ond þone man sceal mid linenan claþe befealdan ond on sceade ahon oððæt he gedriged beon mæge, for ðy hys gecynde is swiþe hat ond slæpbære.

158.1. Gyf hwa mycelne hracan þolige, ond he þone him eaþelice fram bringan ne mæge for ðycnysse ond to hnesce: genime of þysse wyrte wyrtruman ðæs dustes, smæle genucudes, tyn penega gewihte; sylle drincan fæstende on liþon beore—feower scenceas, þry dagas—oþðæt he sy gehæled.

2  Ðam gelice þæt dust þysse sylfan wyrte, on liþon beore geþiged, ðone slep on gelædeþ, ond eac þæra innoþa astyrunge geliþigað.

158.2. Eac swylce þæt dust þysse ylcan wyrte næddrena slitas gelacnaþ.

2  Þæt sylfe gemet þæt we her beforan cwædon þæs dustes ðysse ylcan wyrte *iris Illyrice,* foran mid ecede gemencged ond gedruncen, hyt fremað þam þe his gecyndelice sæd him sylfwylles fram gewiteþ, þone leahtor Grecas *gonorhoeam* nemneþ. Gyf hit þonne soðlice þam ylcan gemete mid wine gemetegud byþ, hit þæra wifa monoðlican astyreð þeah hy ær langæ forlætene wæron.

158.3. Wið cyrnlu ond wið ealle yfele cumulu: genim ðysse ylcan wyrte wyrttruman swa anwealhne, wel gedrigedne ond

because it is thought that, with its colors, it resembles the rainbow in the sky that is called *iris* in Latin, and because it grows best and strongest in the land of Illyria. And it has leaves like sword grass, which the Greeks call *xiphion,* and it has firm roots and a strong smell. It should be folded up in a linen cloth and hung in the shade until it is thoroughly dry, because its nature is very hot and soporific.

158.1. If a person should suffer from a deep cough, and he cannot easily clear it away because it is thick and soft: take powder made from the roots of this same plant, finely pounded, ten pennyweights; give to the patient to drink in light beer while fasting—four cups, for three days—until he is healed.

Likewise the powder of this same plant, when taken in light beer, will induce sleep, and it will also relieve the discomfort of bowel movements. 2

158.2. Likewise the powder of this same plant will heal snakebites.

The same quantity that we said before of the powder of this same plant *iris Illyrica,* mixed in advance with vinegar and given to drink, will benefit a man whose semen ejects spontaneously. This disorder the Greeks call *gonorrhea.* Indeed, if it is prepared with wine in the same quantity, it will stimulate women's menses even if they had long ceased to flow. 2

158.3. For hard swellings and all harmful lumps: take the whole root of this same plant, dried thoroughly and then

siððan gesodenne; cnuca hyne ðonne swa hnescne; wyrc to clyþan; lege to ðam sare. Hyt tofereþ.

158.4. Eac swa some hyt fremað wið ðæs heafodes sare, mid ecede ond mid rosan wose gemencged.

# 159

## *Elleborus albus*

159.1. Wið liferseocnysse: genim þas wyrte, þe man *elleborum album* ond oðrum naman [. . .] nemneþ, gedrigede ond to duste gecnucude; syle drincan on wearmum wætere þæs dustes syx cuculeras fulle. Hit gelacnað þa lifre. Þæt silfe is framigendlic læcedom on wine geþiged ongean ealle attru.

# 160

## *Delfinion*

160.1. Wið þam fefore þe þy feorðan dæge on man becymeþ: genim þysse wyrte seaw, þe man *delfinion* ond oþrum naman

boiled; then pound it until it is soft; make it into a poultice; lay it on the sore. It will cause them to go away.

158.4. Also it likewise does well for headache, when mixed with vinegar and juice of roses.

# Chapter 159

# Smooth rupturewort

159.1. For liver disease: take this plant, which is called *elle borus albus* and by another name [. . .], dried and pounded into a powder; give six spoonfuls of the powder to drink in warm water. It will heal the liver. If drunk in wine, the same plant is an efficacious remedy for all poisons.

# Chapter 160

# Larkspur

160.1. For the fever that comes on a person every fourth day: take the juice of this plant, which is called *delphinion* and by

[. . .] nemneþ, wel gegaderod ond þæt mid pipore gecnucud ond gemencged. Ond ðæra pipercorna sy ofertæl, þæt ys þonne þy forman dæge an ond þrittig, ond þy oðrum dæge seofontyne, ond ðy þryddan dæge þreotyne. Gyf þu him þis syllest toforan þære genealæcincge þæs fefores, wundorlicre hrædnysse he byð alysed.

# 161

## *Acios*

Ðeos wyrt, þe man *æcios* ond oþrum naman [. . .] nemneþ, hafað sæd gelic næddran heafde, ond heo hafað lange leaf ond stiþe, ond heo manega stelan of hyre asendeþ; heo hafað þynne leaf ond ða hwonlice þyrnihte, ond heo hafað betweox þam leafon brune blostman, ond betweonan ðam blostmum heo hafað, swa we ær cwædon, sæd gelic nædran heafde. Ond hyre wyrttruma ys gehwæde ond sweart.

161.1. Wyþ nædrena slitas: genim ðysse ylcan wyrte wyrtruman þe we *æcios* nemdon; syle drincan on wine. Hyt fremað ge ær ðam slite ge æfter. Se sylfa drenc eac swylce þæra lendena sar geliðigað, ond eac drige on breoston meolc gegearwað. Soðlice an miht ys þysse wyrte, ond þæs wyrtruman, ond þæs sædes.

another name [. . .], carefully gathered and pounded and mixed with pepper. And there should be an odd number of peppercorns, that is, thirty-one on the first day, seventeen on the second day, and thirteen on the third day. If you give him this before the onset of the fever, he will be relieved with wonderful rapidity.

# Chapter 161

# Viper's bugloss

This plant, which is called *echios* and by another name [. . .], has seeds that look like a snake's head, and it has long and stiff leaves, and it sends out many stems; it has thin and somewhat thorny leaves, and it has brown flowers in between the leaves, and in between the flowers, as we have said before, it has seeds that look like a snake's head. And its root is small and dark.

161.1. For snakebites: take the root of this same plant that we have called *echios;* give to drink in wine. It helps both before and after the bite. The same drink also relieves pain in the loins, and the dried plant stimulates breast milk. In truth, there is the same power in the plant, in its root, and in its seed.

## 162

### *Centimorbia*

Ðeos wyrt, þe man *centimorbia* ond oðrum naman [. . .] nemneþ, byþ cenned on beganum stowum ond on stænigum,
ond þæt on dunum ond on wynsumum stowum. Ond heo of
anre tyrf manega bogas asendeþ; ond heo is gehwædon lea-
fun ond sinewealton ond toslitenon. Ond heo hafað þas
mihte to lacnunge.

162.1. Gif hors on hricge oððe on þam bogum awyrd sy ond
hyt open sy: genim þas wyrte, ealle gedrigede ond to swyðe
smælon duste gecnucude; gescead to ðam sare. Heo hit gehæleþ; þu wundrast ðære gefremminge.

## 163

### *Scordios*

Ðeos wyrt, ðe man *scordias* ond oðrum naman [. . .] nemneþ,
hafaþ swæc swylce leac, ond heo eac for þy *scordios* gecweden
ys. Þeos wyrt byð cenned on morum; ond heo hafaþ leaf
sinewealte, ond ða bittere on byrgincge; ond heo hafaþ feo-
werecgedne stelan ond fealuwe blostman.

# Chapter 162

# Moneywort

This plant, which is called *centimorbia* and by another name
[. . .], grows in cultivated places and on stony ground,
particularly on hills and in pleasant places. And from one
tussock it sends out many shoots; and it has leaves that are
small and round and are cleft. And it has this power to heal.

162.1. If a horse is injured on the back or the shoulder and
the wound is open, take this plant, dried thoroughly and
pounded into a very fine powder; sprinkle it on the wound.
It will heal it; you will be surprised at its effectiveness.

# Chapter 163

# Water germander

This plant, which is called *scordion* and by another name
[. . .], smells like a leek and for this reason is called *scordion*.
This plant grows on the moors; it has leaves that are round,
and they are bitter to the taste; and it has a four-edged stem
and reddish-yellow flowers.

163.1. Wið þæs migðan astyrunge: genim þas wyrte *scordios,* swa grene, gecnucude ond on wine geþigede, oððe drigge, on wine gewyllede; syle drincan. Heo þone migðan astyreþ.

163.2. Eac þæt sylfe fremað wið nædrena slitas ond wið ealle attru ond wið þæs magan sare, swa we ær cwædon, wið þæs migðan yrmðe.

163.3. Wið þa gerynnincge þæs worsmes ymb ða breost: genim þas ylcan wyrte, tyn penega gewihte mid hunige gemencged; syle þicgean anne cuculere fulne. Þa breost beoð afeormude.

163.4. Wið fotadle: genim þas ylcan wyrte, on ecede gecnucude oððe on wætere; syle drincan. Hyt fremað wel.

163.5. Wið niwe wunda: genim þas ylcan wyrte sylfe, gecnucude; lege to ðam wundum. Heo hy geþeodeþ. Ond eac heo, mid hunige gemencged, ealde wunda afeormaþ ond gehæleþ; ond eac hyre dust wexende flæsc wel gehnæceþ.

# 164

# *Ami*

164.1. Ðeos wyrt, þe man *ami* ond oðrum naman *milvium* nemneþ, ond eac sume men [. . .] hata∂, hafað gecweme sæd to læcedome þæt on wine geseald byð. Wel fremað wið þæs

163.1. To stimulate the urine: take this plant *scordion,* either fresh, pounded and taken in wine, or dried, simmered in wine; give to drink. It will stimulate the urine.

163.2. The same thing helps for snakebites and for all kinds of poison and for stomachache and, as we have just said, for difficulty urinating.

163.3. For an accumulation of foul matter in the chest: take this same plant, ten pennyweights mixed with honey; give one spoonful to eat. The chest will be purged.

163.4. For foot disease: take this same plant, pounded into vinegar or water; give to drink. It will do much good.

163.5. For fresh wounds: take this selfsame plant, pounded; lay it on the wounds. It will close them up. And also, mixed with honey, this plant will purge old wounds and will heal them; and its powder will also inhibit the growth of flesh.

# Chapter 164

# Bishop's-weed

164.1. This plant, which is called *ami* and by another name *milvium,* and also some people call it [. . .], has seed that, when given in wine, is good as a remedy. It is beneficial for

innoðes astyrunge, ond wið earfoðlicnysse ðæs migðan, ond wið wildeora slitas. Ond eac hyt ða monoðlican forð ge-cigeþ.

2    Ond wið wommas þæs lichaman: genim þysse sylfan wyrte sæd, mid hunige gecnucud. Hyt afyrreð þa wommas.

164.2. Wið ablæcnysse ond æhiwnesse þæs lichaman: do þæt sylfe; þæt ys, þæt ðu þone lichaman mid þam ylcan gesmyre, oððe syle drincan. Hyt þa æhiwnesse of genimeð.

# 165

# Banwyrt

Ðeos wyrt, þe man *violam* ond oðrum naman banwyrt nem-neð, ys ðreora cynna; þonne ys an brunbasuw ond oþer hwit; þridde is geoluw. Ðonne is seo geoluwe swaþeah swiþost læceon gecweme.

165.1. Wið þæs cwiðan sare ond wið þone hætan: genim þas ylcan wyrte, gecnucude ond under gelede. Heo hyne ge-lihteþ. Eac swylce heo ða monoðlican forþ gecigeþ.

165.2. Wiþ misenlice leahtras ðæs bæcþearmes þa *ragadas* hatað, þæt is swaþeah swiðost þæs blodes utryne: genim þysse ylcan wyrte leaf, gecnucude ond to clyþan gemenc-gede. Hy þa untrumnysse ealle gehæleþ.

movement of the bowels, and for difficulty urinating, and for the bites of wild animals. It also brings on the menses.

For blemishes on the body: take the seed of this same 2 plant, pounded with honey. It will clear up the spots.

164.2. For paleness and pallor of the body: do the same; that is, you should apply the same thing to the body, or else give it to be drunk. It will take the pallor away.

# Chapter 165

# Wallflower

This plant, which is called *viola* and by another name wallflower, comes in three varieties; one is dark purple and the second one white; the third is yellow. The yellow one, however, is best suited to remedies.

165.1. For pain and inflammation of the womb: take this same plant, pounded and applied to the woman's lower parts. It will soothe the womb. Likewise it will bring on the menses.

165.2. For various disorders of the anus that are called *rhagades,* but especially for a discharge of blood: take leaves of this same plant, pounded and compounded into a poultice. They will heal all these disorders.

165.3. Þysse sylfæn wyrte leaf, mid hunige gecnucude ond gemencgede, þone cancor þæra toða gehæleð of ðam foroft ða
teþ fealleð.

165.4. Wyþ ða monoðlican to astyrigenne: genim þysse ylcan
wyrte sædes, tyn penega gewihte, on wine gecnucud ond gedruncen, oððe mid hunige gecnucud ond to ðam gecyndelican lime geled. Hyt þa monoðlican astyreþ ond þæt tudder
of þam cwiðan gelædeþ.

165.5. Wið miltan sare: genim þysse ylcan wyrte wyrttruman,
on ecede gecnucudne; lege to ðære miltan. Hit fremaþ.


# 166

## *Viola purpurea*


166.1. Wið niwe wundela ond eac wið ealde: genim þysse
wyrte leaf, þe man *viola purpurea* ond oðrum naman [. . .]
nemneþ, ond rysle, ægþres gelice mycel; lege to ðam wundum. Scearplice hyt hy gehæleð, ond eac geswel ond ealle
yfele gegaderunga hyt tolyseð.

166.2. Wiþ ðæs magan heardnysse: genim þysse ylcan wyrte
blostman, on hunige gemencgede ond mid swiðe godon
wine gewesede. Þæs magan heardnys byð geliðigad.

165.3. The leaves of this same plant, pounded and mixed with honey, will heal the canker of the teeth from which all too often the teeth fall out.

165.4. To stimulate the menses: take the seeds of this same plant, ten pennyweights, either pounded in wine and drunk, or pounded in honey and put on the genitals. It will stimulate the menses and will bring forth the fetus from the womb.

165.5. For pain in the spleen: take the roots of this same plant, pounded in vinegar; lay them on the spleen. It will do some good.

# Chapter 166

# Sweet violet

166.1. For fresh wounds and also for old ones: take the leaves of this plant, which is called *viola purpurea* and by another name [. . .], plus some fat, both of them in the same quantity; apply this to the wounds. It will heal them efficaciously, and it will also do away with swellings and all accumulations of diseased matter.

166.2. For hardening of the stomach: take the flowers of this same plant, mixed with honey and soaked in very good wine. The stomach's hardness will be relieved.

# 167

## *Zamalentition*

Ðeos wyrt, þe man *zamalentition* ond oþrum naman [. . .] nemneþ, byð cenned on stænigum stowum ond on dunum.

167.1. Wið ealle wundela: genim þas wyrte *zamalentition,* wel mid rysle gecnucude, butan sealte; lege to ðam wundum. Ealle heo hy gehæleþ.

167.2. Eft, wið cancorwunda: genim þas ylcan wyrte *zamalentition,* gedrigede ond to swyþe smalon duste gecnucude; lege to ðam wundum. Ealne þone bite þæs cancres heo afeormað.

# 168

## *Ancusa*

Ðeos wyrt, ðe man *ancusa* ond oðrum naman [. . .] nemneþ, byð cenned on beganum stowum ond on smeþum; ond ðas wyrte ðu scealt niman on ðam monþe ðe man Martius hateþ. Ðysse wyrte syndon twa cynrenu: an is ðe Affricani *barbatam* nemnað; oðer ys to læcedomum swyþe gecoren, ond ðeos

# Chapter 167

## *Zamalentition*

This plant, which is called *zamalention* and by another name [. . .], grows in stony places and on hills.

167.1. For all wounds: take this plant *zamalention,* pounded well in fat, without salt; lay it on the wounds. It will heal them all.

167.2. Again, for ulcerous wounds: take this same plant *zamalention,* dried and pounded into a very fine powder; lay it on the wounds. It will purge the entire ulcerous infection.

# Chapter 168

## Alkanet

This plant, which is called *anchusa* and by another name [. . .], grows in cultivated places and on flat ground; and you must gather it in the month that is called March. There are two varieties of this plant: there is one that the people of Africa call *barbata;* the other is very suitable for medicines,

byð cenned fyrmest on ðam lande ðe man Persa hateþ. Ond heo ys scearpon leafon þyrnihtum butan stelan.

168.1. Wið forbærnednysse: genim þysse wyrte wyrttruman *ancusa,* on ele gesodene ond wið wex gemencgedne, ðam gemete þe þu plaster oþþe clyþan wyrce; lege to þam bærnytte. Wundorlice hyt gehæleþ.

# 169

## *Psillios*

Ðeos wyrt ys *psillios* gecweden for ðam þe heo hafað sæd swylce flean; þanon hy man eac on Leden *pulicarem* nemneð, ond hy eac sume men [. . .] nemnað. Heo hafað gehwæde leaf ond ruhe, ond heo hafað stelan ond ðone on bogum geþufne. Ond heo ys drige gecynde ond tyddre, ond heo byð cenned on beganum stowum.

169.1. Wið cyrnlu ond wið ealle yfele gegaderunga: genim þysse wyrte sædes, gecnucudes, an elefæt ful, ond twegen bollan fulle wæteres; mencg tosomne; syle drincan. Nim þonne of ðam sylfan sæde; wyrc plaster; lege to ðam sare. Hyt byþ gehæled.

169.2. Wið heafudsare: do þæt sylfe mid rosan wose ond mid wætere gewesed.

and this one is native to the land called Persia. And it has sharp, thorny leaves without stalks.

168.1. For a bad burn: take the roots of this same plant *anchusa*, cooked in wine and mixed with wax, in the quantity that you need to make a plaster or poultice; lay it on the burn. It will heal it wonderfully.

# Chapter 169

# Fleawort

This plant is called *psyllium* because it has seeds like fleas; for this reason it is called *pulicaris* in Latin, and also some people call it [. . .]. It has slight, hairy leaves, and it has a stem that is bushy with branches. Its nature is dry and weak, and it grows in cultivated places.

169.1. For hard swellings and all accumulations of harmful matter: take the seed of this plant, pounded, an oil jar full, and two bowls full of water; mix together; give to drink. Then take some of the same seed; make into a plaster; lay it on the sore. It will be healed.

169.2. For headache: do the same with juice of roses and soaked in water.

## 170

# *Cynosbatus*

Ðeos wyrt, þe man *cynosbatus* ond oðrum naman [. . .] nemneþ, ðonne hy man of ðam stelan genimeþ, heo bið þam goman stið ond wiðerræde for mete geþiged, ac heo swaþeah
ða breost afeormað ond swa hwylce þincg swa syndon afore
oððe bitere. Ðeah hy þam magan derien, hi swaþeah ðære
miltan wel fremað. Þysse ylcan wyrte blostma, gedruncen,
swa þone man gelacnað þæt he þurh ðone innoð ond þurh
ðone migþan forð gelæded bið. Ond he eac blodrynas afeormaþ.

170.1. Eft, wið miltan sare: genim þysse ylcan wyrte wyrttruman, of ðære rinde wel afeormadne; lege to ðære miltan.
Hyt bið hyre nytlic ond fremgendlic; ond se þe þysne læcedom þolaþ, he sceal upweard licgean þy læs he, ungeþyldig,
ða strengþe þysre lacnunge ongite.

# Chapter 170

# Dog rose

This plant, which is called *cynosbatos* and by another name [. . .], when it is picked from its stem, is harsh to the throat and disagreeable if eaten as food, but nevertheless it will purge the chest and all things that are sour or bitter. Even if it hurts the stomach, it nevertheless much benefits the spleen. If drunk, the blossoms of this same plant will heal the patient in such a manner that he is purged through discharge of the bowels and through the urine. And it also staunches bleeding.

170.1. Again, for pain in the spleen: take the roots of this same plant, well cleared of their bark; apply them to the spleen. It will be useful and beneficial to it; and the person who undergoes this treatment should lie facing up, lest, impatient, he feel the strength of this medication.

## 171

# *Aglaofotis*

Ðeos wyrt, ðe man *aglaofotis* ond oðrum naman [. . .] nemneþ, scineð on nihte swa blæse, ond heo mæg wið manega
untrumnyssa.

171.1. Wið þone fefor ðe þy ðriddæn ond ðy feorðan dæge on
man becymeþ: genim þysse ylcan wyrte seaw *aglaofotis,* mid
rosenan ele gemencged; smyre þone seocan. Untweolice þu
hyne alysest.

171.2. Gyf hwa hreohnysse on rewytte þolige: genime ðas
ylcan wyrte, for rycels onælede. Seo hreohnys byð forboden.

171.3. Wiþ hramman ond wið bifunge: genime þas sylfan
wyrte; hæbbe mid him. Gif hy þonne hwa mid him bereþ,
ealle yfelu hyne ondrædað.

# Chapter 171

# Peony

This plant, which is called *aglaophotis* and by another name [. . .], shines at night like a lamp, and it is effective for many infirmities.

171.1. For the fever that comes on a person every third and every fourth day: take the juice of this same plant *aglaophotis*, mixed with oil of roses; apply it to the patient. Undoubtedly you will relieve him.

171.2. If anyone is caught in a storm while rowing: take this same plant, kindled in place of frankincense. It will prevent the storm.

171.3. For cramps and tremors: have the person take this same plant; have him keep it with him. If anyone carries it with him, all things harmful will fear him.

172

# Wudubend

172.1. Wið miltan sare: genim þysse wyrte wyrttruman, þe man *capparis* ond oþrum naman wudubend hateð; cnuca to duste ond gewyrc to clyþan; lege to ðære miltan. He hy adrygeð. Ac swaþeah gewrið þone man þy læs he þurh þæt sar ða lacnunge of him asceace; ond æfter þrim tidum, gelæd hyne to bæþe ond hyne wel gebaþa. He byþ alysed.

173

# *Eryngius*

Ðeos wyrt, þe man *eringius* ond oþrum naman [. . .] nemneþ, hafað hnesce leaf þonne heo ærest acenned byþ, ond ða beoð werede on swæce, ond hi man þigeþ swa oðre wyrta; syððan hy beoð scearpe ond ðyrnihte. Ond heo hafað stelan hwitne oððe grenne, on ðæs heahnysse ufeweardre beoð acennede scearpe ond þyrnyhte pilas; ond heo hafað lancne wyrtruman ond þone uteweardne sweartne; ond se bið godes swæces. Þeos wyrt byþ cenned on feldon ond on wiðerrædon stowum.

## Chapter 172

# Caper

172.1. For pain in the spleen: take the root of this plant, which is called *capparis* and by another name caper; pound into a powder and make into a poultice; lay it on the spleen. It will dry it up. However, you must bind the person lest he shake the medication off him because of the pain; and after three hours, lead him to the bath and bathe him well. He will be cured.

## Chapter 173

# Eryngo

This plant, which is called *eryngium* and by another name [. . .], has tender leaves when it first grows, and they are sweet to the taste, and one eats them just like other plants; later on they are sharp and thorny. And it has a white or a green stalk, and in the tip of its upper part grow sharp and thorny prickles; and it has a long root, black on the outside; and it has a good smell. This plant grows in fields and on unfavorable ground.

173.1. Wið þæs migþan astyrunge: genim þas ylcan wyrte þe we *eringius* nemdun, gecnucude; syle drincan on wine. Na þæt an þæt heo þone migþan astyreþ, ac eac swylce ða monoþlican; ond ðæs innoðes astyrunge ond toðundenysse heo tolyseþ; ond eac wið liferseocnysse ond wið næddrena slitas heo wel fremað.

173.2. Eac swylce wið mænigfealde leahtras þæra innoða heo wel fremað, geþiged mid þære wyrte sæde þe man *olisatrum* nemneþ.

173.3. Wið þæra breosta geswel: genim ðas ylcan wyrte, to clyþan geworhte; lege to ðam breostan. Ealle þa yfelan gegaderungæ onbutan þa breost heo tofereð.

173.4. Wið *scorpiones* stingc ond wið ealra næddercynna slitas, ond eac wið wedehundes slite: genim þas ylcan wyrte; wyrc to plastre; lege to ðære wunde, swa þæt seo wund swaþeah ærest mid iserne geopenud sy, ond syððan þærto geled swa þæt se seoca þone stenc ne ongite.

2     Eac swylce þeos sylfe wyrt wið oman wel fremaþ on þas ylcan wisan gemetegud; ond eac heo fotadle geliðigað gyf hy man æt frymþe to gelegeþ.

173.1. To stimulate the urine: take this same plant that we have called *eryngium,* pounded; give to drink in wine. It will stimulate not only the urine, but also the menses; and it relaxes agitation and swelling of the internal organs; and it also is good for liver disease and snakebite.

173.2. Likewise it is of benefit for many diseases of the abdomen, if eaten with the seed of the plant called *holus atrum.*

173.3. For swelling of the breasts: take this same plant, made into a poultice; lay it on the breasts. It will expel all accumulations of diseased matter in the breasts.

173.4. For a scorpion's sting and for all snakebites, and also for the bite of a mad dog: take this same plant; make it into a plaster; put it on the wound, as long as the wound has first been opened up with an iron blade and the plaster has afterward been applied to it so that the patient does not perceive the bad smell.

This plant also does much good for erysipelas if prepared 2 in the same way; and it will also relieve foot disease if one puts it on at the onset of the disease.

# 174

# *Philantropos*

Ðas wyrte man *philantropos* nemneþ; þæt ys on ure þeode,
menlufigende, for ðy heo wyle hrædlice to ðam men gecly-
fian; ond heo hafað sæd gelic mannes nafolan, þa man eac
oþrum naman clate nemneð. Ond heo of hyre manega bogas
asendeþ, ond þa lange ond feowerecge, ond ys stið on leafon,
ond heo hafað greatne stelan ond hwite blostman; ond heo
hafað heard sæd ond sinewealt ond on middan hol, swa we
ær cwædon, þam gemete þe byð mannes nafla.

174.1. Wið nædrena slitas ond wið þara wyrma ðe man *spalangiones* hateþ: genim þysse wyrte wos, gecnucud on wine; syle
drincan. Hyt fremað.

174.2. Wið earena sare: genim þysse ylcan wyrte wos; drype
on þæt eare. Hyt gehæleþ þæt sar.

# Chapter 174

# Cleavers

This plant is called *philantropos;* that is, people loving in our language, because it will quickly stick to a person; and it has seeds that look like the human navel, thus it is also called by another name, cleavers. And it sends out many branches, which are long and four sided, and its leaves are stiff, and it has a thick stem and white flowers, and it has seeds that are hard and round and are hollow in the middle, as we have said before, so as to resemble a human navel.

174.1. For snakebites and for those venomous creatures we call *spalangiones:* take the juice of this plant, pounded in wine; give to drink. It will be of benefit.

174.2. For earache: take the juice of this same plant; drip it into the ear. It will heal the pain.

# 175

## *Achillea*

Ðeos wyrt, þe man *achillea* ond oðrum naman [. . .] nemneþ, byð cenned on beganum stowum ond neah wætere, ond heo hafað geoluwe blostman ond hwite.

175.1. Wið niwe wunda: genim þysse wyrte croppas, gecnucude; lege to ðam wundum. Heo þæt sar genimð, ond heo ða wunda geðeodeþ ond þone blodryne gewrið.

175.2. Gif wif of ðam gecyndelican limon þone flewsan þæs wætan þoligen: genim þas ylcan wyrte, gesodene; gelege under þam wifon sittendum. Ealne þone wætan of hyre æþme heo gewrið.

175.3. Eac ðeos sylfe wyrt, on wætere gedruncen, wið utsiht wel fremað.

2    Ðeos wyrt is *achillea* gecweden for þam þe is sæd þæt Achilles se ealdorman hyre gelomlice brucan scolde wunda to gelacnigenne.

# Chapter 175

# Yarrow

This plant, which is called *achilleos* and by another name [. . .], grows in cultivated places and near water, and it has yellow and white flowers.

175.1. For fresh wounds: take the flower heads of this plant, pounded; apply to the wound. It will take away the pain, and it will heal the wounds and will stop the bleeding.

175.2. If women are troubled by fluid flowing from the genitals: take this same plant, boiled; put it beneath the woman while she is seated. Its vapor will staunch all the flow.

175.3. In addition this same plant, when drunk in water, does well for diarrhea.

    This plant is called *achilleos* because it is said that Achilles, the chief, would often use it to treat wounds.   2

# 176

# *Ricinus*

176. 1. Wið hagol ond hreohnysse to awendenne: gyf ðu þas wyrte ðe man *ricinum* ond oðrum naman [. . .] nemneð, on þinre æhte hafast, oððe hyre sæd on þin hus ahegst oððe on swa hwilcere stowe swa þu hy hafast oððe hyre sæd, heo awendeð hagoles hreohnysse. Ond gyf þu hy oððe hire sæd on scyp ahehst, to ðam wundorlic heo is þæt heo ælce hreohnysse gesmylteþ. Þas wyrte þu scealt niman þus cweþende:

2    *Herba ricinum, precor uti adsis meis incantationibus et avertas grandines, fulgora, et omnes tempestates, per nomen omnipotentis Dei qui te iussit nasci.*

3    Þæt is ðonne on ure geþeode: "Wyrt *ricinum,* ic bidde þæt þu ætsy minum sangum ond þæt ðu awende hagolas ond ligræsceas ond ealle hreohnyssa, þurh naman ælmihtiges Godes se þe het beon acenned."

4    Ond þu scealt clæne beon þonne þu ðas wyrte nimest.

# Chapter 176

# Castor-oil plant

176.1. To avert hail and storms: if you have in your possession this plant, which is called *ricinus* and by another name [. . .], or if you hang its seeds in your house or in whatever place where you have either the plant or its seeds, the plant will turn away the violence of hail. And if you hang either the plant or its seeds aboard a ship, it is so marvelous that it will calm every storm. You must gather this plant saying these words:

*Herba ricinum, precor uti adsis meis incantationibus et avertas* 2
*grandines, fulgora, et omnes tempestates, per nomen omnipotentis*
*Dei qui te iussit nasci.*

That is in our language: "*Ricinum* plant, I pray you to be 3 present at my singing and to avert hail and lightning and all storms, in the name of the almighty God who bid you to be born."

And you must be free from defilement when you pick the 4 plant.

# 177

## *Polloten*

Ðeos wyrt, ðe man *polloten* ond oþrum naman *porrum nigrum* nemneþ, ond eac sume men [. . .] hatað, ys þyrnihton stelan ond swearton ond rugum, ond bradran leafon þonne leac ond sweartran, ond þa syndon stranges swæces. Ond hyre miht ys scearp.

177.1. Wið hundes slite: genim þysse wyrte leaf, mid sealte gecnucude; lege to þam wundum. Hit hæleþ wundorlice.

177.2. Eft, wið wunda: genim þyssæ ylcan wyrte leaf, mid hunige gecnucude; lege to þam wundum. Ælce wunde hyt gehæleþ.

# 178

## Netele

178.1. Wið forcillede wunda: genim þysse wyrte seaw, þe man *urticam* ond oðrum naman netele nemneþ, mid eledrosnum gemencged ond sumne dæl sealtes ðærto gedon; lege to þære wunde. Binnan þrim dagum heo biþ hal.

# Chapter 177

# Black horehound

This plant, which is called *ballote* and by another name *porrum nigrum,* and also some people call it [. . .], has stems that are thorny and dark and rough, and leaves that are broader than those of a leek and darker, and these have a strong smell. And its power is sharp.

177.1. For the bite of a dog: take the leaves of this plant, pounded with salt; apply to the wounds. It will heal them wonderfully.

177.2. Again, for wounds: take the leaves of this same plant, pounded with honey; apply to the wounds. It will heal every wound.

# Chapter 178

# Nettle

178.1. For wounds caused by cold: take the juice of this plant, which is called *urtica* and by another name nettle, mixed with the dregs of oil and with a quantity of salt added; apply to the wound. Within three days it will be well.

178.2. Wið geswel: do þæt sylfe; þæt ys, þonne þam ylcan gemete lege to þam geswelle. Hyt bið gehæled.

178.3. Gyf ðonne ænig dæl þæs lichaman geslegen sy: genim þas ylcan wyrte *urticam,* gecnucude; lege to þære wunde. Heo byð gehæled.

178.4. Wið liþa sare, gyf hy of hwylcum belimpe, oððe of cyle, oþþe of ænigum þincge gesargade beoð: genim þysse ylcan wyrte seaw, ond eles efenmycel, togædere gewylled; do þonne þærto þær hit swiðost derige. Binnan þrim dagon ðu hyne gehælst.

178.5. Wið fule wunda ond forrotude: genim þas ylcan wyrte *urticam,* gecnucude, ond þærto sumne dæl sealtes; gewrið to þære wunde. Binnan þrym dagon heo biþ hal.

178.6. Wið wifes flewsan: genim þas ylcan wyrte, on mortere wel gepunude oððæt heo wel liþi sy; geyc þonne þærto sumne dæl huniges. Nim syþþan wæte wulle, ond þa wel getæsede; smyre ðonne þa geweald mid þam læcedome, ond syþþan hyne þam wife gesyle þæt heo hyne hyre under gelecge. Þy sylfan dæge hyt þone flewsan beluceð.

178.7. Wið þæt þu cyle ne þolige: genim þas ylcan wyrte *urticam,* on ele gesodene; smyre ðonne þærmid þa handa ond ealne þone lichaman. Ne ongitst ðu þone cile on eallum þinum lichaman.

178.2. For swellings: do the same thing; that is, put it on the swelling in the same manner. It will be healed.

178.3. If any part of the body is struck by a blow: take this same plant nettle, pounded; lay it on the wound. It will be healed.

178.4. For pain in the loins, if they are injured by some accident, or by cold, or by anything else: take the juice of this same plant, plus the same amount of oil, simmered together; apply it where it hurts the most. Within three days you will heal him.

178.5. For foul and putrefied wounds: take this same plant nettle, pounded, and add a quantity of salt; fasten to the wound. Within three days it will be well.

178.6. For a woman's excessive flow of menstrual blood: take this same plant, thoroughly pounded in a mortar until it is very soft; then add to it a quantity of honey. After this take some moist wool, and this must be well teased; then smear the genitals with the medication, and afterward give this medication to the woman so that she can set it under her. The same day this will stop the flow.

178.7. So that you do not suffer from the cold: take this same plant nettle, cooked in oil; then smear it on the hands and on the whole body. You will not feel the cold anywhere in your body.

# 179

## *Priapisci*

179.1. Ðeos wyrt, þe man *priapisci* ond oðrum naman *vicapervica* nemneð, to manegum þingon wel fremað; þæt ys þonne ærest ongean deofolseocnyssa, ond wið nædran, ond wið wildeor, ond wið attru, ond wið gehwylce behatu, ond wið andan, ond wið ogan, ond þæt ðu gife hæbbe. Ond gif ðu þas wyrte mid þe hafast, ðu bist gesælig ond symle gecweme. Ðas wyrte þu scealt niman þus cweþende:

2     *Te precor vicapervica, multis utilitatibus habenda, ut venias ad me hilaris florens cum tuis virtutibus, ut ea mihi prestes, ut tutus et felix sim semper a venenis et ab iracundia inlesus.*

3     Þæt ys þonne on ure geþeode: "Ic bidde þe, *vica pervica*, manegum nytlicnyssum to hæbbenne, þæt ðu glæd to me cume, mid þinum mægenum blowende, þæt ðu me gegearwie þæt ic sy gescyld ond symle gesælig ond ungedered fram attrum ond fram yrsunge."

4     Ðonne ðu þas wyrt niman wylt, ðu scealt beon clæne wið æghwylce unclænnysse. Ond ðu hy scealt niman þonne se mona bið nigon nihta eald, ond endlyfon nihta, ond ðreottyne nyhta, ond ðrittig nihta, ond ðonne he byð anre nihte eald.

# Chapter 179

# Greater periwinkle

179.1. This plant, which is called *priapiscus* and by another name *vinca pervinca,* is beneficial for many purposes; that is, first of all for possession by demons, and for snakes, and for wild animals, and for poisons, and for any threats, and for envy, and for terror, and that you may have good luck. And if you keep this plant with you, you will be happy and always agreeable to others. You must gather this plant saying these words:

*Te precor vinca pervinca, multis utilitatibus habenda, ut venias* 2
*ad me hilaris florens cum tuis virtutibus, ut ea mihi prestes, ut tutus et felix sim semper a venenis et ab iracundia inlesus.*

This is in our language: "I pray you, *vinca pervinca,* you 3
who have many uses, that you come to me happy, with your powers blooming, so that you make me ready that I may be protected and always happy and uninjured by poisons or by anger."

When you want to gather the plant, you must be free 4
from any kind of defilement. And you shall pick it when the moon is nine nights old, and eleven nights, and thirteen nights, and thirty nights, and when it is one night old.

# 180

## *Litosperimon*

Ðeos wyrt, ðe man *litospermon* ond oðrum naman [. . .] nem-neð, byð cenned in Italia, ond seo fyrmeste in Creta. Ond heo hafað maran leaf ðonne rude ond ða rihte; ond on ðære hehnysse heo hafað stanas hwite ond sinewealte—swylce meregrotu—on pysna mycelnysse, ond ða beoð on stanes heardnysse ond eac swylce hy togædere geclifigen, ond hy beoð innan hole, ond ðonne þæt sæd þæron innan.

180.1. Wið þæt stanas on blædran wexen ond wið þæt man gemigan ne mæge: genim of ðysum stanum fif penega ge-wihte; syle drincan on wine. Hit ða stanas tobrycð ond ðone migþan forð gelædeþ.

# 181

## *Stavisagria*

Ðeos wyrt, þe man *stavisagria* ond oðrum naman [. . .] nem-neð, hafað leaf swylce wingeard ond rihte stelan, ond heo hafað sæd on grenum coddum on ðære mycele þe pysan ond

# Chapter 180

# Common gromwell

This plant, which is called *lithospermon* and by another name [. . .], grows in Italy, and it is native to Crete. It has larger leaves than rue, and straight ones; and at its top it has white and round stones—like pearls—the size of peas, and these are as hard as stones and likewise they bunch together, and they are hollow inside, and the plant's seeds are inside them.

180.1. If stones should grow in the bladder, and in case one has difficulty urinating: take five pennyweights of these stones; give to drink in wine. This will break up the kidney stones and will draw out the urine.

# Chapter 181

# Stavesacre

This plant, which is called *staphis agria* and by another name [. . .], has leaves like those of grapevine and a straight stem, and it has seeds in green pods that are of the size of peas and

þæt byð þreohyrne. Ond hyt byþ afor ond sweart; byð swaþeah innan hwit, ond biterre on byrgincge.

181.1. Wið þone yfelan wætan þæs lichaman: genim þysse wyrte sædes fiftyne corn, gecnucude on liðan beore; syle drincan. Hyt þone lichaman ðurh spiwðan afeormað. Ond æfter ðam þe he ðone drenc gedruncan hafað, he sceal gan ond hyne styrian ær ðam þe he hyne aspiwe. Ond þonne he hine spiwan onginneþ, he sceal gelomlice liðne wætan beores þicgean, ði læs seo strengð þære wyrte þa goman bærne ond forðylme.

181.2. Wið scurf ond wið sceb: genim þysse sylfan wyrte sæd ond rosan; cnuca tosomne; lege to ðam scurfe. He byð gehæled.

181.3. Wið toþa sare ond toðreomena: genim þysse ylcan wyrte sæd; seoð on ecede; healde þonne on his muð of ðam ecede lange hwile. Ðæra toða sar, ond ðæra toðreomena, ond ealle þæs muðes forrotudnyssa beoð gelacnude.

# 182

# *Gorgonion*

182.1. Ðeos wyrt, ðe man *gorgonion* ond oðrum naman [. . .] nemneþ, byð cenned on diglon stowum ond on wæton. Be þysse wyrte is sæd þæt hyre wyrttruma sy geanlicud þære

that are triangular. And the seeds are sour and dark; they are white inside, however, and bitter to the taste.

181.1. For the corrupt humor in the body: take fifteen grains of the seed of this same plant, pounded in light beer; give to drink. It will purge the body through vomiting. And after the patient has drunk the beverage, he should walk and bestir himself before he vomits it up. And when he begins to vomit, he should often drink a light drink of beer, lest the strength of this plant burn and choke the throat.

181.2. For scurf and scab: take the seeds of this same plant and some roses; pound together; apply to the scurf. He will be healed.

181.3. For toothache and sore gums: take seeds of this same plant; simmer them in vinegar; have the patient keep this vinegar in his mouth for quite a while. The toothache, the sore gums, and all putrefactions of the mouth will be healed.

# Chapter 182

# Eryngo

182.1. This plant, which is called *gorgonion* and by another name [. . .], grows in remote places and on damp ground. About this plant it is said that its root looks like the head of

nædran heafde ðe man *gorgon* nemneð, ond ða telgran habbað, þæs ðe eac is sæd, ægðer ge eagan ge nosa ge næddrena
hiw.

2 Eac se wyrttruma gehwylcne man him geanlicað, hwilon
of goldes hiwe, hwilon on seolfres; ond þonne ðu þas wyrte
mid hyre wyrttruman niman wylle, ðonne warna þu þæt hy
na sunne bescine ðy læs hyre hiw ond hyre miht sy awend
þurh ðære sunnan beorhtnysse. Forceorf hy þonne mid
anum wogan ond swyþe heardon iserne; ond se þe hy ceorfan wylle, ðonne sy he fram awend, for ðy hit nys alyfed þæt
man hyre wyrtruman anwealhne geseon mote. Se þe þas
wyrte mid him hafað, æghwylce yfele fotswaðu him ongean
cumende he forbugeþ; ge forðon se yfela man hyne forcyr-
reþ oððe him onbugeþ.

# 183

## *Milotis*

183.1. Ðeos wyrt, þe man *milotis* ond oðrum naman [. . .]
nemneð, byð cenned on beganum stowum ond on wætum.
Þas wyrte þu scealt niman on wanigendum monan, on ðam
monþe þe man Augustus hateð. Genim þonne þone wyrttruman þysse wyrte ond gewrið to anum hefelþræde ond
ahoh to ðinum swyran; þy geare ne ongitst þu dymnysse
þinra eagena, oððe gif heo þe belimpeð, heo hrædlice ge-
swiceð ond þu byst hal. Þes læcecræft ys afandud.

the snake that is called Gorgon, and that its shoots—so it is also said—have the eyes and the nose and the color of snakes.

Also this root changes a person so as to make him look like itself, sometimes in the color of gold, sometimes of silver; and when you want to gather this plant, together with its root, be careful that the sun does not shine on it lest its color and its strength be altered by the sun's brightness. Then cut it up with a crooked blade, a very hard one; and whoever intends to cut it, have him turn away from it, because it is not permitted that anyone should see its entire root. Whoever keeps this root with him, he will avoid all evil footsteps coming toward him; indeed, the evil person will either turn away or will yield to him on this account.

# Chapter 183

# Melilot

183.1. This plant, which is called *melilotus* and by another name [. . .], grows in cultivated places and on damp ground. You should pick this plant when the moon is on the wane, in the month that is called August. Then take the root of this plant and fasten it to a weaving thread and hang it around your neck; during this year you will not be affected by dimness of the eyes, or if you are affected by it, it will quickly cease and you will be well. This remedy has been tested.

183.2. Wið sina togunge: genim þysse ylcan wyrte wos; smyre þærmid. Hy beoð geliðegude. Eac ys be þysse wyrte sæd þæt heo on geare twigea blowe.

# 184

## *Bulbus*

Ðeos wyrt, þe man *bulbus* ond oþrum naman [. . .] nemneþ, ys twegea cynna: þonne ys þeos read ond wið þæs magan sare fremgendlic; þonne ys oðer byterre on byrgincge—seo ys *scillodes* gecweden—eac þam magan nytlicre. Ægþer hafað strang mægen, ond hy to mete geþigede mycelon ðone lichaman gestrangiað.

184.1. Wiþ geswel, ond wið fotadle, ond wið gehwylce gederednyssa: genim þas wyrte sylfe, gecnucude oððe mid hunige gemencgede; lege to ðam sare þe man þonne beþurfe.

184.2. Wið wæterseocnysse: genim þas ylcan wyrte swa we ær cwædon, gecnucude; lege to þam innoðe. Eac hy mid hunige gecnucude hunda slitas gelacniað; ond hy eac swylce, mid pipere gemencgede ond to gelede, hy þæs lichaman swat gewriðaþ; ond eac swa some hy þæs magan sare geliðigað.

184.3. Wið wundela þe þurh hy sylfe acennede beoð: genim þyssa wyrta wyrtruman, gecnucude mid ele ond mid

183.2. For sinew spasm: take the juice of this same plant; smear it on. The sinews will feel better. It is also said about this plant that it blooms twice a year.

# Chapter 184

# Tassel hyacinth

This plant, which is called *bulbus* and by another name [. . .], has two types: one is red and is beneficial for stomach pain; the other is more bitter to the taste—it is called *scilliticus*— and is also more beneficial for the stomach. Each has a strong power, and when eaten as food they greatly fortify the body.

184.1. For swellings, and for foot disease, and for any injury: take this same plant, pounded or mixed with honey; lay it on the sore spot as is then needed.

184.2. For dropsy: take this same plant as we said before, pounded; lay it on the abdomen. When mixed with honey, it also cures dog bite; and likewise, mixed with pepper and applied, it restrains the body's sweating; and similarly, it also relieves stomachache.

184.3. For sores that grow of themselves: take the root of this plant, pounded along with oil and with wheat flour and

hwætenan meluwe ond mid sapan, ðam gemete þe þu clyðan
wyrce; lege to ðam wundum. Eac hyt afeormaþ ðone leahtor
þe Grecas *hostopyturas* hatað (þæt ys, scurf þæs heafdes) ond
eac þone þe hy *achoras* nemnað (þæt ys, sceb), se foroft þæt
heafod fexe bereafað. Eæc swylce, mid ecede oððe mid hunige gecnucude, hy of þam andwlitan nebcorn afeormaþ.

184.4. Eac swa some, on ecede geþigede, hy þæra innoða
toþundennesse ond toborstennysse gehæleð. Be þysse wyrte
ys sæd þæt heo of dracan blode acenned beon sceolde, on
ufeweardum muntum on þiccon bearwum.

# 185

## *Colocyntisagria*

Ðeos wyrt þe man *colocynthisagria*—þæt ys, *cucurbita agrestis,*
þe man eac *frigillam* nemneþ—heo ealswa oðer cyrfætte wið
þa eorðann hyre telgran tobrædeþ. Ond heo hafað leaf cucu-
mere gelice ond toslitene, ond heo hafaþ wæstm sinewealtne
ond byterne, se ys to nymenne to þam timan þonne he æfter
his grennysse fealwað.

185.1. Wið innoðes astyrunge: genim þyses wæstmes hnesc-
nysse innewearde, butan þam cyrnlun, twegea penega gewihte, on liðan beore gecnucude; syle drincan. Hyt astyreþ
þone innoð.

with soap, the same way that you would make a poultice; lay it on the sores. It also clears up the disorder that the Greeks call *hostopyturas* (that is, a scurvy scalp) and also the one called *achoras* (that is, scabies), which all too often cause hair loss. Likewise, when pounded with vinegar or honey, it takes spots off the face.

184.4. In like manner, when consumed with vinegar, it also heals swelling and rupture of the internal organs. About this plant it is said that it is generated by dragon's blood, in thick groves in the heights of mountains.

# Chapter 185

## Bitter cucumber

This plant, which is called *colocynthis agria* — that is, *cucurbita agrestis,* which is also called *frigilla* — sends out its shoots over the ground just like other gourds. And it has leaves that are divided and that resemble cucumber, and it has round and bitter fruit, which should be picked at the time when it turns from green to yellow.

185.1. To stimulate the bowels: take the tender inside of this fruit, without the seeds, two pennyweights, pounded in light beer; give to drink. It will move the bowels.

# OLD ENGLISH REMEDIES
# FROM ANIMALS

# I

1.1. Sagað ðæt Ægypta cyning, Idpartus wæs haten, Octaviano þam casere his freonde hælo bodade, þyssum wordum þus cweðende:

2      Monegum bisenum ic eom gewis þinra mægena ond snytro; ond hwæþere ic wene þæt þu næfre to ðus mycles mægnes læcedomum become swylcum swa ic gefregn ða we fram Æscolupio ferdon. Ic þæt þa for ðinre cyððe ond þe weorðne wiste þyses to gewitanne, þæt ys, be wylddeora læcecræftum, swa þæt wel gesæd ys.

1.2. Sum fyþerfete nyten is þæt we nemnað *taxonem;* þæt ys, broc on Englisc. Gefoh þæt deor ond him þonne of cwicum þa teþ ado, þa þe he mæste hæbbe, ond þus cweð: "On naman þæs ælmihtigan Godes ic þe ofslea ond þe þine teþ of abeate." Ond þonne hy syððan on linenum hrægle bewind, ond on golde oþþe on seolfre bewyrc þæt hio ne mægen þinum lice æthrinan. Hafa mid þe, ðonne ne sceþþeð þe ne tungol, ne hagol, ne strang storm, ne yfel man, ne wolberendes awiht, ne þe æniges yfeles onhrine dereþ; oððe gyf hwæt yfeles bið, hraþe hyt byð tosliten, swa wæs Abdias gyrdels þæs witegan.

1.3. Nim þonne þone swyþran fot—þone furðran—ðissum wordum, ond þus cweþ: "On naman þæs lifigendan Godes, ic þe nime to læcedome." Þonne on swa hwylcum geflite

# Chapter 1

1.1. They say that the king of Egypt, who was called Idpartus, sent greetings to his friend the emperor Octavian, saying these words:

From many examples I know about your power and wisdom; however, I expect that you have never come upon such powerful remedies as those I learned that were given by Asclepius. I am giving you this for your information and because I know you are worthy of learning this knowledge, that is, about remedies made from wild animals, as has aptly been said.

1.2. There is a four-footed animal that we call *taxonem;* that is, badger in English. Catch this animal and while it is still alive remove its teeth, the biggest ones, and say this: "In the name of almighty God I am killing you and knocking out your teeth." And after that wrap the teeth in a linen cloth, and cover them over in gold or in silver so that they cannot come in contact with your body. If you keep them with you, then no star, nor hail, nor severe storm, nor evil person, nor anything pestilential will harm you, nor will contact with anything evil harm you; or if there is some evil, it will be quickly destroyed, as was the girdle of the prophet Obadiah.

1.3. Take then its right foot—its forefoot—with these words, and say: "In the name of the living God, I am taking you for medicine." Then, no matter what contest or fight you are

oððe gefeohte swa ðu bist, ðu bist sigefæst; ond þu þæt ge-
digest gif þu ðone fot mid þe hafast.

1.4. Mid his gelynde, smyre þa hors þa þe syn on feofre oþþe
on ænigre adle. Hio him fram ahyldeþ ond lifes tid him ofer
byð. Ond þeah hyt mycel adl sy, hraþe heo onweg gewiteþ.

1.5. Meng hys blod wyþ lytlum sealte horsum ond mulum
ond ælcum fiþerfetum neate þe on wole winnen oþþe on
ænigum yfle. Do þurh horn on muð æfter þæs deores mihte,
ond efne ymb þreo niht hy beoð hale.

1.6. His brægen geseoð on þrim sestrum eles on niwon croccan oðþæt þrydda dæl sy beweallen; fætelsa ond heald hyt.
Gif hwa sy on heafodwræce, æfter bæþe smyre mid on þrim
nyhtum. He byð gehæled, ond swa eac þa fet. Ond þeah man
sy on hwylcre ungewendendlicre adle ond unhalwendlicre,
seo wise hine hæleð, ond lacnað.

1.7. Nim his lifre, todæl, ond bedelf æt þam ymbhwyrftum
þinra landgemæra ond þinra burhstaðola, ond þa heortan æt
þinum burhgeatum behele. Þonne þu ond þine beoð alysde
hale to feranne ond ham to cyrrenne. Eall wol byþ aweg as-
tyred, ond þæt ær gedon wæs naht sceþþeð, ond byþ lytel
frecne fram fyre.

1.8. Cuþ ys eac þæt his hyd is bryce hundum ond eallum
fiþerfetum nytenum wið woles gewinne on to donne. Hafa
þære hyde fellsticceo on þinum sceon; ne gefelest þu gewin
on þinum fotum.

engaged in, you will be victorious; and you will come out of it successfully if you have the foot with you.

1.4. With its fat, rub down horses that are suffering from a fever or from any disease. The disease will be turned aside from them and its time of life will be over. And even if the disease is very severe, it will quickly go away.

1.5. Mix its blood with a little salt for horses and mules and for any four-footed animals that are suffering from pestilence or from any illness. Pour it through a horn into the mouth according to the animal's strength, and within just three days they will be cured.

1.6. Simmer its brain in three sesters of oil in a new earthenware pot until it is reduced by one-third; put it in a vessel and preserve it. If a person suffers from headache, apply it for three nights after he has bathed. The person will be cured, as will the feet. And even if someone suffers from a chronic and incurable disease, this method will cure and heal him.

1.7. Take its liver, divide it, and bury it along the circuit of the boundaries of your property and the base of your walls, and conceal its heart at your gate. Then you and yours will be free to travel and to return home safe and sound. Everything pestilential will be turned away, and whatever was done before will cause no harm, and there will be little danger from fire.

1.8. It is also known that its pelt is useful to put onto dogs and all four-footed animals for suffering due to pestilence. Keep some pieces of its hide in your shoes; you will not feel pain in your feet.

1.9. Ðu halgusta casere, ic wille þæt ðu gelyfe þæt þis wilddeor well fremað gif þu þinum clænsungdagum, þær þu færest geond eorðan ymbhwyrft, hys flæsc gesoden etest ond þigest. Hyt byþ god þe ond þinum weorudum.

1.10. Gif hwam hwæt yfeles gedon bið þæt he ne mæge hys wynlusta brucan, seoðe þonne his sceallan on yrnendum wyllewætere ond on hunige, ond ðicge þonne fæstende þry dagas. Sona he bið gebeted.

## 2

2.1. Wið blodes flewsan þonne eallum mannum: sy seofon-tyne nihta eald mona, æfter sunnan setlgange ær monan upryne, cyme to þam treowe þe man hateþ morbeam, ond of ðam nim æppel mid þinre wynstran handa mid twam fingrum—þæt is, mid þuman ond mid hringfingre—hwitne æppel þe þonne gyt ne readige. Ahefe hyne þonne upp ond upp aris; he bið brice to þam uferan dæle þæs lichaman. Eft do hyne adune ond onlut; he bið behefe to ðam neoðran dæle þæs lichoman. Ær ðon þu þysne æppel nime, cweð þonne þas word:

2   *Aps, aps, aps, sparare rose prospasam emorragiam pantosani opum æmesstanes.*

3   Þonne þu þas word gecweden hæbbe, genim þone æppel, ond hine þonne bewind on weolcreadum godwebbe, ond seoð þonne eft mid sceate oþres godwebbes. Ond beheald

1.9. Most sacred emperor, I want you to believe that this wild animal is very beneficial if you eat and consume its cooked flesh in the days when you are abstinent while traveling the expanse of the earth. It will be good for you and your armies.

1.10. If something evil has been done to a man so that he cannot enjoy sexual pleasure, simmer its testicles in fresh spring water and in honey, and give him this to eat while fasting for three days. He will soon be cured.

# Chapter 2

2.1. For bleeding, for all people: when the moon is seventeen days old, after the setting of the sun and before the rising of the moon, go to the tree that is called the mulberry tree, and pick a berry from it with two fingers of your left hand—that is, with the thumb and the ring finger—a white berry that has not yet begun to turn red. Lift it up high and stand up tall; it will be beneficial for the upper part of the body. Then lower it down and bow low; it will be useful for the lower part of the body. And before you pick this berry, say these words:

*Aps, aps, aps, sparare rose prospasam emorragiam pantosani*  2
*opum æmesstanes.*

When you have said these words, take the berry, and wrap  3
it in a fine purple cloth, and then simmer it inside another piece of fine cloth. And be careful that this medication does

þæt þes læcedom ne hrine ne wæteres ne eorðan. Þonne neadþearf sy ond se ufera dæl þæs lichoman on ænigum sare oððe on earfeþum geswince, wrið on þone andwlitan; gyf hyt sy on þam neoðran dæle, wrið on þa wambe.

2.2. Wið wifes flewsan: genim þone camb þe heo ana hyre heafod mid cemde, ond nænig man ær mid cemde ne æfter cembe; under ðam treowe morbeame cembe þær hyre feax. Þæt þær on þam cambe geþolige gesomnige ond aho on up standende twig þæs morbeames. Ond eft ymb hwyle, clæne, him to gesomnige ond gehealde. Þæt hyre bið læcedom þære ðe hyre heafod þær cembeþ.

2.3. Eft, gif heo wylle þæt ðæt hyre blodryne cyme to: cembe eft hyre heafod under morbeame ond þæt feax þe on þam cambe cleofige somnige, ond do on anne telgran ðe sy adune gecyrred, ond gesamnige eft. Þæt hyre byþ læcedom.

2.4. Gyf ðu wylle þæt wif sy geclænsod þe næfre mihte clene beon: wyrc hyre sealfe of þam feaxe, ond hit æthwego adrig, ond do on hyre lic. Þonne byþ heo geclænsod.

not touch either water or soil. When there is need, if the upper part of the body is in any pain or is suffering from any distress, bind it on the forehead; if the pain is in the lower part, bind it on the abdomen.

2.2. For a woman's excessive flow of menstrual blood: take the comb with which the woman alone has combed her hair, and which no other person has used before nor will use after; have her comb her hair under the mulberry tree. She should gather the hair that has been left in the comb and should hang it on an upright twig of the mulberry tree. Then after a while, she should gather it together, free from impurities, and should keep it. This will be a remedy for the woman who combs her hair there.

2.3. Again, if she wants to have her menstrual flow back: have her comb her head again under a mulberry tree, and have her gather the hair that sticks to the comb, and place it on a twig that is turned downward, and collect it again. This will be her remedy.

2.4. If you want to cleanse a woman who could never previously be cleansed: make her a salve from that hair and let it dry a little, and apply it to her body. Then she will be cleansed.

# 3

## *Medicina de cervo*

3.1. Wiþ nædran slite: heortes horn hafað mægen ælcne wætan to adrigenne. For þam his man bruceþ on eagsealfe.

3.2. Wiþ heafodsare: heortes hornes axan fif penega gewæge drinc. Nim anne sester wines ond twegen wæteres; nim þæs æghwylce dæge scenc fulne ond drince. Þes drenc eac wambe sar gehaþerað.

3.3. Wiþ toþa wagunge: heortes horn, gebærned ond gecnucod, þa teð getrymeþ, gif his man wislice bruceð.

3.4. Wið wifes flewsan: heortes horn, to duste gebeaten, ond drince on wine. Sona him byþ sel.

3.5. Wið wyrmas to cwellenne: heortes horn, gebærnedne; drince on hatum wætere. Þa wyrmas he acwelleð ond ut aweorpeþ.

3.6. Nædran eac to acwellanne: nim þæs hornes acxan ond stred þær hi syn. Hi fleoð sona onweg.

3.7. Wið wifa earfoðnyssum, þas uncyste Grecas hatað *hystem cepnizam:* heortes hornes þæs smælestan dustes bruce; þry dagas on wines drince. Gif he feforig sy, drince þonne on wearmum wætere. Þæt bið god læcecræft.

3.8. Wiþ miltan sare: heortes horn gebærnedne; þicge on

# Chapter 3

# Medicine from the deer

3.1. For snakebite: deer antler has the power to dry up every humor. For this reason it is used as an eye salve.

3.2. For headache: five pennyweights of deer-antler ashes; give it to drink. Take one sester of wine and two of water; take a cupful of this beverage to drink every day. This drink also relieves abdominal pain.

3.3. For loose teeth: deer antler, burned and pounded, firms up the teeth, if one uses it wisely.

3.4. For a woman's excessive flow of menstrual blood: deer antler pounded into a powder, and have her drink it in wine. She will soon be better.

3.5. To kill worms: give burned deer antler to drink in hot water. It will kill and expel the worms.

3.6. To kill snakes as well: take the ashes of this same antler and strew them where they are. They will soon flee away.

3.7. For women's disorders, the malady the Greeks call *hystem cepnizam:* use the finest powder of deer antler; have her drink it in wine for three days. If she is feverish, then have her drink it in warm water. This is a good remedy.

3.8. For pain in the spleen: burned deer antler; partake of it

geswettum drince. He þa miltan adrigeð ond þæt sar onweg afyrreþ.

3.9. Wið teter: heortes horn gebærnedne; meng wið eced; smyre mid þam. Hrædlice him cymeþ bot.

3.10. Eft, wið teter of andwlitan to donne: heortes horn gebærnedne; meng wið ele; smyre, ond þonne þæt bedrugud sy, eft þu hit geniwa. Do þis on sunnan upgange; hrædlice hit hæleþ.

3.11. Eft, wið þam ylcan: heortes horn gebærnedne, nigon penega gewæge, do þærto ond geswyrfes of seolfre syx peninga gewæge; gemeng ond gegnid swiþe wel, ond gewyrc to clyþan, ond smyre mid. Hyt hæleþ wel þæt sar.

3.12. Wið cyrnlu: *patella*—þæt ys heortes heagospind—gif þu hafast mid þe, ne arisað þe cyrnlu; ond þa þe ær arison, mid hys æthrine hy onweg gewitað.

3.13. Wifgemanan to aweccanne: nim heortes sceallan, dryg, wyrc to duste; do hys dæl on wines drinc. Þæt awecceþ wifgemanan lust.

3.14. Wið þæt ilce: nim heortes scytel ond cnoca to duste; do on wines drync. Hit hæleð þæt ilce.

3.15. Wið nædran bite: heortes gecyndlimu; drig to duste ond gedo rosan dust þærto, þreora peninga gewæge on drince, ond þicge on dæge. Scearplice se drenc hæleþ nædran bite.

3.16. Wið stede ond for gebinde: heortes hær beoð swiðe gode mid to smeocanne wifmannum.

in a sweetened drink. It will dry up the spleen and take the pain away.

3.9. For impetigo: burned deer antler; mix with vinegar; rub it on. It will quickly bring recovery.

3.10. Again, to take impetigo off the face: burned deer antler; mix with oil; rub it on, and when this has dried up, renew it again. Do this at sunrise; this will quickly heal it.

3.11. Again, for the same: add to it nine pennyweights of burned deer antler and six pennyweights of silver filings; mix and pound very well, and make a poultice of it and rub it on. It will heal the sore well.

3.12. For hard swellings: if you keep a *patella* with you—that is, a deer's cheek—no hard swellings will affect you; and if they have already arisen, they will go away at its touch.

3.13. To stimulate sexual arousal: take a deer's testicles; dry them, pound them into a powder; put some of this into a drink of wine. It will arouse sexual desire.

3.14. For the same: take a deer's dung and pound it into a powder; put it into a drink of wine. It will heal the same.

3.15. For snakebite: a deer's genitals; dry them to a powder and add powdered roses, three pennyweights, and put into a drink, and have him drink it on the same day. This remedy will heal the snakebite efficaciously.

3.16. For difficulty urinating and for constipation: the hairs of a deer are very good for fumigating women.

3.17. Wið wifes geeacnunge: ban bið funden on heortes heortan, hwilum on hrife; þæt ylce hyt gegearwað. Gif ðu þæt ban on wifmannes earm ahehst (gewriðest scearplice), hræþe heo geeacnað.

3.18. Wið innoþa wræce, ond gif gebind men byþ: heortes mearh gemylted syle him on wearmum wætere. Hrædlice hyt hæleþ.

3.19. Wið nædrena afligenge: heortes mearh, gebærned oðþæt hyt smeoce, oþþe þu hit mid þe hæbbe. Hit afligeþ ða nædran.

3.20. Wið laðum lælum ond wommum: heortes smeoro, gemylted ond mid ostorscillum gecnucud ond gemenged, ond to sealfe gedon ond on geseted. Wundorlice hyt hæleþ.

4

*Medicina de vulpe*

4.1. Wið wifa earfoðnyssum þe on heora inwerdlicum stowum earfeþu þrowiað: foxes leoþu ond his smeoru, mid ealdon ele ond mid tyrwan; wyrc him to sealfe; do on wifa stowe. Hraþe hit þa earfeþu gehæleþ.

4.2. Wið heafodsare: þam gelice þe hyt her bufan gecweden ys, smyre þæt heafod. Hyt hæleþ wundorlice.

3.17. For a woman to conceive: a bone can be found in a deer's heart, or sometimes in its belly; they have the same effect. If you hang this bone on the woman's arm (and bind it firmly), she will quickly conceive.

3.18. For abdominal pain, and if a person is constipated: melted deer's marrow; give it to him in warm water. This will quickly cure it.

3.19. To put snakes to flight: deer marrow, burned until it smokes, or else have it with you. It will put the snakes to flight.

3.20. For harmful bruises and spots: deer's grease, melted and pounded with oyster shells and mixed, and made into a salve and applied to the sores. It heals wonderfully.

# Chapter 4

# Medicine from the fox

4.1. For the pains of women who suffer afflictions in their internal parts: a fox's limbs and its fat, mixed with old oil and tar; work them into a salve; apply it to the womanly parts. It will quickly heal the afflictions.

4.2. For head sores: just as is said above, rub on the head. It will heal wonderfully.

4.3. Wið earena sare, eft: gelice þon þe her bufan gecweden is, genim þa ylcan sealfe, hluttre, drype on þæt eare. Wundorlice hyt hæleþ.

4.4. Wið miltan sare: foxes lungen on hattre æscan gesoden ond ær gecnucud ond to drence gedon. Þa miltan hyt wundorlice gehæleþ. Swa deþ hys lifer þæt ylce.

4.5. Wið weartan: genim foxes sceallan, gegnid swiþe oft þærmid þa weartan. Hraþe hyt hy tobreceþ ond onweg adeþ.

4.6. Wið nearwre sworetunge: foxes lungen gesoden ond on geswettum wine gedon ond geseald. Wundorlice hit hæleþ.

4.7. Wið sare cyrnlu: foxes sceallan genim ond gnid mid gelome. Hraþe hi beoð hale.

4.8. Wið gomena sare: foxes sina genim ond on hunige gewæt, ond gnid mid þa goman swiþe oft. Sona him byþ sel þæs broces.

4.9. Wið heafodece: genim foxes gecynd, ymfoh þæt heafod utan. Hraþe þæs heafodes sar byþ aweg afyrred.

4.10. To wifþingum: foxes tægles se ytemæsta dæl on earm ahangen. Þu gelyfest þæt þis sy to wifþingum on bysmær gedon.

4.11. Wið liþadle: genim cwicenne fox ond seoð þæt þa ban ane beon læfed; astige þærin gelomlice, ond in oþer bæð do he swa swiþe oft. Wundorlice hit hæleþ. Ond æghwylce geare þisne fultum he him sceal gegearwian. Ond ele do þærto ðonne he hine seoðe, ond his þyssum gemete to þearfe bruce.

4.3. For earache, again: just as is said above, take this same salve, unclouded; drip it into the ear. It will heal wonderfully.

4.4. For pain in the spleen: fox's lung, first pounded and then cooked in hot ashes and made into a drink. It will heal the spleen wondrously. It will do the same for the person's liver.

4.5. For warts: take fox's testicles; rub them very often on the warts. It will quickly break them up and will remove them.

4.6. For shortness of breath: fox's lung, cooked and added to sweetened wine and given to drink. This will heal it wonderfully.

4.7. For sore swellings: take fox's testicles and rub the swellings with them often. They will heal quickly.

4.8. For a sore throat: take the kidneys of a fox and moisten them in honey, and rub the throat with this very often. He will soon get over this affliction.

4.9. For headache: take the genitals of a fox; bind to the outside of the head. The headache will be quickly removed.

4.10. For sexual arousal: a fox's tail, the tip of it hung on the arm. You will think that this is done to make a mockery of the intercourse.

4.11. For ailments of the joints: take a live fox and boil it until only the bones are left; immerse frequently, and have him do the same thing very often in a second bath. This will heal it wonderfully. And every year he should prepare this remedy for himself. And add oil to it when cooking it, and use it in this same way when needed.

4.12. Wið earena sare: genim foxes geallan, menc wið ele; drype on þa earan. Hyt wel gehæleþ.

4.13. Wið eagena dymnysse: genim foxes geallan gemencged mid doran hunige, ond on eagan gedon. Hyt hæleþ.

4.14. Wið earena sare: genim foxes gelynde gemylted; drype on þa earan. Him cymð god hæl.

4.15. Wið fotwræce: gif se innera dæl þæs sceos byþ fixen-hyd, ond gyf hit sy fotadl, smyre mid ele þa fet. Hy habbaþ þæs þe leohtran gang.

# 5

## *Medicina de lepore*

5.1. Wið oferslæpe: haran brægen on wine geseald to drence. Wundorlice hyt beteþ.

5.2. Wiþ eagena sare: haran lungen on geseted ond þærto ge-wriþen. Þæt sar byþ gehæled.

5.3. Wið fotswylum ond sceþþum: haran lungen ufan on ond neoþan to gewriþen. Wundorlice þa gongas beoð gehælede.

5.4. Ðam wifum þe him hyra beorðor losie: haran heortan adrige ond wyrc to duste ond þriddan dæl recelses dustes; syle drincan seofon dagas on scirum wine.

4.12. For earache: take fox's gall, mix it with oil; drip it into the ears. It will cure it well.

4.13. For dimness of the eyes: use fox's gall mixed with honey from wild honeybees and applied to the eyes. It works.

4.14. For earache: take fox's fat, melted; drip into the ears. It will bring him good health.

4.15. For pain in the feet: if the lining of the shoe is made with fox hide, and if the ailment is foot disease, rub the feet with oil. This will make walking easier for them.

# Chapter 5

# Medicine from the hare

5.1. For excessive sleeping: hare's brain, given in wine as a beverage. This will make for a wonderful improvement.

5.2. For sore eyes: hare's lungs, put on them and fastened there. The soreness will be healed.

5.3. For swollen feet and foot injuries: hare's lungs, tied onto the top and bottom of the foot. One's footsteps will be wonderfully relieved.

5.4. For women who miscarry: dry a hare's heart and pound it into a powder, along with a third part of powdered frankincense; give to drink in pure wine for seven days.

5.5. Þam þonne þe hyt oft oðfealleþ: þritig daga ge on wine ge on wyrtunge.

5.6. Ðonne þam wifum þe æfter beorþre on sumum stowum swincen: þæt ylce do to drence fæstendum on wearmum wætere. Sona hyt byþ gehæled.

5.7. Wið eagena dymnysse: haran geallan wið hunig gemencged ond mid gesmyred. Þa eagan gebeorhtigeaþ.

5.8. Ðam mannum þe swinglunge þrowiað: haran lungen ond seo lifer somod gemencged, ond feower penega gewæge myrran, ond ðreora befores, ond anes huniges; þis sceal beon awylled on godum ecede, ond syþþan mid geswetton wine gewesed. Ond æfter þam drince, sona hyt hæleþ.

5.9. Wið blædran sare: haran sina, gedrygede ond mid sealte gebrædde ond gehyrste, sceaf on his drinc. Wundorlice hyt hæleþ.

5.10. Wið attorcoppan bite: haran sina gegyre ond him syle þicgan. Eac hyt is æltæwe gyf hi mon hreawe swylgeþ; eac wið wlættan hi beoð gode gesodene.

5.11. Wiþ feallendum feaxe: haran wambe seoð oþþe bræd on pannan on godum ele; smyre þæt feax ond þæt heafod. Þonne nimeþ þæt feax to, ond seo sealf genydeð þæt hyt weaxeþ.

5.12. To þan þæt wif cenne wæpnedcild: haran hrif, gedryged ond gesceafen oððe gegniden on drinc; drincen butu. Gif þæt wif ana hyt drinceþ, ðonne cenð heo *androginem;* ne byþ þæt to nahte, naþer ne wer ne wif.

5.5. The same for those who fall down often: for thirty days, either in wine or in herbal beverages.

5.6. Then for those women who feel pain in certain parts of their body after childbirth: give her the same thing as a beverage, in warm water while fasting. The condition will soon be cured.

5.7. For dimness of the eyes: hare's gall, mixed with honey and smeared around the eyes. The eyesight will improve.

5.8. For those who suffer from giddiness: hare's lungs and liver mixed together, and four pennyweights of myrrh, and three of beaver, and one of honey; this should be simmered in good vinegar, and afterward soaked in sweetened wine. After this drink, it will quickly heal.

5.9. For bladder pain: hare's kidneys, dried and roasted with salt and fried; scrape into the patient's beverage. It will heal wondrously.

5.10. For spider bites: prepare hare's kidneys and give them to the person to eat. It is also efficacious if they are eaten raw; also, when cooked, they are good for nausea.

5.11. For hair loss: simmer or roast a hare's belly in a pan with good oil; smear it on the hair and the head. The hair will stay in place, and the ointment will induce it to grow.

5.12. So that a woman can conceive a male child: hare's womb, dried and either scraped or crushed into a drink; let them both drink. If only the woman drinks it, she will generate an androgynous child; it will be as nothing, neither male nor female.

5.13. Eft, to þam ylcan: haran sceallan wife æfter hyre clæn-sunge; syle on wine drincan. Þonne cenð heo wæpnedcild.

5.14. Wif to geeacnigenne: haran cyslybb, feower penega gewæge, syle on wine drincan, þam wife of wife, ond þam were of were. Ond þonne don hyra gemanan, ond æfter þon hy forhæbben. Þonne hraþe geeacnað heo. Ond for mete heo sceal sume hwyle swamma brucan, ond for bæð smyre-nysse. Wundorlice heo geeacnaþ.

5.15. Wið *scorpiones* bite ond nædran slite: haran cyslyb ge-seald on wines drince. Þæt wel gehæleþ.

5.16. Wið þæt cildum butan sare teð wexen: haran brægen, gesoden, gnid gelome mid þa toðreoman. Hi beoð clæne ond unsare.

5.17. Wið wambe wræce: genim haran helan; ber on þinum hedclaþe. Wundorlice hit hæleð.

5.18. Wið eagena sare: haran lifer gesoden ys god on wine to drincenne, ond mid þam broþe ða eagan to beþianne.

5.19. Ðam manum þe fram þære teoþan tide ne geseoð: þæs ylcan drinces smyc heora eagan onfon ond mid þam broþe recen; ond þa lifre wæten ond gniden ond mid smyrwen.

5.20. Wið blodryne: gebærned haran lifer ond gegniden ond on gestreded. Hraþe hyt gestilleþ.

5.13. Again, for the same: hare's testicles for the woman after her purification; give to drink in wine. She will then give birth to a male child.

5.14. For a woman to conceive: a hare's rennet, four penny-weights. Give it to the patient to drink in wine, to the woman from a female, to the man from a male. Afterward they should have sexual intercourse, and after this they should abstain. She will quickly conceive. And for a while she should take mushrooms for food, and ointments instead of a bath. She will conceive wondrously.

5.15. For the sting of a scorpion and for snakebite: have the person drink hare's rennet in wine. That will heal it well.

5.16. For children to grow teeth painlessly: hare's brain, cooked; rub it often on the gums. They will be unblemished and free from pain.

5.17. For stomach pain: take the heel of a hare; carry it in your belt pouch. It will heal wonderfully.

5.18. For eye pain: simmered hare's liver is good to drink in wine, and it is good to bathe the eyes with its broth.

5.19. For people who cannot see after the tenth hour: let their eyes receive the vapor from this same drink and be steamed with the broth; and have them moisten the liver and massage it and smear this on.

5.20. For bleeding: hare's liver, burned and crushed and strewn on it. It will quickly cease.

# 6

## *Medicina de hirco*

6.1. Wið blodryne of nebbe: firginbuccan—þæt ys, wudu-bucca—oððe gat, þæs lifer gebryted wið ecede ond on næsþyrl bestungen. Wundorlice hraþe hyt ðone blodryne gestilleþ.

6.2. To eagena beorhtnysse: wudubuccan gealla gemencged wið feldbeona hunige ond on gesmyred. Seo beorhtnys him to cymð.

6.3. Þæt ylce mæg wið gomena sare: gemeng þone geallan ond hunig tosomne; hrin þa goman mid. Hyt hælð.

6.4. To eallum uncystum þe on gomum beoð acenned: wudu-gate geallan mid feldbeona hunige gemenged; þær sceal eac gelice awegen myrre ond pipor ond croh. Seoð eall on wine oþþæt hyt sy wel to sealfe geworht. Smyre þonne þa saran goman mid daga gehwylce oðþæt hy haligen.

6.5. Wið eagena dymnesse: wudugate geallan ond lytel wines meng tosomne; smyre mid ðriwa. Þonne beoð hi gehælede.

6.6. Wið dropfagum andwlatan: wudubucan geallan oððe gate gemencged wið wætere ond on gesmyred. Hraþe hit gelacnað.

6.7. Wið nebcorn þe wexað on þam andwlatan: smyre mid gate geallan. Ealle þa nebcorn he of þam andwlitan aclæn-sað, ond ealne þone wom he geðynnað.

# Chapter 6

# Medicine from the wild goat

6.1. For nosebleed: the liver of a mountain buck—that is, a wild buck—or of a goat, pounded in vinegar and pushed into the nostril. It will stop the bleeding amazingly quickly.

6.2. For bright vision: wild buck's gall mixed with the honey from wild honeybees and applied. The brightness will be restored.

6.3. This same remedy is good for a sore throat: mix the gall and the honey together; touch the throat with it. It will heal.

6.4. For all disorders that form inside the mouth: wild goat's gall mixed with honey from wild honeybees; there should also be myrrh and pepper and saffron in an equal weight. Simmer them all in wine until they thicken nicely into a salve. Then rub the sore mouth tissues with this every day until they heal.

6.5. For dimness of the eyes: wild goat's gall and a little wine; mix together; apply three times. Then they will be healed.

6.6. For a spotted face: wild buck's or goat's gall, mixed with water and applied. This will quickly cure it.

6.7. For spots that grow on the face: rub them with goat's gall. It will clear all the spots from the face, and it will reduce the entire blemish.

6.8. Wið earena sare ond swege: wudugate gealla mid neo-wum ele oððe æppeles seawe, wlæc, gemencged ond on þa earan gedon. Hyt hæleþ.

6.9. Wið toþece: wudugate geallan; mencg wið ele; smyre mid swyþe gelome. Þonne beoð hi hale.

6.10. Wið herðbylges sare oððe wunde: fyregate geallan; meng wið hunig; do to þam sare. Hit hæleþ wel.

6.11. To wifes willan: þæs buccan geallan; meng wið recels ond wið netelan sæd; smyre þone teors mid ær foran to þæs restgemanan. Þæt wif onfehð þæs willan on ðam hæmede.

6.12. Þy læs cild sy hreosende—þæt is, fylleseoc—oþþe scin-lac mete: fyregate brægen; teoh þurh gyldenne hring; syle þam cilde swelgan ær þam hyt meolc onbyrge. Hyt byþ ge-hæled.

# 7

## *Medicina de capra*

7.1. Wið homum: nim gate horn ond lege to fyre þæt he byrne on fyrle; do þonne of þa scylle on niwe fæt; cnuca hyt þonne swiþe wið scearpum ecede. Do on þa homan oþþæt hy hale syn.

6.8. For pain and ringing in the ears: wild goat's gall mixed with fresh oil or apple juice, lukewarm, and put into the ears. It will heal.

6.9. For toothache: wild goat's gall; mix with oil; rub it on very often. Then the teeth will be cured.

6.10. For pain or sores in the scrotum: mountain goat's gall; mix with honey; put this on the sore spot. It will heal well.

6.11. For a woman's pleasure: buck's gall; mix with frankincense and nettle seeds; apply to the penis before intercourse. The woman will experience pleasure in intercourse from this.

6.12. To prevent a child from experiencing falling sickness— that is, epilepsy—or from meeting with witchcraft: a mountain goat's brain; draw through a golden ring; give it to the child to swallow before it takes its milk. The child will be cured.

# Chapter 7

# Medicine from the domestic goat

7.1. For erysipelas: take a goat's horn and lay it near the fire so that it burns at a distance; then scrape off the scaly part and put it into a new vessel; then pound it thoroughly in sharp vinegar. Put it on the sores until they are cured.

7.2. To slæpe: gate horn under heafod geled. Weccan he on slæpe gecyrreþ.

7.3. Wið cyrnla sare: smeoc þone man mid gate hærum. Hraþe he byþ þæs sares hal.

7.4. Wið blodryne of nosum: adryg gate blod ond gnid to duste; do on þæt næsþyrl. Hyt wiðstandeþ.

7.5. Wið eagena hætan ond stice: niwe gate cyse ofer geseted mid þa eagbræwas. Him byþ hrædlice bot.

7.6. Wið heafodece: niwe gate cyse þærto gewriþen. Hyt hæleþ.

7.7. Wið fotadle: gate cyse niwe on gelegd þæt sar geliðegað.

7.8. Wið nædran slite: sceaf gate horn on þry scenceas, ond þare ylcan gate meolc wið wine gemencgede on þry siþas drince. Syllice hyt þæt attor tosceadeþ.

7.9. Wið innoðes flewsan: gate horn gesceafen ond wið hunige gemencged ond gegniden, ond æfter þam geþiged. Þære wambe flewsan he forþryceð.

7.10. Wið hreofle ond wið toflogen lic: genim þæt wæter þe innan gæt byþ ond heo hwilum ut geoteð; menge þone wætan wið hunige ond sealte; ond symle on æfenne his heafod ond his lic mid þy þwea ond gnide.

7.11. Wið innoðes heardnysse: swa hwæt swa he ete, menge wið þone wætan, ond þone ylcan drince wið þæs innoðes

7.2. For sleep: a goat's horn laid under the head. The horn will turn waking into sleep.

7.3. For the pain of hard swellings: fumigate the person with goat's hair. He will quickly be cured of the pain.

7.4. For nosebleed: dry goat's blood and pound it into a powder; put this into the nostril. It will act as a preventative.

7.5. For inflammation and sharp pain in the eyes: fresh goat's cheese set over them along with the eyelids. This will quickly be a cure for them.

7.6. For headache: fresh goat's cheese fastened to the head. This will cure it.

7.7. For foot disease: goat's cheese, fresh, laid on the feet will relieve the pain.

7.8. For snakebite: scrape a goat's horn into three cups, and have the person drink the milk of this same goat three times, mixed with wine. It will disperse the poison in an extraordinary manner.

7.9. For intestinal flux: goat's horn, scraped off and mixed with honey and pounded, and afterward eaten. It will stop the belly's flux.

7.10. For severe skin disease and for ulcers on the body: take the fluid that is inside a goat and that sometimes issues from it; mix the fluid with honey and salt; and always in the evening wash and rub the head and the body with this.

7.11. For a hardened abdomen: whatever the person eats, he should mix it with that fluid, and have him drink that same

heardnysse þæt seo getogene wamb sy alysed. Swa he ma drinceð, swa hyt furðor clænsað.

7.12. Wið þone wætan: do him eac to drince gate blod. Wel þæt hyne hæleþ.

7.13. Gif innoð þinde: nim gate blod mid hyre smeorwe, ond berene gryta gemeng, ond on wambe utan gewrið. Wundorlice hyt hælþ.

7.14. Wið ælces cynnes næddran bite: gate smeoro ond hyre tord ond weax; mylt ond gemeng tosomne; wyrc swa hit man gehal forswelgan mæge; onfo se þe him ðearf sy. Þonne bið he gehæled.

7.15. Se man se þe him seo wæteradl: gæten smeoro geþyd to poslum; swelge ond drince mid ceald wæter, ond somod swelge; ond drince æfter þam gate blod. Hym byþ hræd bot.

7.16. Drince eft buccan micgan, ond ete nardes ear ond wælwyrte moran. Selost ys se micga þæt he sy oftost mid feded.

7.17. Wið earena sare: gate micgan do on þæt eare. Þæt sar geliðigað. Gif þær wyrms inne bið, hyt þæt ut awyrpð.

7.18. Wið cyrnlu: gate tord; menge wið hunige; smyre mid. Sona bið sel.

7.19. Wið þeohwræce: gate tord; cned swyþe, þæt hyt sy swylce sealf, ond smyre mid þa þeoh. Sona hy beoð hale.

fluid for a hardened abdomen so that the tight stomach is relieved. The more he drinks, the more fully it will purge him.

7.12. For the corrupt humor: also make him a drink of goat's blood. This will heal him well.

7.13. If the abdomen should swell up: take goat's blood with its fat, and mix with coarse barley meal, and fasten this over the abdomen. It will heal wonderfully.

7.14. For every kind of snakebite: goat's grease, with its dung and some wax; melt them and mix them together; make it such that it can be swallowed whole; let whoever needs it take it. He will then be healed.

7.15. For a person who suffers from dropsy: goat's grease compressed into pills; have him swallow and drink them with cold water, swallowing them together; and after that have him drink goat's blood. This will be a quick cure for him.

7.16. Again, have him drink buck's urine and eat spikenard and dwarf elder roots. The best is the urine of a buck that has often been fed with these.

7.17. For earache: drip goat's urine into the ear. It will relieve the pain. If there is pus inside, it will expel it.

7.18. For hard swellings: goat's dung; mix with honey; apply it. He will soon be well.

7.19. For pain in the thighs: goat's dung; knead it well, so that it is like a salve, and rub the thighs with it. They will soon be well.

7.20. Wið liþa sare: nim gate tord, meng wið scearpum ecede, ond smyre mid. Wel hyt hæleþ. Ond smeoce mid hæþe, ond þæt ylce on wine drince.

7.21. Wið cancre: gate tord gemenged wið hunige ond on þa wunde gedon. Hraþe hyt hæleþ.

7.22. Wið swylas: gate tord; smyre mid þa swylas. Hyt hy todrifð ond gehæleþ ond gedeþ þæt hy eft ne arisað.

7.23. Wið sina getoge: gate tord; meng wið ecede ond smyre mid þæt sar. Hyt hælþ.

7.24. Wið springum: gate tord; meng wið hunige; smyre ond on gelege. Eac þa springas þe beoð on mannes innoðe acenned hyt todrifeþ.

7.25. Gate geallan on wine gedruncen wifa halan him of adeþ ond hi gehæleþ.

# 8

## Medicina de ariete

8.1. Wiþ wearras ond wið swylas: blacu rammes wul on wætere gedyfed ond æfter þam on ele, ond syþþan aled on þa saran stowe. Þæt sar heo onweg afyrreþ. Ond gyf hyt bið mid gereced, þa toslitenan wunda heo forþrycceþ.

7.20. For pain in the joints: take goat's dung, mix it with strong vinegar, and apply it. This will heal it well. And fumigate the patient with heather, and have him drink the same mixture in wine.

7.21. For an ulcerous sore: goat's dung mixed with honey and put on the sore. It will quickly heal.

7.22. For swellings: goat's dung; apply to the swellings. It will reduce them and heal them and ensure that they do not erupt again.

7.23. For a sinew spasm: goat's dung; mix with vinegar and apply to the sore area. It will heal.

7.24. For carbuncles: goat's dung; mix with honey; apply it and leave it on. It will also expel those carbuncles that have grown in a person's abdomen.

7.25. Goat's gall drunk in wine will remove women's afterbirth from them and will heal them.

# Chapter 8

# Medicine from the ram

8.1. For calluses and swellings: black ram's wool dipped in water and after that in oil, and afterward laid on the sore place. It will take the pain away. And if the sore spot is fumigated with this, the wool will close lacerated wounds.

8.2. Þa wearras ond ða swylas þe beoð on mannes handum oððe on oþrum limum oððe ymb þone utgang: smyre mid þam wætan þe drype of healfsodenre rammes lungenne. Hraþe heo hy onweg afyrreð.

8.3. Wið wundspringum on anwlatan: rammes lungen, smel tocorfen ond to þam sare geled. Sona hyt gehælþ.

8.4. Wið scurfum: rammes smeoru, ond meng ðærto sot ond sealt ond sand, ond hyt wulla onweg; ond æfter smyre. Hyt byþ eft liðre.

# 9

## *Medicina de apro*

9.1. Wið ælc sar: bares brægen gesoden ond to drence geworht on wine. Ealle sar hyt geliðegaþ.

9.2. Wið hærþena sare ond teorses: bares brægen; meng wið hunig ond wrið on. Wundorlice hyt hæleþ.

9.3. Wið nædran bite: bares brægen, gesoden ond gemencged wið hunig. Wundorlice hyt gehæleþ.

9.4. Eft, wið sarum ond gewundedum fotum: bares lungen gebeaten swiðe smale ond wið hunig gemenged ond to sealfe gedon. Hraþe heo þæt sar gehæleþ.

8.2. For calluses and swellings that are on a person's hands or on other limbs or round about the anus: rub them with the liquid that drips from half-boiled ram's lung. It will quickly remove them.

8.3. For ulcerous wounds on the face: ram's lung, cut up into small pieces and laid on the sore. It will soon heal them.

8.4. For scurfy skin: ram's grease, and mix soot and salt and sand with it, and wipe it away with wool; then apply it. It will become smoother again.

# Chapter 9

# Medicine from the boar

9.1. For every kind of pain: boar's brain, cooked and made into a drink with wine. It will relieve all pains.

9.2. For pain in the testicles and the penis: boar's brain; mix with honey and fasten it on. It will heal wonderfully.

9.3. For snakebite: boar's brain, cooked and mixed with honey. It will heal wonderfully.

9.4. Again, for sore and injured feet: boar's lungs, crushed very thin and mixed with honey and made into a salve. It will heal the soreness quickly.

9.5. Wið innoðes flewsan: niwe bares lifre; wyrc to drence on wine, ond þonne drince. Sona him bið sel.

9.6. Oras onweg to adonne: nim bares lifre ond swetre apuldre rinde; wyl tosomne on wine gemenged ond drince. Hraðe hy fleoð onweg fram him.

9.7. Gif earan syn innan sare ond þær wyrms sy: on do þa ylcan sealfe. Heo ys swiþe god to þam.

9.8. Weres wylla to gefremmanne: nime bares geallan ond smyre mid þone teors ond þa hærþan. Þonne hafað he mycelne lust.

9.9. Wið fylleseocum men: bares sceallan; wyrc to drence on wine oððe on wætere. Se drync hyne gehæleþ.

9.10. Wið spiwþan ond wlættan ond hnappunge: genim bares gelynde ond seoð on þrim sestrum wæteres oþþæt se dridda dæl sy beweallen; do þærto bares fam, ond drince. He byþ hal, ond he sylf wundrað ond weneþ þæt hyt sy oþer læcedom þæt he dranc.

9.11. Wið stede ond wið blæddran sare: genim eoferes blædran mid þam micgan, ahefe upp ond abid oþþæt se wæta of aflogen sy; seoð syððan, ond syle etan þam þe earfoþo þrowie. Wundorlice hit gehæleþ.

9.12. Þam þe under hy migað: bares blædre, gebræded ond geseald to etanne. Þa unhæle heo gehælþ.

9.13. Wið homum: bares scearn ond swefel gegniden on wine, ond gelome drince. Þa homan hyt beteþ.

9.5. For intestinal flux: fresh boar's liver; make it into a drink with wine, and then have the patient drink it. He will be better straightway.

9.6. To remove phlegm: take boar's liver and the bark of a sweet apple tree; simmer them together, mixed in wine, and have the person drink it. The phlegm will quickly leave him.

9.7. If ears are internally sore and there is pus inside, apply the same salve. It is very good for that.

9.8. To increase a man's desire: take boar's gall and apply it to the penis and testicles. Then he will feel great pleasure.

9.9. For a person with falling sickness: boar's testicles; make a drink out of them in wine or in water. This drink will heal him.

9.10. For vomit and nausea and drowsiness: take boar's fat and cook it in three sesters of water until it is reduced by one-third; add to it boar's saliva, and have the person drink it. He will be restored to health, and he will be surprised and think it is some other medicine that he drank.

9.11. For difficulty urinating and bladder pain: take a wild boar's bladder with its urine, lift it up and wait until the liquid has drained out; then cook it, and give it to the person to eat who suffers from this troubling condition. It will heal it wonderfully.

9.12. For those who suffer from leakage of urine: boar's bladder, roasted and given to eat. It will heal those who are sick.

9.13. For erysipelas: boar's dung and sulfur, crushed in wine, and have him drink this often. It will cure erysipelas.

# 10

## *Medicina de lupo et de canibus*

10.1. Wiþ deofulseocnysse ond wið yfelre gesihðe: wulfes flæsc, wel getawod ond gesoden; syle etan ðam þe þearf sy. Þa scinlac þe him ær ætywdon ne geunstillað hy hine.

10.2. To slæpe: wulfes heafod; lege under þone pyle. Se unhala slæpeþ.

10.3. Gif þu gesyxt wulfes spor ær þonne hyne, ne gesceþþeð he þe. Gif ðu hafast mid þe wulfes hrycghær ond tæglhær þa ytemæstan on siðfæte, butan fyrhtu þu ðone sið gefremest, ac se wulf sorgað ymbe sið.

10.4. Eagwræc onweg to donne: genim wulfes swyþre eage ond hyt tosting ond gewrið to ðam eagon. Hit gewanad þæt sar gyf hyt gelomlice þærmid gesmyred byþ.

10.5. Wið miltwræce: cwices hundes milte abred of; wyrc to drence on wine; syle drincan. Hyt hæleþ. Sume nimað hwelpes inylfe ond wriðaþ on.

10.6. Wið wiþerweard hær onweg to adonne: gif þu nimest wulfes mearh ond smyrest mid hraðe ða stowe þe þa hær beoð of apullod, ne geþafað seo smyrung þæt hy eft wexen.

# Chapter 10

# Medicine from the wolf and from dogs

10.1. For one who is possessed and for an evil vision: wolf's flesh, well prepared and cooked; give it to be eaten by one who has need of it. The phantoms that previously appeared to him will not disturb him.

10.2. For sleep: a wolf's head; lay under the pillow. The restless one will sleep.

10.3. If you see the tracks of a wolf before seeing the wolf itself, it will not harm you. If you keep hair from a wolf's back and hair from the tip of its tail with you while on a journey, you will undertake it without fear, but the wolf will be sorry about its lot.

10.4. To take away eye pain: take a wolf's right eye and prick it to pieces and fasten it on the eye. It will lessen the pain if the eye is often smeared with it.

10.5. For pain in the spleen: remove the spleen from a live dog; work it up as a drink with wine; give it to the patient to drink. It will be healed. Some people take the entrails of a whelp and fasten them on.

10.6. To remove ingrown hair: if you take wolf's marrow and rub it briskly on the place from which the hairs have been pulled, the ointment will not allow them to grow back.

10.7. Se wifman se þe hæbbe dead bearn on innoðe: gif he drinceð wylfene meolc mid wine ond hunige gemenged gelice efne, sona hyt hælð.

10.8. Biccean meolc: gif ðu gelome cilda toðreoman mid smyrest ond æthrinest, butan sare hy wexað.

10.9. Wearras ond weartan onweg to donne: nim wulle ond wæt mid biccean hlonde; wrið on þa weartan ond on þa wearras. Hraþe hi beoð awege.

10.10. Þam mannum þe magon hwon gehyran: hundes gelynde, ond wermodes seaw mid ealdum ele gemylt; dryp on þæt eare. Hyt þa deafan gebeteþ.

10.11. Wið wedehundes slite: nim þa wyrmas þe beoð under wedehundes tungan; snið onweg; ymb læd utan fictreow; syle þam þe tosliten sy. He bið sona hal.

10.12. Wið fefore: nim blæces hundes deades þone swyþran fotensceancan, hoh on earm. He tosceaceð þone fefor.

10.13. Warna ðe þæt ðu ne mige þær se hund gemah. Sume men secgað þæt þær oncyrre mannes lichama, þæt he ne mæge, þonne he cymeþ to his wife, hyre mid gerestan.

10.14. Scinseocum men: wyrc drenc of hwites hundes þoste on bitere lege. Wundorlice hyt hæleþ.

10.15. Hnite ond wyrmas onweg to donne ðe on cildum beoð: bærn hundes ðost ond gnid smale; menge wið hunige ond smyre mid. Seo sealf adeþ ða wyrmas onweg. Nim eac þæt græs þær hund gedriteþ; cnuca, wrið on. Hraðe hyt hælð.

10.7. For a woman who has a dead fetus inside her: if she drinks she-wolf's milk, along with wine and honey mixed in equal quantities, it will soon heal her.

10.8. Bitch's milk: if you touch and rub the gums of children with it often, the teeth will grow painlessly.

10.9. To remove calluses and warts: take some wool and wet it with bitch's urine; fasten it on the warts and the calluses. They will soon be gone.

10.10. For people who cannot hear well: dog's fat, and wormwood juice melted in old oil; drip this into the ear. It will cure the deaf.

10.11. For the bite of a mad dog: take the worms that are under the tongue of a mad dog; cut them away; carry them around a fig tree; give them to the person who is wounded. He will quickly be cured.

10.12. For a fever: take the right shank of a dead black dog; hang it on the arm. It will expel the fever.

10.13. Beware not to urinate where a dog has urinated. Some people say that there the body of a man changes so that, when he goes to his wife, he cannot lie with her.

10.14. For those who are possessed: make a drink from the excrement of a white dog in bitter lye. This will heal it wonderfully.

10.15. To remove nits and ringworms that are on children: burn dog's excrement and pound it finely; mix with honey and smear it on. The ointment will expel the worms. Also take the grass on which a dog has defecated; pound it, fasten it on. It will quickly heal.

10.16. Wið wæteradle: nim drigne hundes þost, wyrc to drence. He hæleð wæterseoce.

10.17. Dweorg onweg to donne: hwites hundes þost gecnucadne to duste, ond gemænged wið meolowe, ond to cicle abacen. Syle etan þam untruman men ær þær tide hys tocymes, swa on dæge swa on nihte, swæþer hyt sy. His togang bið ðearle strang, ond æfter þam he lytlað ond onweg gewiteþ.

10.18. Wið wæteradle: hundes spiwþan lege ond wrið on þam innoðe. Þurh þone utgang seo wæteradl ut afloweð.

## II

11.1. Ða þe scinlac þrowien: etan leon flæsc. Ne þrowiað hy ofer þæt ænig scinlac.

11.2. Wið earena sare: nim leon gelynde, mylt on scylle; drype on þæt eare. Sona him byþ sel.

11.3. Wið ælcum sare: gemylted leon gelynde ond þærmid gesmyred. Ælc sar hyt geliðigað.

11.4. Wið sina ond wið cneowa leoða sarum: nim leon gelynde ond heortes mearg, mylt ond gemeng tosomne; smyre mid þæt sar ðæs lichoman. Sona hyt byþ hal.

10.16. For dropsy: take dried dog's excrement, work it up into a drink. It will heal a person who is water-sick.

10.17. To remove a fever accompanied by delirium: white dog's excrement, pounded into a powder and mixed with meal and baked into a small cake. Give this to the person to eat before the time of the fever's onset, whether by day or night, whichever it may be. Its onset is exceedingly strong, and after that it diminishes and goes away.

10.18. For dropsy: dog's vomit; lay it on the abdomen and fasten it there. The liquid from the dropsy will flow out of the anus.

# Chapter 11

11.1. For those who are troubled by specters: let them eat lion's flesh. They will never suffer from any specters.

11.2. For earache: take lion's fat; melt it in a shell; drip into the ear. He will soon be better.

11.3. For any pain: melted lion's fat applied to the sore spot. It will relieve every pain.

11.4. For pains of the sinews and knee joints: take lion's fat and deer's marrow; melt them and mix them together; rub the body with it where it is sore. It will soon be well.

# 12

12.1. Wið næddrena eardunge ond aflygennysse: fearres horn gebærnedne to acsan; stred þær nædran eardien. Hy fleoð onweg.

12.2. Wommas of andwlatan to donne: smyre mid fearres blode. Ealle þa wommas hyt of genimeþ.

12.3. Fearres geallan, wið eagena þystru ond genipe; meng wið feldbeona hunig; do on þa eagan. Wundorlice hyt gehæleþ.

12.4. Wambe to astyrigenne: nim fearres geallan, somna on wulle, wrið under þæt setl neoðan. Sona he þa wambe onlyseþ. Do þæt ylce cildum ofer ðone nafolan. He weorpeþ ut þa wyrmas.

12.5. Wið earena sare: fearres geallan; meng wið hunige ond drype on ða earan. Sona him byþ sel.

12.6. Wið cyrnlu ðe beoþ on mannes andwlatan: smyre mid fearres geallan; sona he byþ clæne.

12.7. Wið apan bite oððe mannes: smyre mid fearres geallan. Sona heo bið hal.

12.8. Wið ælce heardnysse: fearres smeru; mylt wið tyrwan ond lege on. Ealle þa sar ond þæt hearde hyt geliðigað ond gehnesceaþ.

12.9. Wiþ fortogonysse: fearremearg on gehættum wine; drince. Þæt beteþ.

# Chapter 12

12.1. For snakes' nests and to expel them: bull's horn, burned to ashes; strew where the snakes have their home. They will flee.

12.2. To remove stains from the face: apply bull's blood. It will remove all the stains.

12.3. Bull's gall, for dimness and mistiness of the eyes; mix with honey from wild honeybees; apply to the eyes. This will heal it wonderfully.

12.4. To stimulate the bowels: take bull's gall, collect it in wool, fasten it underneath the person's rear. It will soon loosen the belly. For children, do the same over the navel. It will expel worms.

12.5. For earache: bull's gall; mix with honey and drip it into the ears. They will soon be better.

12.6. For hard swellings on a person's face: apply bull's gall. It will soon be without blemish.

12.7. For the bite of an ape or of a person: apply bull's gall. It will soon be well.

12.8. For all accumulation of hard matter: melt bull's fat with resin and lay it on. It will relieve the pain and soften the hardness.

12.9. For abdominal cramps: bull's marrow in warmed wine; drink. That will make it better.

12.10. Wið ælcum sare: drince fearres gor on hatum wætere. Sona hyt hælþ.

12.11. Wiþ bryce: fearres gor, wearm; lege on þone bryce. Syþþan him bið sel.

12.12. Wið wæteres bryne oððe fyres: bærn fearres gor ond scead þæron.

12.13. Gyf þu wylle don beorhtne andwlitan: nim fearres scytel, cnuca ond bryt, ond gnid swiðe smale on eced; smyre mid þone andwlatan. Ðonne byð he beorht.

12.14. Wifgemanan to donne: nim drige fearres sceallan; wyrc to duste oððe elcor gnid on win, ond drince gelome. He bið þy gearwra of wifþingum.

# 13

13.1. Wið gehwylce wommas of lichoman onweg to nimenne: genim ylpenban mid hunige gecnucud ond to geled. Wundorlice hyt þa wommas of genimeð.

13.2. Eft, wið wommas of andwlatan to donne: gyf wifman mid þam sylfan duste dæghwamlice hyre andwlatan smyreð, heo þa wommas afeormaþ.

12.10. For every kind of pain: have the person drink bull's dung in hot water. This will soon cure it.

12.11. For an injury: bull's dung, warm; lay it on the injury. The person will afterward be better.

12.12. For a burn from water or fire: burn bull's dung and scatter it over the spot.

12.13. If you want to make a face look splendid: take bull's dung, pound it and break it up, and crush it very fine in vinegar; apply this to the face. Afterward it will look splendid.

12.14. For sexual arousal: take dried bull's testicles; make into a powder or otherwise crush them into wine, and have him drink this often. He will be readier for sexual intercourse.

# Chapter 13

13.1. To remove all kinds of blemishes from the body: take ivory, pounded with honey and applied to the spot. It will remove the blemishes wonderfully.

13.2. Again, to remove blemishes from the face: if a woman applies this same powder to her face daily, she will clear the blemishes.

# 14

## *Medicina de canibus*

14.1. Wið ealle sar: gyf þu on foreweardon sumera þigest hwylcne hwelp na þonne gyt geseondne, ne ongitest þu ænig sar.

14.2. Wið fortogenysse: drince hundes blod. Hyt hæleþ wundorlice.

14.3. Wið geswel þæra gecyndlima: hundes heafodpanne, gecnucud ond to gelegd. Wundorlice heo hæleþ.

14.4. Wið cynelice adle: wedehundes heafod, gecnucud ond mid wine gemenged to drence. Hyt hæleþ.

14.5. Wið cancorwund: hundes heafod, to acxan gebærned ond on gestreded. Hit þa cancorwunda gehæleþ.

14.6. Wið scurfedum næglum: gebærned hundes heafod, ond seo acxe þæron gedon. Þa ungerisnu hyt onweg afyrreþ.

14.7. Wid wedehundes slite: hundes heafod gebærned to acxan ond þæron gedon. Eall þæt attor ond þa fulnysse hyt ut awyrpeð ond þa wedendan bitas gehæleþ.

14.8. Eft: wedehundes heafod ond his lifer, gesoden ond geseald to etanne þam þe tosliten bið. Wundorlice hyt hyne gehæleþ.

# Chapter 14

# Medicine from the dog

14.1. For all pains: if in the early summer you eat a whelp whose eyes have not yet opened, you will not perceive any pain.

14.2. For abdominal cramps: drink a dog's blood. This will cure them wonderfully.

14.3. For swelling of the genitals: a dog's skull, pounded and applied. This will heal it wonderfully.

14.4. For the king's disease: the head of a mad dog, pounded and mixed with wine so as to make a beverage. This will heal it.

14.5. For a cancerous wound: a dog's head, burned to ashes and sprinkled on the wound. It will heal the infected wounds.

14.6. For scabby nails: a burned dog's head, and its ashes laid on. It will remove the condition.

14.7. For a mad dog's bite: a dog's head burned to ashes and laid on. This will dispel all the poison and the foulness and will heal the mad dog's bite.

14.8. Again: the head of a mad dog and its liver, cooked and given to the bitten person to eat. This will heal him wonderfully.

14.9. To gehwylcum bryce: hundes brægen aled on wulle ond on þæt tobrocene to gewriþen feowertyne dagas. Þonne byþ hyt fæste gebatod. Ond þær byð þearf to fæstere gewriðennysse.

14.10. Wið eagwræce ond stice: tobrec hundes heafod; gif þæt swyþre eage ace nim þæt swyþre eage, gif þæt wynstre eage ace nim þæt wynstre, ond wrið utan on. Hyt hæleþ wel.

14.11. Wið toþwræce: hundes tuxas bærn to acxan, hæt scenc fulne wines, do þæt dust on, ond drince. Ond do swa gelome; þa teþ beoð hale.

14.12. Wið toþreomena geswelle: hundes tux, gebærned ond gegniden ond seted on. He wel hæleþ.

14.13. Wið þæt teþ wexen buton sare: hundes tux, gebærned ond smale gegniden ond on gedon. Toþreomena swylas gedwæsceað.

14.14. Wið hunda reðnysse ond wiðerrædnysse: se þe hafað hundes heortan mid him, ne beoð ongean hine hundas cene.

14.9. For any fracture: the brain of a dog, laid in wool and fastened onto the broken bone for fourteen days. It will then be firmly healed. And a quite firm binding will be needed.

14.10. For aching and for sharp pain affecting the eyes: break up a dog's head; if the right eye aches take the right eye, if the left eye aches take the left, and bind it on externally. This will heal it well.

14.11. For toothache: burn a dog's canine teeth to ashes, heat a cupful of wine, add the powder, and have the person drink. And do this often; the teeth will heal.

14.12. For swelling of the gums: a dog's canine tooth, burned and pounded into a powder and applied. This will heal it well.

14.13. To make teeth grow painlessly: a dog's canine tooth, burned and pounded into a fine powder and applied. This will remove the swelling of the gums.

14.14. For a dog's fierceness and hostility: if a person keeps a dog's heart with him, vicious dogs will never molest him.

# LACNUNGA

# Lacnunga

1. Wið heafod-wræce.

Genim hamorwyrt ond efenlastan nyðowearde. Cnuca; lege on clað. Gnid in wæter; gnid swiðe þæt heo sy eall geleðred. Þweah mid þy leaðre þæt heafod gelome.

2. Wið heafod-wræce.

Hindhæleða ond grundeswylgean ond fæncyrsan ond giðrifan. Wyl in wætere. Læt reocan in þa eagan þa hwile hy hate synd, ond ymb ða eagan gnid mid þam wyrtum swa hatum.

3. Wið heafod-wærce.

Betan wyrtruman. Cnuca mid hunige; awring. Do þæt seaw on þæt neb. Gelicge upweard wið hatre sunnan, ond ahoh þæt heafod nyþerweard oððæt seo ex sy gesoht. Hæbbe him ær on muðe buteran oððe ele. Asitte þonne uplang. Hnige þonne forð; læte flowan of þæn nebbe þa gilstre. Do þæt gelome oððæt hyt clæne sy.

4. To heafod-sealfe ond to eh-sealfe.

Aluwan. Gegnid in eced. Smyre þæt heafod mid, ond in þa eagan do.

5. Eah-sealf.

Win ond piper. Do in horn, ond in þa eagan þonne þu ðe restan wille.

# Lacnunga

1. For headache.

Take hammerwort and the lower part of dog's mercury. Pound; lay in a cloth. Rub in water; rub vigorously until it is all lathered. Bathe the head with the lather frequently.

2. For headache.

Hindheal and groundsel and watercress and cockle. Boil in water. Let the steam penetrate into the eyes while the plants are hot, and rub the plants around the eyes while still hot.

3. For headache.

Beetroots. Mash with honey; strain. Put the juice into the nose. Have him lie on his back facing the warmth of the sun, and have him hang his head back until the juice reaches the brain. See that he keeps butter or oil in his mouth beforehand. Then have him sit upright. Then have him lean forward; let the mucus run from his nose. Do this often until it runs clear.

4. To make a head salve and an eye salve.

Aloes. Rub them in vinegar. Smear the head with them, and put them in the eyes.

5. An eye salve.

Wine and pepper. Put them into a horn vessel, and into the eyes when you want to lie down.

6. Eah-sealf.

Genim streawberian nyþeweardan ond pipor. Do in clað; bebind; lege on gesweted win. Drype of þan claðe ænne dropan in ægðer eage.

7. Gif eagan forsetene beoð.

Genim hræfnes geallan ond hwitmæringc, wudulehtric ond leaxes geallan. Do tosomne. Dryp on þæt eage þurh linhæwenne clað, ond gehwæde arodes woses. Þonne wacað þæt eage.

8. Þis is seo seleste eah-sealf.

Nim doran hunig ond foxes smero ond rah-deores mearh. Mæng tosomne.

9. Gif poc sy on eagan.

Nim mærcsapan ond hinde meolc. Mæng tosomne ond swingc; læt standan oð hit sy hluttor. Nim þonne þæt hluttre; do on ða eagan. Mid Godes fultume he sceal aweg.

10. Þis is seo æðeleste eah-sealf wið eah-wyrce, ond wið miste, ond wið wænne, ond wið weormum, ond wið gicðan, ond wið tyrendum eagan, ond wið ælcum uncuðum geswelle.

Genim feferfugian blosman ond ðunorclæfran blosman ond dyles blosman ond hamorwyrte blosman, ond twegra cynna wyrmod, ond pollegian ond neoðowearde lilian, ond hæwene-hnydelan ond lufestice ond dolhrunan, ond geporta ða wyrta tosomne, ond awyll on heortes mearge oððe on his smerwe, ond menge. Do ðonne on tela micel in ða eagan ond smere utan, ond wyrm to fyre. Ond ðeos sealf deah wið æghwylcum geswelle, to ðicganne ond to smergenne, on swa hwylcum lime swa hit on bið.

6. An eye salve.

Take the lower part of wild strawberry and some pepper. Put into a cloth; tie it up; steep in sweetened wine. Drip a single drop from the cloth into each eye.

7. If the eyes are stopped up.

Take crab's gall and cowslip, prickly lettuce and salmon's gall. Mix together. Drip into the eye through a linen cloth, along with a little arum juice. Then the eye will wake up.

8. This is the best eye salve.

Take honey from a wild honeybee and fox's fat and a roe deer's bone marrow. Mix together.

9. If there is a stye in the eye.

Take bone-marrow soap and doe's milk. Mix together and whip; let stand until it is clear. Then take the clear liquid; put it into the eye. With God's help the stye will go away.

10. This is the finest eye salve for eye pain, and for clouded vision, and for a stye, and for mites, and for itching, and for teary eyes, and for every kind of inflammation of an unknown kind.

Take common centaury blossoms and bugle blossoms and dill blossoms and hammerwort blossoms, and two kinds of wormwood, and pennyroyal and the lower part of lily, and scurvy grass and lovage and pellitory-of-the-wall, and pound the plants together, and boil in deer marrow or deer grease, and stir. Then apply a generous portion to the eyes and daub it outside the eyes, and warm it at the fire. And this salve is effective against any kind of swelling, whether one takes it internally or applies it as a salve on whatever part of the body is affected.

11. Wið hwostan.

Nim huniges tear, ond merces sæd ond diles sæd. Cnuca þæt sæd smale. Mæng ðicce wið ðone tear, ond pipera swiðe. Nim ðry sticcan fulle on nihstig.

12. Wið eagena dymnesse.

Nim wulfescamb neoðeweardne ond lege on hunig ðreo niht. Nim þonne ond wipa þæt hunig of. Cnuca þonne an sticce ðære wyrt. Wring þonne ðurh linhæwenne clað on þæt eage.

13. Gif eagan tyran.

Genim grene rudan. Cnuca smale, ond wes mid doran hunige oððe mid dunhunige. Wring þurh linenne clað on þæt eage swa lange swa him ðearf sy.

14. Se man se-ðe biþ on heals-oman.

Nime healswyrt ond wudamerce ond wudafillan, ond streawbergean wisan ond eoforþrotan ond garclifan, ond isenheardan butan ælcan isene genumen, ond æðelferðþin-cwyrt ond cneowholen, ond brad bisceopwyrt ond brun-wyrt. Gesomnige ealle þas wyrta togædere þrim nihtan ær sumor on tun ga, ælcre efenmicel, ond gewyrce to drænce on wyliscan ealaþ. Ond þonne oniht þonne sumor on tun gæð on mergen, þonne sceal se man wacyan ealle þa niht þe ðone drenc drincan wile. Ond þone coccas crawan forman syðe, þonne drince he æne; oþre siðe þonne dæg ond niht scade; þriddan siðe þonne sunne upga. And reste hine syþþan.

11. For a cough.

Take honey that drips from the comb, and wild celery seed and dill seed. Pound the seeds finely. Blend with the honey into a thick mix, and pepper it well. Take three spoonfuls on an empty stomach.

12. For loss of vision.

Take the lower part of wild teasel and lay it in honey for three days. Then take it and wipe the honey off. Then mash one piece of the plant. Strain it into the eye through a linen cloth.

13. If there is discharge from the eye.

Take fresh rue. Pound finely, and saturate it with honey from a wild honeybee or with mountain honey. Strain it into the eye through a linen cloth as long as the patient needs it.

14. For a person suffering from acute inflammation on the neck.

Take throatwort and wild celery and wild thyme, and wild strawberry runners and carline thistle and agrimony, and knapweed picked without use of an iron tool, and stitchwort and butcher's-broom, and broad-leaved mallow and figwort. Gather all these plants together three days before the first day of summer, an equal amount of each, and make a concoction out of them in Welsh ale. And then the night before the first day of summer, the person who is to drink this concoction must stay awake all night long. And he should take a first drink at first cockcrow; a second drink when day and night separate; and a third drink when the sun rises. And after that he should go to bed.

15. Þis is seo grene sealf.

Betonica, rude, lufestice, finol, salvie, æðelferþincwyrt, savine, helde, galluces moran, slarige, merce, cearfille, hræmnesfot, mugwyrt, organa, melde, quinquefolium, valeriane, clate, medewyrt, dweorgedwoslan, pipeneale, solsequium, biscupwyrt, hæsel, quice, hegeclive, grundeswylie, brocminte ond oþre mintan, cicenamete, gagel, hegehymele, cost, eorðnafala, hnutbeames leaf, lauberge, cymen, ele, weax.

16. Wið adle.

Nim þre leaf gageles on gewylledre mealtre meolce. Syle þry morhgenas drincan.

17. Wið heafod-ece.

Rude ond dweorgedwosle ond betan more ond wudurove. Nim ealra evenmicel swa ðu mæge mid þinan scitefingre to þinum ðuman befon. Cnuca hy smale. Ond mylt buteran ond do of eall þæt fule, ond do on clæne pannan, ond awyl ða wyrta þæron wel, ond wring ðurh claeð. Do ele to, gif ðu begytan mæge, ond smyre his heafod mid þær hit acy.

18. Sealf wið fleogendum attre ond færspryngum.

Nim hamorwyrte handfulle ond mægeðan handfulle ond wegbrædan handfulle, ond eadoccan moran—sece ða þe fleotan wille, þære ðeah læst—ond clænes huniges ane ægscylle fulle. Nim þonne clæne buteran. Þrywa gemylte ðe þa sealfe mid weorcean wile. Singe man ane mæssan ofer ðam wyrtum ær man hy tosomne do ond þa sealfe wyrce.

15. This is the green salve.

Betony, rue, lovage, fennel, sage, stitchwort, savine, tansy, comfrey root, clary sage, wild celery, chervil, crowfoot, mugwort, marjoram, common orache, creeping cinquefoil, valerian, burdock, meadowsweet, pennyroyal, burnet saxifrage, deadly nightshade, marshmallow, hazel, couch grass, hedge cleavers, groundsel, water mint and other mints, chickweed, sweet gale, hop plant, costmary, wild asparagus, nut-tree leaves, laurel berries, cumin, oil, wax.

16. For disease.

Take three leaves of sweet gale in sour milk that has been brought to a boil. Give to drink for three mornings.

17. For headache.

Rue and pennyroyal and beetroot and woodruff. Take an equal amount of each, as much as you can hold between your index finger and your thumb. Pound them finely. And melt some butter and remove all the scum, and put it in a clean pan, and simmer the herbs well in it, and strain through a cloth. Add oil, if you can get it, and daub his head with it where it aches.

18. A salve for flying venom and for sudden rashes.

Take a handful of hammerwort and a handful of chamomile and a handful of common plantain, and some water-lily roots—look for the kind that floats, though this should be the smallest ingredient—and one eggshell-full of pure honey. Then take some pure butter. Whoever is to make the salve should melt the butter three times. Have a Mass sung over the plants before you put them together and make the salve.

19. Wið ðone bledende fic.

Nim murran ða wyrt ond ceorf of nygan penegas, ond do on ælcne hunig. Ond ðige ða on æfen, ond eft oðre nygan on mergen. Ond do swa nigon dagas ond nigon niht, butan ðe raðor bot cume.

20. Oleo roseo sic facis.

Oleo libram unam, flos roseo viride uncium unum. Commiscis in ampulla vitria sub gipsos, et suspendis ad solem dies quadraginta ut virtus eius erit stiptica et frigida. Facis eum ad plurimas passiones, maxime ad dolorem capitis quod Grecae *æncausius* vocant, hoc est emigraneum capitis.

21. *Cardiacus* hatte seo adl ðe man swiðe swæteð. On hy man sceal wyrcean utyrnende drænceas, ond him wyrcean cliðan toforan his heafde ond to his breostan.

2      Genim grene rudan leaf. Scearfa smale ond cnuca swiðe, ond berenmeala gesyft do ðærto ond swetedne eced. Wyrc to cliðan, ond do on þicne cla\ð, ond bind on þreo niht ond þry dagas. Do eft niwne to. Ond drince seoca of bræmelberian gewrungene oft.

22. Sing ðis wið toð-ece syððan sunne beo on setle, swiðe oft.

2      "*Caio laio, quaque voaque,* ofer *sæloficia,* sleah manna wyrm!"

3      Nemne her þone man ond his fæder; cweð þonne:

19. For hemorrhoids.

Take the plant named myrrh and cut off nine penny-weight bits, and put honey on each one. And eat them in the evening, and another nine bits in the morning. And do this for nine days and nine nights, unless you are well before then.

20. This is how you make rose oil.

One pound-weight of oil, one ounce of fresh rose petals. Mix them in a glass flask covered by a plaster made of lime, and hang this in the sun for forty days so that it will have astringent and cooling properties. Make it for many ailments, especially for that kind of headache that the Greeks call *enkausios,* that is, migraine.

21. *Cardiacus* is what the disease is called when someone sweats excessively. To treat it one must make purgative drinks, and one must make poultices for the front part of the person's head and for his chest.

Take fresh rue leaves. Shred them into little pieces and pound them well, and add sifted barley meal and sweetened vinegar. Make a poultice from it, and attach to a thick cloth, and tie on for three nights and three days. Afterward apply a fresh one. And the sick person should take frequent drinks of strained brambleberry juice.

22. Sing this for toothache after the sun has set, repeating it very often:

"*Caio laio, quaque voaque,* over *sæloficia,* kill the *wyrm* of men!"

At this point name the person and his father; then say:

4    "*Lilumenne.* Æceð þæt ofer eall þonne alið; coliað þonne hit on eorðan hatost byrneð. Finit, amen."

23. Wið ðone dropan.

Ive ond fifleafe, nædderwyrt ond hlædderwyrt ond eorð-geallan. Wyrc ðas wyrta on hærfeste, ond scearfa hy smale ond drige hy, ond heald hy ofer winter, ond nytta hy þonne ðe ðearf sy. Wylle hy on ealað.

24. Wið geswel.

Genim lilian moran ond ellenes spryttinge ond porleaces leaf, ond scearfa swiðe smale, ond cnuca swiðe. Ond do on ðicne clað ond bind on.

25. Sing ðis gebed on ða blacan blegene nigon syðan; ærest "Pater noster."

2    "*Tigað, tigað, tigað. Calicet aclu cluel sedes adclocles acre earcre arnem nonabiuð ær ærnem niðren arcum cunað arcum arctua fligara uflen binchi cuterii nicuparam rafafð egal uflen. Arta, arta, arta; trauncula, trauncula.*

3    "Querite et invenietis. Adiuro te per Patrem et Filium et Spiritum Sanctum non amplius crescas sed arescas. Super aspidem et basilliscum ambulabis, et conculcabis leonem et draconem. Crux Matheus, crux Marcus, crux Lucas, crux Iohannes."

"*Lilumenne*. It aches more than ever when it is subsiding;  4
when the ground burns hottest it is growing cool. It is done,
amen."

23. For gout.

Ground pine and cinquefoil, bistort and Jacob's ladder
and yellowwort. Prepare these plants in the autumn, and
shred them into small pieces and dry them, and preserve
them over the winter, and use them when you have need.
Simmer them in ale.

24. For a swelling.

Take bulbs of lilies and new shoots from an elder tree and
leaves of the common leek, and shred them very finely, and
pound them well. And put them into a thick cloth and tie it
on.

25. Sing this prayer on the black boils nine times; but first of
all the "Our Father":

"*Tigað, tigað, tigað. Calicet aclu cluel sedes adclocles acre earcre*  2
*arnem nonabiuð ær ærnem niðren arcum cunað arcum arctua*
*fligara uflen binchi cuterii nicuparam rafafð egal uflen. Arta, arta,*
*arta; trauncula, trauncula.*

"Seek and you shall find. I adjure you by the Father and  3
the Son and the Holy Spirit to grow no bigger, but to wither
away. Thou shalt walk upon the asp and the basilisk, and
thou shalt trample underfoot the lion and the dragon. The
cross of Matthew, the cross of Mark, the cross of Luke, the
cross of John."

26. Wið ðon þe mon oððe nyten wyrm gedrince.

2    Gyf hyt sy wæpnedcynnes, sing ðis leoð in þæt swiðre eare þe heræfter awriten is; gif hit sy wifcynnes, sing in þæt wynstre eare.

3    *Gono mil orgo mil marbu mil marb sair amum tofeð tengo do cuillo biran cuiðær cæfmiil scuiht cuillo scuiht cuib duill marb sir amum.*

4    Sing nygon siðan in þæt eare þis galdor, ond "Pater noster" æne.

27. Þis ylce galdor mæg mon singan wið smeogan wyrme.

Sing gelome on ða dolh ond mid ðinan spatle smyre. Ond genim grene curmeallan, cnuca, lege on þæt dolh, ond beðe mid hattre cu-micgan.

28. Wið ðon ðe mon attor gedrince.

Nim marubian sæd; mængc wið wine. Syle drincan.

29. Þis is se halga drænc wið ælfsidene ond wið eallum feondes costungum.

2    Writ on husldisce "In principio erat verbum" usque "non comprehenderunt," et plura "Et circumibat Ihesus totam Galileam docens" usque "et secuti sunt eum turbe multe"; "Deus in nomine tuo" usque in finem; "Deus misereatur nobis" usque in finem; "Domine Deus in adiutorium" usque in finem.

3    Nim cristallan ond disman ond sidewaran ond cassuc ond finol, ond nim sester fulne gehalgodes wines, ond hat unmælne mon gefeccean swigende ongean streame healfne sester yrnendes wæteres. Nim þonne ond lege ða wyrta ealle

26. In the event that a person or a domestic beast drinks a *wyrm*.

If the person is of male sex, sing in his right ear the chant   2
that is written below; if of female sex, sing in the left ear.

*Gono mil orgo mil marbu mil marb sair amum tofeð tengo do*   3
*cuillo biran cuiðær cæfmiil scuiht cuillo scuiht cuib duill marb sir*
*amum.*

Sing this spell into the ear nine times, and the "Our Fa-   4
ther" once.

27. This same incantation can be sung in case a *wyrm* has
burrowed into the skin.

Sing it repeatedly upon the wounds and daub them with
your spittle. And take fresh knapweed, pound it, lay it on
the wound, and bathe the area with warm cow's urine.

28. In the event that someone drinks poison.

Take seed of white horehound; mix with wine. Give to
drink.

29. This is the holy drink for elf-wrought magic and for all
afflictions caused by the devil.

Write on a paten "In the beginning was the word" as far as   2
"they understood it not," and also "and Jesus went all about
Galilee teaching" as far as "and a great multitude followed
him"; also "God, in thy name" to the end, and "God have
mercy on us" to the end, and "Lord God in our aid" to the
end.

Take cristalle and musk and zedoary and hassock grass   3
and fennel, and take a pitcher full of consecrated wine, and
have a virgin fetch a half pitcher of running water from a
stream while maintaining silence. Then take the plants and

in þæt water, ond þweah þæt gewrit of ðan husldisce þærin swiðe clæne. Geot þonne þæt gehalgade win ufon on ðæt oþer.

4    Ber þonne to ciricean. Læt singan mæssan ofer, ane "Omnibus," oðre "Contra tribulatione," þriddan "Sancta Marian." Sing ðas gebedsealmas: "Misere mei Deus," "Deus in nomine tuo," "Deus misereatur nobis," "Domine Deus," "Incline Domine," ond "Credo," ond "Gloria in excelsis Deo," ond letanias, ond "Pater noster." Ond bletsa georne in ælmihtiges Drihtnes naman ond cweð: "In nomine Patris et Filii et Spiritus Sancti, sit benedictum." Bruc syþþan.

30. To wen-sealfe.

Nim elenan ond rædic ond cyrfillan ond hræmnesfot, Ængliscne næp ond finul ond salvia ond suþernewuda, ond cnuca tosomne. Ond nim garleaces godne dæl. Cnuca, ond wring þurh claõ on gemered hunig. Þonne hit swiðe gesoden sy, þonne do ðu pipor ond sideware, gallengar ond gingifre ond rinde ond lawerbergean ond pyretran, godne dæl ælces be ðære mæðe. Ond syððan hit swa gemænged sy—þara wyrta wos ond þæt hunig—þonne seoð ðu hit twa swa swiðe swa hit ær wæs. Þonne hæfst þu gode sealfe wið wennas ond wið nyrwet.

31. To godre ban-sealfe þe mæg wið heafod-ece ond wið ealra lyma tyddernysse.

2    Rude, rædic, ond ampre, vane, feverfuge, æscðrote, eoforðrote, cilðenige, bete ond betonican, ribbe ond reade hofe, elene, alexandrian moran, clufðung ond clate, liðwyrt ond lambes cerse, hylwyrt, hæsel, cwice, wudurofe ond

lay them fully in the water, and wash the writing completely off the paten into the solution. Then pour the consecrated wine down onto the other liquid.

Then take it to church. Have Masses sung over it, first 4 "To all saints," second "Against tribulation," third "To Saint Mary." Sing these precatory psalms: "God have mercy on me," "God in thy name," "God have mercy on us," "To the Lord God," and "Listen O Lord," plus then the Creed, "Glory to God in the highest," and litanies, and the "Our Father." And earnestly bless it in the name of the almighty Lord and say: "In the name of the Father and the Son and the Holy Spirit, be blessed." Then partake of it.

30. For a wen salve.

Take elecampane and radish and chervil and crowfoot, turnip and fennel and sage and southernwood, and pound together. And take a good deal of garlic. Pound, and wring through a cloth into clarified honey. Once it is well boiled, then add pepper and zedoary, galingale and ginger and cinnamon bark and laurel berries and common centaury, a good deal of each according to good measure. And after it is mixed together like this—the juice of the plants and the honey—then cook it twice as much as before. Then you will have a good salve for wens and for constriction of breath.

31. For a good bone salve that works for headache and for infirmity in any limb.

Rue, radish, and dock, iris, common centaury, vervain, 2 carline thistle, greater celandine, beet and betony, hound's-tongue and red ground ivy, elecampane, roots of alexanders, celery-leaved crowfoot and burdock, limbwort and shepherd's purse, pennyroyal, hazel, couch grass, woodruff and

wrættes ciŏ, springwyrt, sperewyrt, wegbræde ond wermod, ealhtran ond hæferŏan, hegeclife ond hymelan, gearwan ond geaces suran, belenan ond bradeleac.

3    Nim ealra ŏyssa wyrta efenfela; do on mortere. Cnuca eall tosomne, ond do ŏærto ifig croppas. Ond nim æsc-rinde ond weliges twiga ond ac-rinde ond wir-rinde ond surre apoldre rinde ond seales rinde ond wudubindan leaf. Þas ealle sculan beon genumene on neŏoweardan ond on easteweardan þan treowan. Scearfige ealle ŏas rinda togædere ond wylle on haligwætere oŏŏæt hy wel hnexian. Do þonne to þan wyrtum on mortære. Cnuca eall tosomne.

4    Nim þonne heortes smera ond hæferes smera ond eald morod, ond fearres smeru ond bares smeru ond rammes smeru. Mylte mon ealle tosomne, ond geote to trindan. Somnige mon þonne ealle þa ban tosomne ŏe man gegaderian mæge, ond cnocie man þa ban mid æxse yre, ond seoŏe ond fleote þæt smeru. Wyrce to trindan. Nime þonne ealde buteran ond wylle þa wyrta ond þa rinda. Don eall tosomne. Donne hit beo æne awylled, sette þonne. Scearfa þonne eall þæt smera on pannan, swa micel swa þu sealfe haban wille ond þu getyrwan mæge. Sete ofer fyr. Læt socian, næs to swiŏe weallan oŏŏæt hyo genoh sy. Seoh ŏurh claŏ; sete eft ofer fyr.

5    Nim þonne nygon clufa garleaces gehalgodes. Cnuca on wine; wring þurh claŏ. Scaf on myrran þa wyrt, ond fanthalig wex ond brunne stor ond hwitne rycels. Geot þonne innan

sprouts of madder, springwort, spearwort, common plantain and wormwood, lupin and greater stitchwort, cleavers and hop plant, yarrow and wood sorrel, henbane and common leek.

Take an equal amount of all these plants; put into a mortar. Mash them all together, and add clusters of ivy berries. And take ash-tree bark and willow twigs and oak-tree bark and myrtle bark and crab-apple bark and willow-tree bark and woodbine leaves. These must all be taken from the lower part of the tree and from its east-facing side. Shred all these barks together and boil them in holy water until they are good and soft. Then add them to the plants in the mortar. Pound them all together.  3

Then take hart's fat and he-goat's fat and mulled wine, and bull's fat and boar's fat and ram's fat. Melt them all together, and pour it into a mold. Then have someone collect all the bones that can be gathered together, and break up the bones with the back of an ax, and simmer them and skim off the fat. Work it into a mold. Then take some aged butter and simmer the plants and the barks. Put them all together. As soon as this has been brought to a boil, let it sit. Then scrape all the grease into a frying pan, as much as you wish to have as a salve and as you can reduce to a tar. Set it over the fire. Let it soak, but not boil too much, until it is cooked enough. Strain through a cloth; set it back over the fire.  4

Then take nine cloves of consecrated garlic. Pound them in wine; wring through a cloth. Scrape the plant named myrrh into them, and consecrated wax and brown storax and white frankincense. Then pour this into the salve, as  5

ða sealfe, swa micel þæt sy þreo ægscylla gewyrðe. Nim
þonne ealde sapan ond ealdes oxsan mearh ond earnes
mearh. Do þonne ða tyrwan, ond mæng þonne mid cwic
beamenum sticcan oð heo brun sy.

6      Sing þonne þærofer "Benedictus Dominus, Deus meus,"
ond þone oþerne, "Benedictus Dominus, Deus Israel," ond
"Magnificat" ond "Credo in unum" ond þæt gebed "Matheus,
Marcus, Lucas, Iohannes." Sy þæt sar þær hit sy, smite mon
ða sealfe ærest on þæt heafod.

32. Gif poc sy on eagan.

Nim mærcsapan ond hinde meoluc. Mæng tosomne ond
swyng; læt standan oð hit sy hluttor. Nim þonne þæt hlutre;
do on ða eagan. Mid Godes fultume heo sceal aweg.

33. Nim clatan moran. Cnuca swiðe, ond wyl on beore. Syle
drincan wel wearm þonne ðu geseo þæt hy ut slean. Mid
Godes fultume ne wyrð him nan orne.

34. Þas wyrte sculon to lungen-sealfe.

Banwyrt ond brunwyrt, betonican ond streawberian wise,
suþernewuda ond isopo, salvie ond savine ond rude, garclife
ond hæsel, cwice, medewyrt, dolhrune.

35. Wið heafod-ece.

Wyl in wætere pollegian ond leac, mintan, fen mintan,
ond þæt ðridde cyn mintan þæt bloweð hwite. Þweah þæt
heafod mid þys wose gelome.

36. Wið hreofum lice.

Adelf ampron ond gelodwyrt; teoh ut lange. Cnuca ealle
wel; wyll in buteran; do hwon sealtes in. Þæt bið god sealf

much as three eggshells worth. Then take aged soap and aged ox marrow and eagle's marrow. Then add the tar, and stir it with rowan sticks until it is brown.

Then sing over it "Blessed be the Lord, my God," and a 6 second prayer, "Blessed be the Lord, the God of Israel," and the Magnificat and the Creed and the prayer "Matthew, Mark, Luke, John." No matter where the sore is located, daub the salve on the patient's head first of all.

32. If there is a stye in the eye.

Take bone-marrow soap and doe's milk. Mix together and whip; let it stand until it is clear. Then take the clear liquid; put it in the eye. With God's help the stye will go away.

33. Take roots of burdock. Pound well, and simmer in beer. Give to drink, quite warm, when you see that the pustules are breaking out. With God's help no great harm will befall him.

34. These plants are to be used for a salve for lung disease.

Wallflower and figwort, betony and wild strawberry runner, southernwood and hyssop, sage and savine and rue, agrimony and hazel, couch grass, meadowsweet, pellitory-of-the-wall.

35. For headache.

Boil in water pennyroyal and leek, mint, fen mint, and the third kind of mint, the one with white blossoms. Wash the head with this juice frequently.

36. For a leprous body.

Dig up dock and silverweed; draw out the whole length of the plant. Pound everything well; simmer in butter; add a bit

wið hreofum lice. Þweah þone man mid hate, ond mid ðare sealfe smyre.

37. Wið cneo-wærce.

Genim weodewisan ond hegerifan. Gecnuca well tosomne, ond do mela. Læt standan nyhternum on þæm wyrtum. Syle drincan.

38. To eah-sealfe.

Nim aluwan ond sidewaran, lawerberian ond pipor. Gescaf smale, ond cu-buteran fersce lege on wæter. Nim þonne hwetstan bradne, ond gnid ða buteran on ðæm hwetstane mid copore þæt heo beo wel toh. Do þonne sumne dæl þara wyrta þærto. Clæm ðonne on ar-fæt. Læt standan nygon niht; wende man ælce dæge. Mylte siþþan on ðæm ar-fæte sylfan. Aseoh þurh clað. Do syþðan on swylc fætels swylce ðu wille. Nyttige þonne þe þearf sy. Þes sealf mæg wið ælces cynnes untrumnysse ðe eagan eigliað.

39. Wið utsihte.

Genim hænne æg. Lege twa niht on eced. Gif hit ne tocine, tosleah hwon. Lege eft in ðone eced nyhterne. Gesleah þonne in buteran; lege in ele; ado þonne hwon ofer fyr. Syle etan.

40. Eft wið þonne.

Hunig ond hwætesmedman ond unsylt smeoru ond wex. Wyl eall tosomne. Syle etan gelome.

41. Wyll wið ðon miclan eorðnafolan ond fifleafan ond gyðhrofan ond gearwan ond hæferþon ond eoforfearn ond moldcorn ond medewyrt neoðewearde. Drinc gelome.

of salt. This is a good salve for a leprous body. Give the person a steam bath, and smear the salve on.

37. For knee pain.

Take weed runners and goose grass. Pound them well together, and add meal. Let the meal remain on the plants overnight. Give to drink.

38. For an eye salve.

Take aloe and zedoary, laurel berries and pepper. Shave finely, and lay fresh cow butter in water. Then take a broad whetstone, and rub the butter on the whetstone with a copper implement until it is good and firm. Then add a portion of the plants. Then pack it into a bronze vessel. Let it stand for nine nights; stir the contents every day. Afterward melt it in the same bronze vessel. Strain through a cloth. Afterward put it into any vessel you want. Make use of it when you have need. This salve is good for every kind of infirmity that affects the eyes.

39. For diarrhea.

Take a hen's egg. Steep in vinegar for two days. If it does not crack, knock it lightly. Steep it in the vinegar again overnight. Then mash it in butter; lay it in oil; then put it over the fire for a bit. Give to eat.

40. Another cure for it.

Honey and fine wheaten flour and unsalted fat and wax. Boil them all together. Give to eat frequently.

41. For the same ailment, boil the larger type of wild asparagus and cinquefoil and cockle and yarrow and greater stitchwort and common polypody and saxifrage seeds and the lower part of meadowsweet. Drink it often.

42. Scæf ifig wið þonne bol in meolc, ond þige wærlice. Ond seoð ealle ða in meolce, ond hwilum þa meolc geren mid cyslybbe, ond ðige hy.

43. Wyrc ut-yrnendne drænc.

Genim fif ond hund-eahtatig lybcorna ond neogon pipor-corn ond fiftene sundcorn wel berended. Cnuca smale; do sealt in ond wyrm-melo. Mæng tosomne ond gnid swiðe þæt hit sy þæt smælste, geworht to duste. Genim scæncbollan fulne leohtes beores, oððe hluttor eala wel gesweted, oððe gesweted win. Mængc ða wyrta þærwið geornlice; læt ston-dan nihterne. Hrer hine eft on mergen þonne he hine drin-can scyle swiðe wel, ond ða wyrte geornlice wið þone wætan gemengce. Drince þonne. Gif he sy to unswið, wyl merce in wætere; syle drincan. Gif he to swið sy, wyl curmeallan.

44. Oþer ut-yrnynde drænc.

Genim medmicle moran glædenon, fædme longe ond swa greate swa ðin þuma. Ond swylc tu hamwyrte ond celðenian moran ond heleleafes moran ond ellen-rinde neoðewearde, ond wæsc ða moran ealle swiðe wel, ond bescæf utan swiðe clæne ða moran ond ða rinde. Gecnuca ealle ða wyrte swiðe; ado in hluttor eala. Berend ond gegnid feowertig lybcorna ond ado þonne in ðæm wyrtum. Læt standan þreo niht. Syle drincan ær uhton lytelne scænc fulne, þæt se drænc sy ðe ær geleored.

42. Shave ivy, close by the stem, into milk, and taste carefully. And simmer all the shavings in milk, and curdle the milk from time to time with rennet, and partake of it.

43. Make a purgative drink.

Take eighty-five caper-spurge seeds and nine peppercorns and fifteen well-husked saxifrage seeds. Pound finely; add salt and meal made of small insects. Mix together and grate vigorously so that it is as fine as possible, worked into a powder. Take a drinking bowl full of light beer, or else clear ale that has been well sweetened, or sweetened wine. Make a thorough blend of this with the plants; let stand overnight. Stir it again very well in the morning whenever the patient is to drink it, and thoroughly blend the plants with the concoction. Then have him drink. If it is not strong enough, boil wild celery in water; give to drink. If it is too strong, boil knapweed instead.

44. Another purgative drink.

Take medium-sized roots of iris, a cubit long and as thick as your thumb. And likewise two houseleeks and roots of greater celandine and oleaster roots, and the lower part of elder-tree bark, and wash all the roots very well, and scrape the surface of the roots and the bark until they are free of all dirt. Pound all the plants well; put into clear ale. Husk and grate forty caper-spurge seeds and add them to the plants. Let stand for three days. Give a small cupful to drink before dawn, so that the drink is absorbed more quickly.

45. Þridde ut-yrnende drænc.

Wyl secg ond glædenan neoðewearde in suran ealað; asih þonne. Lege eft in niwe; læt ane niht inne beon. Syle drincan.

46. Wyrc spiw-drænc.

Wyl hwerhwettan in wætere; læt weallan lange. Asih þonne healfne bollan. Gegnid hundeahtatig libcorna in þonne drænc.

47. Wyrc oðerne of beore ond of feowertig lybcorna. Ado seofontene pipercorn, gif ðu wille.

48. Spiw-drænc.

Ado in beor oððe in win finul. Læt standan ane niht; syle drincan.

49. Wyrc sealfe wið heafod-wærce ond wið lið-wyrce ond wið eah-wyrce ond wið wenne ond wið ðeore.

2    Genim eolonon ond rædic, wermod ond bisceopwyrt, cropleac, garleac, ond holleac, ealra efenfela. Gecnuca; wyl in buteran; ond celleðenian ond reade netelan. Ado in æren fæt; læt ðærin oþþæt hit hæwen sy. Asih ðurh clað. Smyre mid þæt heafod ond ða leome þær hit sar sy.

50. Wið sid-wærce.

Betonican, bisceopwyrt, eolonan, rædic, ompran ða ðe swymman, marufian, grundeswylie, cropleac, garleac, rude, hæleðe, ealhtre, hune. Seoð in buteran; smyre mid ða sidan. Him bið sel.

45. A third purgative drink.

Boil sedge and the lower part of iris in sour ale; then strain. Put back into fresh ale; let it stand overnight. Give to drink.

46. Make up a drink to induce vomiting.

Boil cucumbers in water; let them boil for a long time. Then strain off half a bowlful. Grate eighty caper-spurge seeds into the drink.

47. Make up another drink from beer and from forty caper-spurge seeds. Add seventeen peppercorns, if you wish.

48. A drink to induce vomiting.

Put fennel in beer or wine. Let stand for a night; give to drink.

49. Make up a salve for headache and for pain in the joints and for eye ache and for wens and for theor-disease.

Take elecampane and radish, wormwood and marsh-mallow, crow garlic, domestic garlic, and shallot, an equal amount of each. Pound; simmer in butter; and take greater celandine and red dead nettle. Put into a bronze vessel; leave it there until it is a tawny brown. Strain through a cloth. Smear the head and the limbs with it where there is pain. 2

50. For pain in the sides.

Betony, marshmallow, elecampane, radish, water lilies, white horehound, groundsel, crow garlic, domestic garlic, rue, hindheal, lupin, black horehound. Simmer in butter; smear it on the person's sides. He will be better.

51. Wyrc briw wið lungen-adle.

Wyll in buteran þas wyrte, ond scearfa smale: cropleac ærest, wyl hwile; ado ðonne hrædic in, ond eolonan ond berenmela ond hwites sealtes fela. Wyl loncge, ond hatne ete.

52. Wyrc oðerne.

Wyl in buteran gið hrofan, attorlaðan, betonican. Mænc ealle tosomne, ado syððan ofer fyr.

53. Wyrc þriddan briw.

Wyl in buteran merce, eolonan, rædic, þa clufehton wen wyrt, hoc, wermod læst. Cnuca ealle swiðe wel. Syle wearm etan, ond on ufan drincon þriwa on dæg ær þonne he ete.

54. Feorða briw.

Wyl in hunige beton oððe marubian; syle etan wearme. Wyrc ær drænc of ðære beton anre. Wyll in wine oððe on ealað; he drince ær he ðone briw ete.

55. Drænc wið lungen-adle.

Wyl marubian in wine oððe in ealað; geswet hwon mid hunige. Syle drincan, wearme, on niht-nicstig. Ond þonne licge on ða swiðran sidan gode hwile æfter ðæm drænce, ond þænne þone swiðran earm, swa he swiþast mæge.

56. Genim betan; seoð on buteran. Syle hate etan, mid ðære buteran. A bið swa selre swa he fættron mete ete, ond gif he mæge gedrincan hwilum ðære buteran.

51. Make up a porridge for disease of the lungs.

Simmer the following plants in butter, and cut them up finely: first of all, crow garlic, simmer for a good while; then put radish in, and elecampane and barley meal and a lot of white salt. Simmer for a long time, and eat it hot.

52. Make up another.

Simmer cockle, cockspur grass, and betony in butter. Mix them all together, then put it on the fire.

53. Make up a third porridge.

Simmer in butter wild celery, elecampane, radish, cloved figwort, common mallow, and (in the smallest measure) wormwood. Pound them all very well. Give to eat while warm, and also give to drink three times a day before he eats.

54. A fourth porridge.

Boil beets or white horehound in honey; give to eat warm. Beforehand make a drink of the beets alone. Boil them in wine or in ale; the patient should drink this before eating the porridge.

55. A drink for lung disease.

Boil white horehound in wine or in ale; sweeten it a bit with honey. Give it to the person to drink, warm, before he has any food in the morning. And then have him lie on his right side for a good while after that drink, and then on the right arm, as long as he can.

56. Take beets; cook them in butter. Give to eat hot, along with the butter. The fatter the food the person eats, the better he will always be, likewise if he can drink some of the butter from time to time.

57. Eft drænc.

Genim marubian ond þa lancge cliton ond wermod ond boðen, gearwan, ond betonican godne dæl. Do ealle in eala; syle drincan on nyht-nicstig.

58. Genim feldmoran; gecnuca swiðe. Lege in win oððe in eala; læt standan a niht oððe twa. Syle drincan on niht-nicstig.

59. Eft wið þon.

Genim gagel ond marubian ond acrimonian. Wyl in ealað. Geswet mid hunige.

60. Wyrc briw.

Wyll ysopon in buteran, ond rædic ond eolonan, ond berenmela mest. Wel longe; syle wearm etan.

61. Briw.

Seoð in buteran ond in hunige beton swiðe oððæt he swa ðicce sy swa briw. Ete on niht-nicstig ðreo snæda swa hates.

62. Slæp-drænc.

Rædic, hymlic, wermod, belone. Cnuca ealle þa wyrte; do in ealað. Læt standan ane niht; drince ðonne.

63. To haligre sealfe sceal betonican ond benedicte ond hindhæleðe ond hænep ond hindebrer, isenhearde, salfige ond safine, bisceopwyrt ond boðen, finul ond fifleafe, heals-wyrt ond hune, mucwyrt, medewyrt ond mergelle, agri-monia ond æðelferðingwyrt, rædic ond ribbe ond seo reade

57. Another drink.

Take white horehound and long cleavers and wormwood and rosemary, yarrow, and a good deal of betony. Put them all into ale; give this to the person to drink before he has any food in the morning.

58. Take wild carrot; pound vigorously. Soak in wine or ale; let stand a day or two. Give to drink before he has any food in the morning.

59. Another drink for that same illness.

Take sweet gale and white horehound and agrimony. Boil in ale. Sweeten with honey.

60. Make up a porridge.

Simmer hyssop in butter, and radish and elecampane, and, in the greatest proportion, barley meal. Simmer for a long time; give to eat while warm.

61. A porridge.

Cook beets thoroughly in butter and in honey until they are as thick as a paste. Have the person eat three servings, piping hot, before he has any other food in the morning.

62. A drink to help someone sleep.

Radish, hemlock, wormwood, henbane. Pound all the plants; put into ale. Let stand overnight; then drink.

63. For a holy salve use betony and herb bennet and hindheal and hemp and raspberry, knapweed, sage and savine, marsh-mallow and rosemary, fennel and cinquefoil, throatwort and black horehound, mugwort, meadowsweet and marsh gen-tian, agrimony and stitchwort, radish and hound's-tongue

gearuwe, dile, oportanie, dracanse, cassoc ond cawlic, cyle-
ðenie, wyir-rind, weax, wudorofe ond wrættes cið, saturege
ond sigelhweorfa, brunewyrt ond rude ond berbene, streawberian wise ond blæces snegles dust, ealhtre, fanan, merce,
pollegian, attorlaðe, haranspicel, wudufille, wermod, eofor-
þrote, Æncglisc cost, hæwene-hnydele, vica pervica, feverfuge, hofe, cymen ond lilige, levastica, alehsandrie, petre-
silige, grundeswylige. Þysra feor wyrta man sceal mæst don
to, ond eallra oðra ælcre efenfela.

2 Ond ðus man sceal ða buteran gewyrcean to ðære haligan
sealfe. Æt anes heowes cy, þæt heo sy eall reod oððe hwit,
ond unmæle, mon ða buteran aðwere; ond gif ðu næbbe
buteran genoge, awæsc swiðe clæne, mængc oðre wið. Ond
ða wyrta ealle gescearfa swiðe smale tosomne, ond wæter gehalga font-halgunge, ond do ceac innan in ða buteran.

3 Genim þonne ænne sticcan ond gewyrc hine feðorbyrste.
Writ onforan ðas halgan naman: Matheus, Marcus, Lucas,
Iohannes. Styre þonne mid ðy sticcan ða buteran. Eal þæt
fæt ðu sing ofer ðas sealmas: "Beati inmaculati" . . . ælcne
ðriwa ofer, ond "Gloria in excelsis Deo" ond "Credo in
Deum Patrem," ond letanias arime ofer—þæt is, ðara haligra
naman—ond "Deus meus et Pater," et "In principio," ond
þæt wyrm-gealdor. Ond þis gealdor singe ofer:

4 *"Acre arcre arnem nona ærnem beoðor ærnem nidren arcun
cunað ele harassan fidine."*

and red yarrow, dill, southernwood, dragonwort, hassock grass and cabbage, greater celandine, myrtle bark, wax, woodruff and sprouts of madder, wild basil and heliotrope, figwort and rue and gypsywort, wild strawberry runner and powder of black snails, lupin, iris, wild celery, pennyroyal, cockspur grass, viper's bugloss, wild thyme, wormwood, carline thistle, tansy, scurvy grass, greater periwinkle, common centaury, ground ivy, cumin and lily, lovage, alexanders, rock parsley, groundsel. These last four plants should be added in the greatest amount, and an equal amount of each of all the others.

And this is how to make the butter for the holy salve. 2 Have the butter churned from a cow of a single color, one that is either all ruddy or all white, and unblemished; and if you do not have enough of this butter, mix some other butter in, cleansing it of all impurities. And shave all the plants very finely when you combine them, and consecrate some water at the baptismal font, and work a bowlful into the midst of the butter.

Then take a stick and square off its four sides. Write on 3 the faces these holy names: Matthew, Mark, Luke, John. Then stir the butter with this stick. Sing these psalms over the entire vessel: "Blessed are the undefiled" . . . singing each one three times over it, and "Glory to God in the highest" and "I believe in God the Father," and recite litanies over it—that is, the names of the saints—and "My God and Father," and "In the beginning," and the *wyrm* charm. And sing this charm over it:

*"Acre arcre arnem nona ærnem beoðor ærnem nidren arcun* 4 *cunað ele harassan fidine."*

5 . Sing ðis nygon siðan, ond do ðin spatl on, ond blaw on, ond lege ða wyrta be ðæm ceace, ond gehalga hy syððan mæssepreost. Singe ðas orationes ofer:

6 "Domine, sancte Pater, omnipotens eterne Deus, per inpositionem manum mearum refugiat inimicus diabolus a capillis, a capite, ab oculis, a naribus, a labis, a linguis, a sublinguis, a collo, a pectore, a pedibus, a calcaneis, ab universis confaginibus membrorum eis, ut non habeat potestatem diabolus, nec loquendi, nec tacendi, nec dormiendi, nec resurgendi, nec in die, nec in nocte, nec in tangendo, nec in somno, nec ingressu, nec in visu, nec in risu, nec in legendo; sed in nomine Domini Ihesu Cristi, qui nos suo sancto sanguine redemit, qui cum Patre vivit et regnat Deus in secula seculorum. Amen.

7 "Domine mi rogo te, Pater te deprecor, Fili obsecro te, Domine et Spiritus Sanctus, ex totis viribus, sancta Trinitas, ut deleas omnia opera diaboli ab isto homine. Invoco sanctam Trinitatem in adminiculum meum, id est Patrem et Filium et Spiritum Sanctum. Converte Domine istius hominis cogitationes et cor eius, ut confiteatur omnia mala sua et omnes iniquitates eius que habet, ut venit omnia bona sua et . . . voluntatem eius. Unde ergo, maledicte, recognosce sententiam tuam et da honorem Deo, et recede ab hoc famulo Dei, ut pura mente deserviat consecutus gratiam.

8 "Domine, sancte Pater, omnipotens eterne Deus, tu fecisti celum et terram et omnes ornatus eorum et omnes sanctos spiritus et angelorum exercitus; tu fecisti solem et

Sing this nine times, and put your spittle on it, and blow 5
on it, and lay the plants by the bowl, and afterward have a
priest consecrate them. Have him sing these prayers over
them:

"Lord, holy Father, omnipotent, eternal God, by the ap- 6
plication of my hands may the enemy, the devil, flee from
the hair, from the head, from the eyes, from the nose, from
the lips, from the tongue, from the epiglottis, from the
neck, from the breast, from the feet, from the heels, from
the whole framework of his limbs, so that the devil may
have no power over him, neither in speaking, nor in being
silent, nor when going to sleep, nor when getting up, nor by
day, nor by night, nor in touching, nor in slumber, nor at entry, nor in appearance, nor in laughter, nor in reading; but in
the name of the Lord Jesus Christ, who redeemed us by his
holy blood, who lives and reigns with the Father, God forever and ever. Amen.

"My Lord I ask you, Father I entreat you, Son I implore 7
you, Lord and Holy Spirit, by all your powers, holy Trinity,
that you expunge all the works of the devil from this man. I
invoke the holy Trinity to my aid, that is, the Father and the
Son and the Holy Spirit. Turn, O Lord, the thoughts of this
man and his heart, so that he may confess all his ill deeds
and all the iniquities that he has wrought, so that all his good
deeds may come and . . . his will. Wherefore, accursed one,
recognize your sentence and give honor to God, and retreat
from this servant of God, so that with a pure mind he may
serve zealously in the path of grace.

"Lord, holy Father, omnipotent, eternal God, you made 8
heaven and earth and all their adornments and all holy spirits and the hosts of angels; you made the sun and the moon

lunam et omnia astra celi; tu fecisti Adam de limo terre et dedisti ei adiutorium Evam uxorem suam, id est mater vivo- rum. Tu Domine vivificasti nos super nomen sanctum tuum et liberasti nos a periculis malis super nomen filii Ihesu Christi Domini nostri. Libera Domine animam famuli tui NOMEN et redde sanitatem corpori famuli tui NOMEN per nomen sanctum tuum.

9 "Domine, sancte Pater, omnipotens eterne Deus, roga- mus te, Domine Deus noster, propter magnam misericor- diam tuam, ut liberas famulum tuum et da honorem nomini tui, Domine, in secula seculorum. Amen.

10 "[Sanctificata omnia atque benedicta, depulsis atque abi- ectis vetusti hostis atque primi facinoris incentoris insidiis, salubriter ex huius diei anniversaria solemnitate, de univer- sis terre edendis germinibus sumamus, qui vivis et regnat in secula. Amen.]

11 "Sanctifica, Domine, hunc fructum arborum ut qui ex eo utuntur simus sanctificate, per."

64. Et circumibat Ihesus totam Galileam, docens in sinago- gis eorum et predicans evangelium regni et sanans omnem languorem et omnem infirmitatem in populo.

2 "Sanat te Deus, Pater omnipotens, qui te creavit; sanat te fides tua, qui te liberavit ab omni periculo. Criste adiuva nos. Deus meus, et Pater et Filius et Spiritus Sanctus."

3 Þas gebedu þriwa man sceal singan, ælc þriwa, on þysne drænc, ond þæs mannes oruð eallinga on þone wætan þa

and all the stars of the sky; you made Adam from the mud of the earth and gave him his wife Eve—that is, the mother of living creatures—as a help to him. You, Lord, gave us life through your holy name and liberated us from evil dangers through the name of your son Jesus Christ, our Lord. Free, O Lord, the soul of your servant NAME and restore bodily health to your servant NAME through your holy name.

"Lord, holy Father, omnipotent, eternal God, we pray 9 you, Lord our God, on account of your great mercy, to free your servant and give honor to your name, Lord, forever and ever. Amen.

"[We pray you that—the wiles of the ancient enemy and 10 the first inciter of crime having been driven away and cast down—we may take all things sanctified and blessed, in good health from the annual festival of this day, from the edible fruits of the whole earth, you who live and reign forever. Amen.]

"Sanctify, O Lord, this fruit of the trees so that we who 11 make use of it will be blessed, by your holy name."

64. And Jesus went about all Galilee, teaching in their synagogues and preaching the gospel of the kingdom and healing all manner of sickness and all manner of disease among the people.

"God is healing you, the omnipotent Father, who created 2 you; your faith is healing you, which has delivered you from every danger. May Christ help us. My God, and Father and Son and Holy Spirit."

These prayers are to be sung three times onto this drink, 3 each one three times, and the singer's breath should fall

hwile þe he hit singe. Gif se mon sy innan forswollen þæt he
ne mæge þone wætan þicgean, sinc him on þone muþ innan:

4    "Tunc beatus Iohannes, iacentibus mortuis his qui vene-
num biberant, intrepidus et constans accipiens calicem et
signaculum crucis faciens in eo dixit:

5    "'Deus meus et Pater et Filius et Spiritus Sanctus, cui omnia subiecta sunt, cui omnis creatura deservit et omnis
potestas subiecta est et metuit et expavescit cum nos te
ad auxilium invocamus; cuius audito nomine serpens con-
quiescit et draco fugit, silet vipera, et rubita illa que dicitur
"rana" quieta extorpescit, scorpius extinguitur et regulus
vincitur et spalagias nihil noxium operatur, et omnia vene-
nata et adhuc ferociora repentia et animalia noxia tenebran-
tur et omnes adverse salutis humanæ radices arescunt: tu,
Domine, extingue hoc venenatum virus, extingue operationes eius mortiferas, et vires quas in se habet et evacua; et
da in conspectu tuo omnibus quos tu creasti oculos ut vide-
ant, aures ut audiant, cor ut magnitudinem tuam intelligant.'

6    "Et cum hoc dixisset, totum semetipsum armavit crucis
signo, et bibit totum quod erat in calice. Et postea quam
bibit dixit:

7    "'Peto ut propter quos bibi convertantur ad te, Domine,
et ad salutem que apud te est, te inluminante meriantur. Per
eundem.'"

wholly onto the liquid while he is singing it. If the person is so badly swollen internally that he cannot partake of the water, sing into his mouth:

"Then when those who had drunk the poison were lying dead, the blessed John, undaunted and steadfast, taking up the cup and making the sign of the cross on it, said: 4

"'My God, and Father and Son and Holy Spirit, to whom all things are subject, whom all creatures serve and to whom all power is subject and whom all creatures fear and dread when we call upon you for help; upon hearing whose name the serpent becomes quiescent and the dragon flees, the viper grows silent, and the venomous toad that is called *rana* becomes quietly torpid, the scorpion is destroyed and the venomous snake is conquered and the poisonous spider works no harm at all, and all venomous and hitherto most ferocious reptiles and noxious animals are covered in darkness and all roots that are counter to the health of humankind dry up: you, Lord, obliterate this venomous poison, extinguish its deadly workings, annihilate the powers that inhere in it, and grant, in your countenance, to all those whom you have created, eyes that they may see, ears that they may hear, a heart that they may perceive your greatness.' 5

"And when he had said this, he armed his whole body with the sign of the cross, and he drank all that was in the cup. And after he had drunk it, he said: 6

"'I pray that those on behalf of whom I have drunk be converted to you, O Lord, and that, by your enlightening, they may merit the health that pertains to you. By our Lord Jesus Christ.'" 7

## 65. The Lorica of Laidcenn

Suffragare Trinitas Unitas,
unitatis miserere Trinitas.
Suffragare queso mihi posito
maris magni velut in periculo,
5   ut non secum trahat me mortalitas
huius anni neque mundi vanitas.
Et hoc idem peto a sublimibus
caelestis militiae virtutibus:
ne me linquant lacerandum hostibus
10   sed defendant iam armis fortibus,
ut me illi procedant in acie,
caelestis exercitus militiae:
cheruphin et seraphin cum milibus,
Michael et Gabriel similibus.
15   Opto thronos, viventes archangelos,
principatus et potestates, angelos,
ut me denso defendentes agmine
inimicos valeam prosternere.
Tum deinde ceteros agonithetas,
20   patriarchas, quattuor quater prophetas,
et apostolos duodecim, navis Christi proretas,
et martyres omnes peto athletas Dei,
ut me per illos salus eterna sepiat
atque omne malum a me pereat.
25   Christus mecum pactum firmum fereat;
timor tremor tetras turbas terreat.
Deus, inpenetrabile tutela,
undique me defende potentia tua.
Mea gibre pernas omnes libera,

65. The Lorica of Laidcenn

Help me, O Trinity, O Unity,
have mercy on me, O Trinity of unity.
Help me, I ask—placed
in danger as if from a vast sea—
so that neither the mortality of the present year          5
nor earthly vanity may pull me down with it.
And this too I ask of the sublime
powers of the celestial army:
that they not abandon me to be ripped to shreds by foes
but that they defend me now with weapons of might          10
so that they advance before me in the vanguard,
the army of celestial soldiers:
cherubim and seraphim in their thousands,
Michael and Gabriel with similar numbers.
I beg of thrones, living archangels,          15
principalities and powers, angels,
that—shielding me in a dense formation,
I may lay low my enemies.
I ask the other champions as well,
patriarchs, four fours of prophets,          20
and the twelve apostles, helmsmen of Christ's ship,
and all the martyrs, those athletes of God—
that through them eternal well-being may envelop me
while every evil may pass from me.
May Christ make a firm covenant with me;          25
may fear and trembling strike the ghastly throngs.
God, with your impenetrable defenses,
protect me with your might from every quarter.
Deliver my body's thighs,

30   tuta pelta protegente singula,
      ut non tetri daemones in latera
      mea librent, ut solent, iacula;
      gigram, cephale cum iaris, et conas,
      patham, liganam, sennas, atque michinas,
35   cladum, crassum, madianum, taleas
      bathma, exugiam, atque binas idumas.
      Meo ergo cum capillis vertici
      galea salutis esto capite,
      fronte, oculis, et cerebro triformi,
40   rostro, labio, faciei, timpori,
      mento, barbae, supercilis, auribus,
      genis, buccis, internasso, naribus,
      pupillis, rotis, palpebris, tautonibus,
45   gingis, anele, maxillis et faucibus,
      dentibus, linguae, ori, ubae, gutturi,
      gurgulioni et sublingue cervici,
      capitali centro, cartilagini;
      collo, clemens, adesto tutamine.
      Deinde esto mihi lorica tutissima,
50   erga viscera mea, erga membra mea,
      ut retundas a me invisibiles
      sudum clavos quos fingunt odibiles.
      Tege ergo Deus forti lorica
      cum scapulis humeros et brachia;
55   tege ulnas cum cubitis et manibus,
      pugnos, palmas, digitos cum ungibus;
      tege spinam et costas cum artibus,
      terga, dorsum, nervosque cum ossibus;
      tege cutem, sanguinem cum renibus,
60   catacrines, nates cum femoribus;

guard every single part with your protective shield,          30
so that no ghastly demons hurl their missiles
into my sides, as they so often do:
my skull, my head with its hair, my eyes,
my mouth, tongue, teeth, and nostrils,
my neck, breast, sides, limbs,          35
my joints, my internal fat, and both my hands.
May you be a helmet of safety for my head,
up to its crown, with my head of hair,
for my forehead, eyes and triform brain,
for my nose, lips, face, and temples,          40
for my chin, beard, eyebrows, ears,
for my cheeks, jowls, septum, and nostrils,
for my pupils, irises, eyelids, and all else,
for my gums, breath, jaws, and gullet,
for my teeth, tongue, mouth, uvula, and throat,          45
for the larynx and epiglottis of my throat,
for the inner core of my head, for its cartilage,
for my neck: in your mercy, come as my protector.
Be as well an invincible corselet for me
covering my internal organs, covering my limbs,          50
so that you ward off from me the invisible points
of the darts that the odious ones are forging.
O God, shield with a mighty breastplate
my shoulders with their shoulder blades and arms;
protect my elbows with their forearms and hands,          55
my fists, my palms, my fingers with their fingernails;
defend my spine and my ribs with their vertebrae,
my trunk, my back, my sinews with their bones;
defend my skin, my blood with its kidneys,
my hip and my buttocks along with the thighbones;          60

tege gambas, surras, femoralia
cum genuclis, poplites, et genua;
tege ramos concrescentes decies,
cum mentagris unges binos quinquies;
65   tege talos cum tibiis et calcibus,
crura, pedes plantarum cum bassibus;
tege pectus, iugulam, pectusculum,
mamillas, stomachum, et umbilicum;
tege ventrem, lumbos, genitalia,
70   et aluum et cordis vitalia;
tege trifidum iacor et ilia,
marsem, reniculos, fibrem cum obligio;
tege toliam, toracem cum pulmone,
venas, fibras, fel cum bucliamine;
75   tege carnem, inguinem cum medullis,
splenem cum tortuosis intestinis;
tege vesicam, adipem, et pantes
compaginum innumeros ordines;
tege pilos adque membra reliqua
80   quorum forte preteribi nomina;
tege totum me cum quinque sensibus,
et cum decem foribus fabrefactis,
ut a plantis usque ad verticem
nullo membro meo foris intus egrotem,
85   ne de meo possit vitam trudere
pestis, langor, dolor corpore,
donec iam Deo dante seneam
et peccata mea bonis deleam,
ut de carne iens imis caream

defend my hams, my calves, my femurs
with their knee joints, hocks, and knees;
defend my ten branches growing in unison
along with the toe tips, two groups of five toenails,
defend my ankles with their shins and heels,                    65
my shanks, my feet with the bottoms of their soles;
defend my chest, my windpipe, and my collarbone,
my nipples, stomach, and navel;
defend my belly, my loins, my genitals
and my abdomen and the vital parts of my heart;              70
defend my three-cornered liver and my ilium,
my scrotum, my testicles, my colon with its fold;
defend my tonsils, my chest cavity with its lungs,
my veins, my entrails, my gallbladder with its secretions,
defend my flesh, my groin with its inner parts,             75
my spleen with its twisted entrails;
defend my bladder, fatty tissue and belly,
the innumerable sets of connecting tissues;
defend the hairs on my head and my other body parts
of which I have perhaps neglected to speak the names;   80
protect me in my entirety, with my five senses
and with ten skillfully fashioned orifices
so that from the soles of my feet to the tip of my head
in not one part of my body may I fall ill, without or
    within,
so that neither pestilence, weakness, or bodily illness     85
may expel the life from me
until, God willing, I grow old
and extinguish my sins through good deeds
so that, released from the flesh, I may avoid the infernal
    depths

90   et ad alta evolare valeam
     et miserto Deo ad aetheria
     laetus vehar regni refrigeria. Amen.

66. Wið færlicre adle.

Sie clufehte wenwyrt, clate, bisceopwyrt, finul, rædic. Wyl in ealað; syle drincan.

67. Wið lænden-wyrce.

Finol-sæd, betonican leaf grene, acrimonian nyoðe-wearde. Gnid to duste; wes mid geswettan ealað; gewlece. Syle hat drincan in stalle; stonde gode hwile.

68. Wið þeore.

Genim cwic-rinde ond æsc-rinde ond berehalm. Wel in wætere. Genim alo-malt mid ðy wætere. Gebreow mid gryt cumb fulne ealað mid ðy wætere. Geclænsa, ðonne læt standan ane niht, gesweted mid hunige. Drince nygon mor-genas, ond ete secgleac ond cropleac ond cymen tosomne, ond nænigne oþerne wætan ne ðige.

69. Gif ðeor sy in men, wyrc drænc.

Nim þas wyrte nyoþowearde: finul ond bisceopwyrt ond æscðrote, ealra efenmicel; þyssa twiga mæst ufonwearde, ru-dan ond betonican. Ofgeot mid þreowum mædrum ealoð, ond gesinge þreo mæssan ofer. Drince ymbe twa niht þæs ðe hy ofgoten sie. Syle drincan ær his mete ond æfter.

and may succeed in flying to the heights                                      90
and, by the mercy of God, may be borne, full of joy,
to the heavenly solace of his kingdom. Amen.

66. For the sudden onset of a disease.

Bulbed figwort, burdock, marshmallow, fennel, radish.
Boil in ale; give to drink.

67. For pain in the loins.

Fennel seed, fresh betony leaves, the lower part of agrimony. Grind to a powder; steep in sweet ale; make lukewarm. Give to drink, warm, while the patient is standing
upright; he should remain standing for a good while.

68. For theor-disease.

Take rowan-tree bark and ash-tree bark and barley stalk.
Boil in water. In addition to that liquid, take some malt that
is used for brewing ale. Using that same liquid, brew a vessel
full of ale made from bran. Strain, then let stand for a night,
sweetened with honey. Drink for nine mornings, and eat
chives and crow garlic and cumin mixed together, and partake of no other liquid.

69. If a person suffers from theor-disease, make a drink.

Take the lower part of these plants: fennel and marshmallow and vervain, an equal amount of each; and take mostly
the upper parts of these two others, rue and betony. And
pour three pitchers of ale over them, and have three Masses
sung over them. Drink within two days of when they are
soaked. Give to drink both before and after the person has a
meal.

70. Drænc wið ðeore.

Nim ðas wyrte neoðowearde: ceasteræsc, ontre neoðo-
weart; ðas ufonwearde: betonican, rude, wermod, acremo-
nia, felterre, wuduþistel, feferfuge, æþelferðingcwyrt. Ofgeot mid ealað; læt stondan ane niht. Drince nigon morgenas
lytle bollan fulle, swiðe ær, ond ete sealtne mete ond nowiht
fersces.

71. Wyrc ðeor-drænc godne.

Genim wermod ond boðen, acrimonian, pollegan, ða
smalan wenwyrt, feltere, ægwyrt, ðyorwyrt, ceasteraxsan
twa snada, eofolan þreo snada, cammuces feower, wudu-
weaxan godne dæl, ond curmeallan. Gescearfa ða wyrta in
god hluttor eala oððe in god wylisc eala; læt standan þreowe
niht bewrogen. Syle drincan scænc fulne tide ær oþrum
mete.

72. Wið þeore ond wið sceotendum wenne.

Genim boðen ond gearwan ond weoduweaxan ond hræfnes fot; do in god eala. Syle drincan on dæge þreowe drænceas.

73. Gif ðeor sy gewunad in anre stowwe, wyrc gode beðingce.

Genim ifig ðe on stane wyxð on eorþan, ond gearwan ond
wudubindan leaf ond cuslyppan ond oxsanslyppan. Gecnuca
hy ealle swiðe wel. Lege on hatne stan in troge; do hwon
wæteres in. Læt reocan on þæt lic swa him ðearf sy oððæt
col sy. Do oþerne hatne stan in; beþe gelome. Sona him bið
sel.

70. A drink for theor-disease.

Take the lower part of the following plants: black helle-bore, the lower part of radish. Take the upper part of these: betony, rue, wormwood, agrimonia, centaury, sow thistle, common centaury, stitchwort. Steep in ale; let stand over-night. Have the patient drink a small bowlful for nine morn-ings, very early, and have him eat salted food but nothing fresh.

71. Make a good drink for theor-disease.

Take wormwood and rosemary, agrimony, pennyroyal, the slender type of figwort, centaury, eyebright, plowman's-spikenard, two slices of black hellebore, three slices of dwarf elder, four of hog's fennel, a good deal of dyer's greenwood, and knapweed. Shave the plants into good clear ale or good Welsh ale; let stand, covered, for three days. Give him a cup-ful to drink an hour before any other food.

72. For theor-disease and for wens that cause shooting pain.

Take rosemary and yarrow and dyer's greenwood and crowfoot; put into good ale. Give him three drinks a day to drink.

73. If theor-disease is persistent in a particular spot, make a good medicinal hot bath.

Take ivy that grows on rocky outcrops, and yarrow and woodbine leaves and cowslip and oxlip. Pound them all very well. Lay them onto a heated stone in a trough; pour some water on them. Let the steam rise onto the body, according to the person's need, until it cools. Add another hot stone; steam him frequently. He will be better straightway.

74. Wið ðeore.

Ealhtre, wælwyrt, weoduweaxe, æsc-rind in eorþan, cneowholen, wermod se hara, rædic, ceasteræsc, lytel savinan.

75. Gif se fic weorðe on mannes setle geseten, þonne nim ðu clatan moran þa greatan, þreo oððe feower, ond berec hy on hate æmergean. Ond ateoh þonne ða ane of ðan heorðe ond cnuca, ond wyrc swylc an lytel cicel, ond lege to þæm setle swa ðu hatost forberan mæge. Þonne se cicel colige, þonne wyrc þu ma ond lege to, ond beo on stilnesse dæg oððe twegen. Þonne þu þis do—hit is afandad læcecræft—ne delfe hy nan man þa moran mid isene, ond mid wætere ne þwea, ac strice hy mid claðe clæne. Ond do swiþe þynne clað betweonan þæt setl ond ðone cicel.

76. The Nine Herbs Charm

✚ Gemyne ðu, Mucg-wyrt,    hwæt þu ameldodest,
hwæt þu renadest    æt Regenmelde.
"Una" þu hattest,    yldost wyrta.
Ðu miht wið þrie    ond wið þritig,
5    þu miht wiþ attre    ond wið onflyge,
þu miht wiþ þam laþan    ðe geond lond færð.

✚ Ond þu, Weg-brade,    wyrta modor,
eastan opene,    innan mihtigu;
ofer ðy cræte curran,    ofer ðy cwene reodan,

74. For theor-disease.

Lupin, dwarf elder, dyer's greenwood, ash-tree bark taken from beneath ground level, butcher's-broom, the gray type of wormwood, radish, black hellebore, a little savine.

75. If hemorrhoids persist in a person's buttocks, then take three or four burdock roots—the greater burdock—and smoke them on hot embers. And then take one off the hearth and pound it, and shape it like a little cake, and apply it to the person's buttocks, the hottest he can stand. When that poultice cools, then make more of them and apply them, and have him rest quietly, for a day or two. When you do this—it is a tried and tested medical practice—no one should dig up any of the roots with an iron tool, nor should they be washed with water, but rather use a cloth to wipe them clean. And place a very thin cloth between the person's buttocks and the poultice.

76. The Nine Herbs Charm

✚ Remember, Mugwort, what you made known,
what you determined at the Place of Proclamation.
You are called "Una," the oldest of plants.
You have power against three and against thirty,
you have power against venom and against infection,     5
you have power against the enemy who journeys
    throughout the land.

✚ And you, Waybroad, mother of plants,
open from the east, mighty within;
over you chariots have rumbled, over you queens have
    ridden,

10  ofer ðy bryde breodwedon,    ofer þy fearras fnærdon.
Eallum þu þon wiðstode    ond wiðstunedest.
Swa ðu wiðstonde    attre ond onflyge
ond þæm laðan    þe geond lond fereð.

"Stune" hætte þeos wyrt;    heo on stane geweox;
15  stond heo wið attre,    stunað heo wærce.
"Stiðe" heo hatte;    wiðstunað heo attre,
wreceð heo wraðan,    weorpeð ut attor.

✠ Þis is seo wyrt    seo wiþ Wyrm gefeaht.
Þeos mæg wið attre,    heo mæg wið onflyge,
20  heo mæg wið ðam laþan    ðe geond lond fereþ.
Fleoh þu nu, Attor-laðe!    Seo læsse ða maran,
seo mare þa læssan,    oððæt him beigra bot sy.

Gemyne þu, Mægðe,    hwæt þu ameldodest,
hwæt ðu geændadest    æt Alorforda:
25  þæt næfre for gefloge    feorh ne gesealde
syþðan him mon mægðan    to mete gegyrede.

Þis is seo wyrt    ðe "Wergulu" hatte;
ðas onsænde Seolh    ofer sæs hrygc
ondan attres,    oþres to bote.

30  Ðas nygon magon    wið nygon attrum.

✠ Wyrm com snican;    toslat he nan.
Ða genam Woden    nigon wuldor-tanas,
sloh ða þa næddran    þæt heo on nigon tofleah.

over you brides have trampled, over you oxen have     10
   snorted.
You withstood them all then and dashed them down.
Just so may you withstand venom and infection
and the enemy who journeys throughout the land.

"Striker" is the name of this plant; she grew up on rocky
   ground;
she stands fast against venom, she strikes down pain.    15
"Unyielding" she is called; she strikes down poison,
she drives out the wrathful one, casts out venom.

✚  This is the plant that battled the Wyrm.
She has power against venom, she has power against
   infection,
she has power against the enemy who journeys    20
   throughout the land.
Disperse them now, Attorlathe! The lesser the greater,
the greater the lesser, until he is healed of both.

Remember, Maythe, what you made known
the pact you concluded at Alorford:
that never for infection would a life be lost    25
after chamomile was prepared for his food.

This is the plant called "Wergulu";
Seal sent it over the sea's back
as a torment to venom, but a cure for someone else.

These nine have power against nine poisons.    30

✚  The Wyrm came crawling; it killed no one.
Then Woden took up nine wonder-twigs;
he struck the snake so that it flew into nine parts.

Þær geændade    Æppel ond Attor,
35  þæt heo næfre ne wolde    on hus bugan.

✠ Fille ond finule,    fela-mihtigu twa,
þa wyrte gesceop    witig Drihten,
halig on heofonum,    þa he hongode;
sette ond sænde    on siofon worulde
40  earmum ond eadigum,    eallum to bote.

Stond heo wið wærce,    stunað heo wið attre,
seo mæg wið þrie    ond wið þritig,
wið feondes hond    ond wið fær-bregde,
wið malscrunge    manra wihta.

45  ✠ Nu magon þas nygon wyrta    wið nygon wuldor-
                                geflogenum,
wið nigon attrum    ond wið nygon onflygnum,
wið ðy readan attre,    wið ða runlan attre,
wið ðy hwitan attre,    wið ðy wedenan attre,
wið ðy geolwan attre,    wið ðy grenan attre,
50  wið ðy wonnan attre,    wið ðy wedenan attre,
wið ðy brunan attre,    wið ðy basewan attre,
wið wyrm-geblæd,    wið wæter-geblæd,
wið þorn-geblæd,    wið þystel-geblæd,
wið ys-geblæd,    wið attor-geblæd,
55  gif ænig attor cume    eastan fleogan
oððe ænig    norðan cume
oððe ænig westan    ofer wer-ðeode.

There Apple and Attor made a pact
that it would never make its home in a house.                          35

✠ Fille and Finule, two herbs of great might—
these are plants that the wise Lord created,
the holy one in heaven, when he hung on the cross;
he set them down and sent them into the seven worlds
for poor and for rich, as a remedy for all.                          40

She stands firm against pain, she strikes down venom,
she has power against three and against thirty,
against the devil's hand and against sudden acts of
    cunning,
against the spells of abhorrent creatures.

✠ These nine plants, then, have power against nine          45
    fugitives from glory,
against nine venoms and nine infections,
against red venom, against foul venom,
against white venom, against deep blue venom,
against yellow venom, against green venom,
against blue-black venom, against deep blue venom,          50
against brown venom, against crimson venom,
against snake blister, against water blister,
against thorn blister, against thistle blister,
against ice blister, against poison blister,
if any venom should come flying from the east                        55
or if any should come from the north
or any from the west, over the world of men.

✚ Crist stod ofer adle     ængan cundes.

Ic ana wat     ea rinnende
60  þær þa nygon     nædran behealdað.
Motan ealle weoda nu     wyrtum aspringan,
sæs toslupan,     eal sealt wæter,
ðonne ic þis attor     of ðe geblawe.

2     Mugcwyrt, wegbrade þe eastan open sy, lombes cyrse, attorlaðan, mægðan, netelan, wudusuræppel, fille ond finul, ealde sapan. Gewyrc ða wyrta to duste; mængc wiþ þa sapan ond wiþ þæs æpples gor. Wyrc slypan of wætere ond of axsan. Genim finol, wyl on þære slyppan. And beþe mid æg-gemongc þonne he þa sealfe on do, ge ær ge æfter.

3     Sing þæt galdor on ælcre þara wyrta, þriwa ær he hy wyrce, ond on þone æppel ealswa; ond singe þon men in þone muð ond in þa earan buta, ond on ða wunde þæt ilce gealdor ær he þa sealfe on do.

77. Gif se wyrm sy nyþergewend, oððe se bledenda fic.

Bedelf ænne wrid cileþenigan moran, ond nim mid þi-num twam handum upweardes, ond sing þærofer nigon Pa-ternostra. Æt þam nigeðan, æt "libera nos a malo," bred hy þonne up. Ond nim of þam ciðe, ond of oþrum þæt þær sy, an lytel cuppeful, ond drinc hy þonne, ond beðige hine mon to wearman fyre. Him bið sona sel.

✛ Christ prevailed over diseases of any kind.

I alone know a flowing river
where the nine serpents are on the watch.    60
May all woodlands now be filled with plants,
may all seas disperse, all salt water,
when I blow this venom from you.

Then take mugwort, common plantain that is open from    2
the east, shepherd's purse, cockspur grass, chamomile, nettle, wild crab apple, chervil and fennel, old soap. Work the
plants into a powder; mix with the soap and with the apple's
pulp. Make a cream out of water and ashes. Take fennel; simmer it in the cream. And bathe the person with an egg mixture when the healer applies the salve, both before and after.

Sing this incantation on each of the plants, three times    3
before they are worked into a powder, and sing it on the apple in like manner; and sing into the person's mouth and into
both ears, and sing the same spell onto the wound before
the healer applies the salve.

77. If the wyrm has turned downward, or for bleeding hemorrhoids.

Dig out a clump of shoots growing from the root of
greater celandine, and hold the clump with your two hands
directed upward, and sing nine "Our Fathers" over it. At the
ninth, at "deliver us from evil," hold your hands up high. And
from the shoot, and from whatever other shoots are there,
take a little cupful, and have the person then drink it, and
give him a bath at a warm fire. He will soon be better.

78. Eft wið þon ylcan.

Læt niman ænne greatne cwurnstan, ond hætan hine, ond lecgan hine under þone man. Ond niman wælwyrt ond leomucan ond mugcwyrt, ond lecgan uppan þone stan ond onunder, ond do þærto ceald wæter. Ond læt reocan þone bræð upon þone man, swa hat swa he hatust forberan mæge.

79. Gif fot oððe cneow oððe scancan swellan.

Nim neoðewearde betonican oððe elehtran. Cnuca hy swiþe; mængc wiþ smale hwætenan meoluwe. Clæme on þæt geswel.

80. Wið micclum lice ond bringc-adle, wyrce sealfe.

Wyll in buteran þas wyrta: elenan moran, ond hegerifan ufewearde, ond savinan ond curmeallan ond feferfugean, ond dolhrunan ond brunwyrt. Awringc ðurh claðٍ. Hafa þonne gegniden ond gebærned sealt, ond an penigweorð swefles.

81. Writ þis ondlang ða earmas wið dweorh:

2    ✚ T ✚ P ✚ T ✚ N ✚ ω ✚ T ✚ UI ✚ M ✚ ω̄ Ā

3   Ond gnid cyleþenigean on ealað. Sanctus Macutus, sancte Victorici.

82. Wið wennas æt mannes heortan.

Nim hwerhwettan ond rædic ond smælne tunnæp ond garleac ond suþernewuda ond fifleafan ond pipor on un-sodenan hunige. Ond wring ðurh claðٍ, ond pipera þonne, ond wylle þonne swiðe.

78. Again, for the same ailment.

Have a large millstone taken up, and heat it, and place it beneath the patient. And take dwarf elder and brooklime and mugwort, and place them on top of the stone and beneath it, and add cold water. And let the hot fume steam upon the person, as hot as he can bear.

79. If a foot or a knee or the shanks should swell up.

Take the lower parts of betony or lupin. Pound them well; mix with finely ground wheatmeal. Plaster it onto the swelling.

80. For a distended body and for brinch-disease, make a salve.

Simmer these plants in butter: roots of elecampane, and the upper part of goose grass, and savine and knapweed and common centaury, and pellitory-of-the-wall and figwort. Strain through a cloth. Then obtain salt that has been ground and burned, and one pennyweight of sulfur.

81. Write this along the length of the arms for dwarf-driven fever.

✚ T ✚ P ✚ T ✚ N ✚ ω ✚ T ✚ UI ✚ M ✚ ω̄ Ā ₂

And grate greater celandine into ale. Saint Macutus, Saint ₃ Victoricus.

82. For cysts at a person's heart.

Take cucumber and radish and small garden rape and garlic and southernwood and cinquefoil and pepper in uncooked honey. And wring through a cloth, and then pepper it, and then boil vigorously.

83. Þis gebed man sceal singan on ða blacan blegene nigon siðum: *Tigað.*

2    Ond wyrc þonne godne cliðan. Genim anes æges gewyrðe greates sealtes, ond bærn on anan claðe þæt hit si þurh-burnen. Gegnid hit þonne to duste, ond nim þonne þreora ægra geolcan ond gemængc to þam duste, þæt hit sy swa stið þæt hit wille wel clyfian. Ond geopenige mon þonne þone dott, ond binde þone cliðan to þan swyle ðe þearf sy.

3    Wyrc him þonne sealfe ðæt hit halige. Genim æðel-ferðingcwyrt ond elehtran ond reade fillan ond merce. Ge-cnuca ealle tosomne, ond wyll on ferscre buteran.

84. Gif men eglað seo blace blegen, þonne nime man great sealt. Bærne on linenum claðe swa micel swa an æg; grinde þonne þæt sealt swiþe smæl. Nime þonne þreora ægra geol-can, swinge hit swiðe togædere, ond lege hit siex niht þærto. Nim þonne eorðnafelan ond grundeswylian ond cawel-leaf ond eald smera. Cnuca þæt eal tosomne, ond lege hit þreo niht þærto. Nim þonne gearwan ond grundeswylian ond bræmbel-leaf ond clæne spic. Cnuca togædere, ond lege þærto—him bið sona sel—oððæt hit hal sy. Ond ne cume þæræt nan wæta butan of þan wyrtan sylfan.

85. Gif þin heorte ace.

Nim ribban ond wyl on meolc. Drinc nygon morgenas; þe bið sona sel.

83. This prayer is to be sung on the black boils nine times: *Tigað.*

And then make up a good poultice. Take an eggcup-full of coarse salt, and burn it in a cloth until it is burned through. Then grate it into a powder, and then take three egg yolks and mix them with the powder until it is firm enough that it will adhere well. And then lance the head of the boil, and bind the poultice to the swelling for as long as is needed. 2

Then make up a salve for him so that it may heal. Take stitchwort and lupin and red chervil and wild celery. Pound them all together, and simmer in fresh butter. 3

84. If black boils are afflicting a person, then take some coarse salt. Burn as much as an eggcup of it in a linen cloth; then grind the salt very fine. Then take three egg yolks, vigorously whip them together with the salt, and apply this to the boil for six days. Then take wild asparagus and groundsel and cabbage leaves and aged fat. Pound them all together, and apply this to the boil for three days. Then take yarrow and groundsel and bramble leaf and pure lard. Pound together, and apply this to the boil—he will soon be better— until it is cured. And see that no liquid comes into contact with it apart from the plant extracts themselves.

85. If you suffer from heart pains.

Take hound's-tongue and simmer in milk. Drink for nine mornings; you will soon be better.

86. ✚  Wið dweorh.

2    Man sceal niman seofon lytle oflætan swylce man mid ofrað ond writtan þas naman on ælcre oflætan: Maximianus, Malchus, Iohannes, Martinianus, Dionisius, Constantinus, Serafion. Þænne eft þæt galdor þæt her æfter cweð man sceal singan, ærest on þæt wynstre eare, þænne on þæt swiðre eare, þænne bufan þæs mannes moldan. And ga þænne an mædenman to ond ho hit on his sweoran, ond do man swa þry dagas; him bið sona sel.

3    Her com ingangan an spider-wiht.
     Hæfde him his haman on handa;
     cwæð þæt þu his hæncgest wære;
     legde þe his teage an sweoran.
   5 Ongunnan him of þæm lande liþan;
     sona swa hy of þæm lande coman
     þa ongunnan him ða liþu colian.
     Þa com ingangan deores sweostar;
     þa geændade heo ond aðas swor
  10 ðæt næfre þis ðæm adlegan derian ne moste,
     ne þæm þe þis galdor begytan mihte,
     oððe þe þis galdor ongalan cuþe.
     Amen. Fiat.

87. Her syndon læcedomas wið ælces cynnes omum ond on-feallum ond ban-coþum, eahta ond twentige.

Grenes merces leaf gecnucude mid æges þæt hwite ond ecedes dræstan. Smyre on þa stowe þær þæt sar sy.

86. ✠ For dwarf-driven fever.

One must take seven little wafers of the kind used at ₂
Communion, and write these names on each wafer: Maximianus, Malchus, John, Martimianus, Dionysius, Constantinus, Serafion. Then in turn one must sing the spell that is
related hereafter, first into the left ear, then into the right
ear, then above the crown of the person's head. And have a
virgin go and hang it on his neck, and have this done for
three days; he will soon be better.

Here a spider-creature came walking in. ₃
He had his skin-coat in his hand;
he said that you were his steed;
he laid his strings on your neck.
They began to speed from the land; ₅
as soon as they came away from that land
their limbs began to cool.
Then the creature's sister came walking in;
then she made a pact and swore oaths
that this could never harm the sick one ₁₀
nor anyone who might get hold of this charm
nor anyone who knew how to voice this charm.
Amen. May it be so.

87. Here are remedies for every kind of skin eruption and
sudden onset of disease and bone disease, twenty-eight in
number.

Fresh leaves of wild celery pounded with the white of an
egg and the residue of vinegar. Smear it on the place where
there is pain.

88. Wið omum ond blegnum.

Cristus natus, *a aius,* sanctus, *a.* Cristus passus, *a aius a.* Cristus resurrexit a mortuis, *a aius,* sanctus, *aa.* Superare potens.

89. Wið omum ond ablegnedum.

Sur meolc. Wyrce cealre, ond beþe mid cealre.

90. Eft.

Genim beordræstan, ond sapan ond æges þæt hwite, ond ealde grut. Lege on wið omena geswelle.

91. Eft, wið omena geberste.

Sitte on cealdum wætere oððæt hit adeadad sy; teoh þonne up. Sleah þonne feower scearpan ymb þa poccas utan, ond læt yrnan þa hwile þe he wille. Ond wyrc þa sealfe: brunewyrt, merscmergyllan, ond reade netlan. Wel on buteran, smyre mid, ond beþe mid þam wyrtum.

92. Eft.

Angeltwæccan. Gegnid swiþe. Do eced to, ond onbind ond smyre mid.

93. Eft.

Safinan. Gegnid to duste ond mængc wiþ hunige, ond smyre mid.

94. Eft wið þonne ylcan.

Genim gebrædde ægru; meng wið ele. Lege on, ond besweþe mid betan leafum.

88. For skin eruptions and boils.

Christ is born, *a aius,* holy, *a.* Christ has suffered, *a aius a.* Christ has risen from the dead, *a aius,* holy, *aa.* The power to conquer.

89. For skin eruptions and inflamed sores.

Sour milk. Make curds of it, and bathe with the curds.

90. Another.

Take the sediments from beer, and soap and the white of an egg, and some old draff. Apply it for skin eruptions.

91. Another, for skin eruptions that have burst open.

Soak in a cold-water bath until the area is numbed; then take out. Then cut four incisions around the outside of the pustules, and let the pus run out as long as it will. And make up this salve: figwort, marsh gentian, and red dead nettle. Simmer in butter, daub on, and bathe the sore with those same plants.

92. Another.

Earthworms. Crush thoroughly. Add vinegar, and break them up and daub on.

93. Another.

Savine. Pound to a powder and mix with honey, and daub on.

94. Another for the same ailment.

Take roasted eggs; mix with oil. Apply, and wrap up with beet leaves.

95. Eft.

Cealfes scearn, oððe ealdes hryþeres. Wearm ond lege on.

96. Eft.

Genim heoretes sceafeþan of felle—ascafen mid pumice—ond wese mid ecede, ond smyre mid.

97. Eft.

Genim eofores geallan oððe oþeres swynes, ond smyre mid þær hit sar si.

98. Wið þon ylcan.

Genim swolwan nest ond gebræc mid ealle, ond gebærne mid scearne mid ealle. Ond gegnid to duste, ond mæng wiþ eced, ond smyre mid.

99. Eft.

Gehæt ceald wæter mid isene, ond beþe mid gelome.

100. Wið hwostan ond neorunyse.

Wyl sealvian ond finol on geswettum ealoð, ond sup hat. Do swa swa oft swa þe þearf sie.

101. Wið morgen-wlætunga.

Wyl on wætre eorþgeallan; swet mid hunige; sele him godne bollan fulne on morgenne.

102. Wið þon þe mon blode wealle þurh his muð.

Genim betonican, þreora trymess gewæge, ond cole gate meoloc, þreo cuppan fulle, ond drince. Þonne bið he sona hal.

95. Another.

Calf's dung, or dung of a mature bullock. Warm it and apply.

96. Another.

Take shavings from a deer hide—ones that have been scrubbed off with pumice—and soak in vinegar, and daub on.

97. Another.

Take the gall of a boar or other pig, and daub on where it is sore.

98. For the same ailment.

Take a swallow's nest and break it all to pieces, and burn it up with its filth and all. And pound to a powder, and mix with vinegar, and daub on.

99. Another.

Heat cold water with iron, and bathe the sores frequently.

100. For coughing and congestion.

Boil sage and fennel in sweet ale, and eat while hot. Do this as often as you need to.

101. For morning nausea.

Boil yellowwort in water; sweeten with honey; give him a good bowlful in the morning.

102. If blood wells up through a person's mouth.

Take betony, the weight of three small coins, and cool goat's milk, three cupfuls, and have him drink it. Then he will be well straightway.

103. Wið ælces monnes tydernesse innewearde.

Genime wegbrædan. Do on win. Sup þæt wos, ond ete þa wyrta; þonne deah hit wið æghwylcre innancundre unhælo.

104. Gif man sceorpe on þone innað.

Galluc hatte. Delf þa moran; do to duste. Do godne cucelere fulne, ægscylle fulle wines oððe godes ealað, ond hunig. Syle drincan ær on mergen.

105. Wið eagena teara.

Heortes hornes axan. Do on geswet win.

106. Wið earon, æþele drænc.

Genim hrædic nyþeweardne ond elenan, þa bradan biscopwyrt ond cassuc-leaf, rudan ond rosan, safenan, fefer-fuigan. Gebeat ealle tosamne. Ofgeat mid ænne sester fulne ealoð ær þu mete þicge.

107. Wið lungen-adle ond breost-wræce.

Genim merces sæd ond diles sæd. Gnid, wyl, ond gemæng wið huniges teare. Do sumne dæl pipores, ond do him þreo snæda on niht-nyhstig.

108. Wið heals-omena.

Smyra hy sona mid hryþeres geallan, ond swiþost mid oxan. Him bið sona sel.

109. Wið lænden-ece.

Genim betonican, teon pænega gewæge. Do þær geswet-tes wines to, twegen bolan fulle; mæng wið hat wæter. Syle hit nistigum drincan.

103. For anyone suffering from internal weakness.

Take common plantain. Put into wine. Drink the juice and have him eat the plants; this will do well for every kind of internal ailment.

104. If a person is scratching at his belly.

The plant called comfrey. Dig up the roots; make into a powder. Combine a good spoonful, an eggshell-full of wine or good ale, and honey. Give to drink early in the morning.

105. For discharge from the eyes.

The ashes of a deer's antlers. Put into sweetened wine.

106. For the ears, an excellent drink.

Take the lower part of radish and elecampane, broad-leaved mallow and blades of hassock grass, rue and rose, savine, common centaury. Beat them all together. Steep in a pitcher-full of ale before you have any food.

107. For lung disease and chest pain.

Take wild celery seed and dill seed. Crush, boil, and mix with drippings of honey from the comb. Add a bit of pepper, and give the person three servings before he has any other food in the morning.

108. For skin eruptions at the neck.

Smear them straightway with bullock's gall, and especially with ox's gall. The person will be better straightway.

109. For pain in the loins.

Take betony, ten pennyweights. Add sweetened wine, two bowls full; mix with hot water. Give it to him to drink on an empty stomach.

110. Wið utsihte.

Genim lemocan. Wyl hy ongemetlice mid smale hwætenan melowe. Do hryþeres smera to oððe sceapes. Syle him
etan wearm.

111. Gif hors gescoten sy, oððe oþer neat.

Nim ompran sæd ond scyttisc wex. Gesinge mæssepreost
twelf mæssan ofer, ond do halig wæter on; ond do þonne on
þæt hors, oððe on swa hwylc neat swa hit sie. Hafa þe þa
wyrta symle mid.

112. Gif men synd wænnas gewunod on þæt heafod foran,
oððe on ða eagan.

Wring neoþewearde cuslyppan ond holleac in ða næs-
þyrlo; læt licgan upweard gode hwile. Þis is gewis læcedom.

113. To monnes stæmne.

Nim cyrfillan ond wudu cyrfillan, biscopwyrt, ontran,
grundeswyligean. Wyrc to drænce on hluttrum ealað. Nim
þreo snada buteran, gemængce wið hwæten meola, ond ge-
sylte. Þyge mid ðy drænce. Do swa neogan morgnas, ma gyf
þe þearf sy.

114. Wið angc-breoste.

Wyll holen-rinde on gate-meolce, ond sup weram,
nyhstig.

115. Wið ðone swiman.

Nim rudan ond salfian ond finul ond eorðifig, bettonican
ond lilian. Cnuca ealle þas wyrta tosomne. Do on ænne
pohchan; ofgeot mid wætere. Gnid swyðe; læt sigan ut on
sum fæt. Nim þone wætan ond wyrm, ond lafa þin heafod
mid. Do swa oft swa þe þearf sy.

110. For diarrhea.

Take brooklime. Boil it very well, along with fine wheaten meal. Add the fat of a bullock or a sheep. Give it to him to eat while warm.

111. If a horse is shot, or other livestock.

Take dock seed and Irish wax. Have a priest sing twelve Masses over it, and sprinkle holy water on it; and then apply it to the horse, or to whatever type of livestock it is. Always keep that plant with you.

112. If wens have settled onto a person's forehead or in his eyes.

Wring the lower parts of cowslip and shallots into the nostrils; have him lie on his back for a good while. This is a surefire cure.

113. For hoarseness.

Take garden chervil and asparagus, marshmallow, radish, groundsel. Make a drink out of them using clear ale. Take three pieces of butter, mix with wheaten meal, and salt it. Eat it along with the drink. Do so for nine mornings, more than that if you have need.

114. For tightness of the chest.

Boil holly-tree bark in goat's milk, and eat warm, on an empty stomach.

115. For dizziness.

Take rue and sage and fennel and ivy that grows on the ground, betony and lily. Mash all these plants together. Put them into a bag; steep in water. Knead thoroughly; let it flow out into a vessel. Take the liquid and warm it, and bathe your head with it. Do this as often as you need.

116. Wyrc godne drenc wið sid-ece.

Wyl betonican ond pollegan in aldum wine. Do seofon-
ond-twentig piporcorna gegrundenra. Syle him on niht-
nyhstig godne scenc fulne, wearmes, ond gereste gode hwile
æfter ðæm drence on ða saran sidan.

117. Wið ðon ylcan.

Wyll in ealaþ þa haran hunan ond rudan; geswet mid
hunige. Syle drincan on mergene on niht-nihstig godne bollan fulne, ond oðerne þonne he restan wille; ond symle reste
ærest on ða saran sidan oððæt he hal sy.

118. Eft wið sid-ece.

Genim hoclæf grene; cnuca swiðe; mængc wið ele þæt hit
sy swylce clam. Clæm ðonne on ða sidan þær se sy mæst, ond
wrið mid claðe. Læt swa gewriðen þreo niht; þonne bið se
man hal.

119. Wið fot-adle.

Genim betonican; wyl in wætere. Bewyll þriddan dæl;
syle þonne drincan, ond ða wyrt gecnuca. Lege on. Wundorlice hraðe þæt sar gelyhteð, þæs ðe gelærede læceas secgeað.

120. Wið ðære miclan siendan fot-adle, þære ðe læceas hatað
*podagre,* þær seo adl bið aswollen ond heo sihð wursme ond
gilstre, ond seonuwa fortogene, ond ða tan scrinceð up.

2       Genim grundeswyligean, ða ðe on ærenu wexeð, ond þa
readan wudufillan, bega efenfela. Cnuca wið ealdum swines

116. Make a good drink for pain in the side.

Simmer betony and pennyroyal in aged wine. Add twenty-seven ground peppercorns. Give the patient a good bowlful, warm, before he takes any food in the morning, and after he drinks it have him lie down for a good while on the side that is sore.

117. For the same ailment.

Boil in ale the gray kind of horehound and rue; sweeten with honey. Give the person a good bowlful to drink in the morning before he has anything else to eat, and another one when he goes to lie down; and have him always lie down on the side that is sore until he is well.

118. Another cure for pain in the side.

Take common mallow, fresh; pound it thoroughly; mix with oil until it is like a poultice. Then fasten it to the side where the pain is greatest, and wrap with a cloth. Leave it wrapped up like that for three days; the person will then be well.

119. For foot disease.

Take betony; boil in water. Boil away a third of it. Give it to him to drink, and mash the plant. Apply it. The pain will diminish remarkably quickly, according to what learned physicians say.

120. For foot disease with copious discharge, the ailment that doctors call "podagra," where the disease is marked by swelling and it oozes matter and pus, and the sinews are twisted, and the toes contract.

Take groundsel, the type that grows among people's 2 houses, and the red type of wild thyme, an equal amount of

rysle. Wyrc to clame; do on ða fet. Wrið mid claðe on niht, ond ðweah eft on morgen, ond dryg mid claðe. Smyre mid henne æges þe hwitan. Do eft nyowne clam. Do swa seofon niht; þonne bið ða seonuwa rihte ond ða fet hale.

121. Wyrc drænc wiþ þonne ylcan.

Genim ða ylcan grundeswyligean ond hindheoloðan ond ða smalan cliðwyrt ond wuduhrofan ond pollegian, ealra efenfela. Do in win oððe on wylisc eala. Syle drincan godne scænc-fulne on niht-nihstig. Þes drænc is god wið end-werce ond wið þeor-werce ond wið fot-swilum.

122. Wið giccendre wombe.

Wyll pollegian on wætere. Syle supan swa he hatost mæge aræfnan. Ðam men bið sona se gicða læssa.

123. Wyrc sealfe wið lusum.

Wyll in buteran nyoðeweardne hymlic, ond wyrmod oððe boðen. Smyre mid þæt heafod. Seo sealf gedeð þæt þær bið þara lusa læs.

124. Wyrc godne drænc wið lusum.

Genim lufestice ond wyrmod ond hymlic; doo in eala. Syle drincan on niht-nihstig godne bollan fulne.

125. Wið innoðes hefignese.

Syle etan rædic mid sealte, ond eced supan. Sona bið þæt mod leohtre.

both. Mash with old pig's lard. Make into a poultice; put it onto the feet. Bind them with a cloth at night, and wash them again in the morning, and dry them with a cloth. Smear them with the white of a hen's egg. Make up a fresh poultice. Do this for seven days; then the sinews will be straight and the feet will be well.

121. Make a drink for the same ailment.

Take the same groundsel and hindheal and the small kind of cleavers and woodruff and pennyroyal, an equal amount of all. Put into wine or Welsh ale. Give the person a good cupful before he has any food in the morning. This drink is good for pain in the buttocks and for pain from theordisease and for swollen feet.

122. For an itchy belly.

Boil pennyroyal in water. Give it to the person to sip, as hot as he or she can stand. The itching will soon diminish.

123. Make a salve for lice.

Simmer in butter the lower part of hemlock, and wormwood or rosemary. Smear it on the head. This salve will reduce the number of lice.

124. Make a good drink for lice.

Take lovage and wormwood and hemlock; put into ale. Give the person a good bowlful to drink before he has any food in the morning.

125. For heaviness of the internal organs.

Give him radish to eat with salt, and vinegar to sip. His mood will soon be brighter.

126. Wið fleogendan attre.

2     Asleah feower scearpan on feower healfa, mid æcenan brande. Geblodga ðone brand; weorp on weg. Sing ðis on þreowa:

3     ✚ Matheus me ducat ✚ Marcus me conservat ✚ Lucas me liberet ✚ Iohannes me adiuvet semper. Amen.

4     Contere Deus omnem malum et nequitiam, per virtutem Patris et Filii et Spiritus Sancti. Sanctifica me Emanuhel Ihesus Christus; libera me ab omnibus insidiis inimici. Benedictio Domini super caput meum. Potens Deus in omni tempore. Amen.

127. Wið færstice.

2     Feferfuige, ond seo reade netele ðe þurh ærn inwyxð, ond wegbrade. Wyll in buteran.

3     Hlude wæran hy, la, hlude,    ða hy ofer þone hlæw ridan,
wæran anmode    ða hy ofer land ridan.
Scyld ðu ðe nu! Þu ðysne nið    genesan mote.
Ut, lytel spere,    gif herinne sie!

5    Stod under linde,    under leohtum scylde,
þær ða mihtigan wif    hyra mægen beræddon
ond hy gyllende    garas sændan.
Ic him oðerne    eft wille sændan,
fleogende flane,    forane togeanes.

10    Ut, lytel spere,    gif hit herinne sy!
Sæt smið,    sloh seax lytel,
iserna,    wundrum swiðe.
Ut, lytel spere,    gif herinne sy!

126. For flying venom.

Cut four incisions on four sides, using an oaken imple-   2
ment. Bloody the implement; cast it away. Sing this on the
wound three times:

✚ May Matthew lead me ✚ May Mark keep me   3
safe ✚ May Luke deliver me ✚ May John help me always.
Amen.

Annihilate, O God, all evil and vileness, by the power of   4
the Father and the Son and the Holy Spirit. Sanctify me,
Emmanuel Jesus Christ; deliver me from all the wiles of the
enemy. The blessing of the Lord be upon my head. God
powerful in eternity. Amen.

127. For a stabbing pain.

Common centaury, and red nettle of the kind that grows   2
round about the house, and common plantain. Simmer in
butter.

Loud were they, lo, loud, when they rode over the       3
    mound,
they were fierce when they rode over the land.
Shield yourself now! You can survive this attack.
Out, little spear, if you are within!
I stood under the linden wood, under a bright shield,    5
where the mighty women gathered their powers
and hurled screaming spears.
I will send another one back at them,
a flying shaft, right back in return.
Out, little spear, if it is within!              10
A smith sat, forged a little knife,
out of iron bits, very wondrously.
Out, little spear, if you be within!

Syx smiðas sætan, wæl-spera worhtan.
15 Ut, spere! Næs in, spere!
Gif herinne sy isenes dæl,
hægtessan geweorc, hit sceal gemyltan.
Gif ðu wære on fell scoten oððe wære on flæsc scoten
oððe wære on blod scoten oððe wære on lið scoten,
20 næfre ne sy ðin lif atæsed.
Gif hit wære esa gescot oððe hit wære ylfa gescot
oððe hit wære hægtessan gescot, nu ic wille ðin helpan.
Þis ðe to bote esa gescotes, ðis ðe to bote ylfa gescotes,
ðis ðe to bote hægtessan gescotes; ic ðin wille helpan.
25 Fleoh þær on fyrgen-heafde!
Hal westu! Helpe ðin Drihten!

4      Nim þonne þæt seax; ado on wætan.

128. Wið lusan, an sealf.
Commuc, clofðung, rædic, wermod, ealra efenfela. Gecnuca to duste; gecned wið ele; smyre mid ealne ðone lichoman.

129. Nim eac meldon ða wyrt; gewyrc to duste swiðe smale.
Do in hat wæter; syle drincan. Sona ða lys ond oðre lytle
wyrmas swyltað.

130. Nim eac wermod ond marufian ond wyl gelice micel
ealra; wyll in wine oððe on geswettum wætere. Gedo þriwa
on þone nafolan; þonne swylteð ða lys ond oðre lytle wyrmas.

131. Nim eac cylendran wið ðon. Wyll in eala swiðe; smyre
mid þæt heafod.

Six smiths sat, they made deadly spears.
Out, spear! Not in, spear!                                               15
If here within be any piece of iron,
a witch's work, it shall dissolve.
If you were shot in the skin or were shot in the flesh
or were shot in the blood or were shot in a limb,
may your life never be threatened.                                       20
If it were shot by gods or were shot by elves
or were shot by a witch, now I will help you.
*This* to cure you of gods' shot, *this* to cure you of elf shot,
*this* to cure you of witch's shot; I will help you.
Fly away there to the mountain head!                                     25
Be well! May the Lord help you!

Then take the knife; put it into the liquid.                              4

128. For lice, a salve.

Hog's fennel, celery-leaved crowfoot, radish, wormwood, each in the same amount. Pound to a powder; knead with oil; smear the body all over.

129. Also take the plant called common orache; work into a very fine powder. Put into hot water; give to drink. Straightway the lice and other little insects will die.

130. Also take wormwood and white horehound and boil an equal amount of each; boil in wine or in sweetened water. Apply it to the navel three times; then the lice and other little insects will die.

131. Also take coriander for this same condition. Boil thoroughly in ale; smear the head with it.

132. Gif hryðera steorfan.

Do in halig wæter grundeswyligean ond springcwyrt ond attorlaðan neoðewearde ond cliðwyrt. Geot on ðone muð. Sona hy batigeað.

133. Wyþ lungen-adle hriðerum.

2    Þa wyrt þe weaxaþ on worðigum—heo bið gelic hundes micgean ðære wyrte—þær wexeð blacobergean eal swa micele swa oðre pys-beana. Gecnuca, do in haligwæter; do þonne on muð þæm hryþerum.

3    Genim þa ylcan wyrte. Do in glede, ond finol ond cassuc ond godeweb ond recels. Bærn eal tosomne on ða healfe ðe se wind sy. Læt reocan on ðone ceap.

4    Weorc Criste-mæl of cassuce, fifo. Sete on feower healfe þæs ceapes ond an to middes. Sing ymb þone ceap: "Benedicam Dominum in omni tempore," usque in finem, ond "Benedicite" ond letanias ond "Pater noster." Stred on haligwæter; bærn ymb recels ond godeweb, ond geeahtige mon ðone ceap. Syle þone teoþan pænig for Gode; læt syþðan beotigean. Do ðus þriwa.

134. Gif sceap sy abrocen, ond wið fær-steorfan.

Cæsteræsc, elehtre, wulfescamb, finol, stancrop. Wyrc to duste. Do in haligwæter. Geot in þæt abrocyne sceap, ond stred on ða oþur þriwa.

132. If cattle are dying.

Put groundsel and springwort and the lower part of cockspur grass and cleavers into holy water. Pour into the mouth. Straightway they will recover.

133. For lung disease in cattle.

The plants that grow in open spaces in settled areas — it looks like the plant called "dog's piss" — where brambles grow just as big as the peas of other kinds of plants. Pound; put into holy water; then put into the mouth of the cattle. [2]

Take the same plant. Put onto glowing coals, along with fennel and hassock grass and precious cloth and frankincense. Burn everything together on the side from which the wind is blowing. Have the smoke fumigate the cattle. [3]

Make signs of the cross out of blades of hassock grass, five of them. Set them to four sides of the cattle and one in their midst. Sing round about the cattle, "I will bless the Lord at all times," all the way to the end, and the "Benedicite" and litanies and the "Our Father." Sprinkle holy water on them; burn frankincense and precious cloth round about, and have the cattle appraised. Give the tenth penny to God; afterward leave them to recover. Do this three times. [4]

134. If a sheep is afflicted with illness, and to prevent sudden death.

Black hellebore, lupin, wild teasel, fennel, stonecrop. Make into a powder. Put into holy water. Pour into the diseased sheep, and sprinkle three times on the others.

135. Wið poccum ond sceapa hreoflan.

Elehtre ond eoforfearn neoðeweard, sperewyrt ufan-
wearde agrundene; greate beane. Cnuca ealle tosomne swiðe
smale in hunig ond in haligwæter, ond gemengc well tosomne. Do in muð mid cucylere ane snade, þreo symle ymb
ane niht; nigon siðum gif micel þearf sy.

136. Wið swina fær-steorfan, do a in heora mete.

Seoð clitan; syle etan. Nim eac elehtran, bisceopwyrt ond
cassuc, ðefeþorn, hegerifan, haranspicel. Sing ofer feower
mæssan; drif on fald. Hoh ða wyrte on feower healfe ond on
þan dore. Bærn; do recels to; læt yrnan ofer þone rec.

137. Wið þeofentum.

*Luben luben niga efð niga efð fel ceid feldelf fel cumer orcggæi*
*ceufor dard giug farig pidig delou delupih.*

138. Wið hondwyrmmum.

Scip-teron, swefl, pipor, hwit sealt. Mængc tosomne;
smyre mid.

139. Eft.

Wex, swefl, ond sealt. Mængc; smyre mid.

140. Gif nægl of honda weorðe.

Nim hwætene corn; gecnuca; mængc wið hunig. Lege on
ðone finger. Wyll slahþorn rinde; þweah mid ðy drænce.

141. Wið hwostan.

Wyll curmeallan wyrtruman; wyrc to duste. Syle him on
wine drincan. Sona se hwosta blinneð.

135. For pustules and scabbiness of sheep.

Lupin and the lower part of common polypody; the upper part of spearwort, ground up; broad beans. Pound them all together very finely in honey and in holy water, and blend them well. Using a spoon, put one portion into the mouth, always three times a day; nine times if there is great need.

136. For the sudden death of swine, always put this into their food.

Cook cleavers; give to eat. Also take lupin, marshmallow and hassock grass, buckthorn, goose grass, viper's bugloss. Sing four Masses over them; drive them into the pen. Hang the plants on the four sides of the pen and on the gate. Burn them; add frankincense; let the smoke fumigate them.

137. For thefts.

*Luben luben niga efið niga efið fel ceid feldelf fel cumer orcggæi ceufor dard giug farig pidig delou delupih.*

138. For itch mites.

Ship's tar, sulfur, pepper, white salt. Mix together; smear it on.

139. Another.

Wax, sulfur, and salt. Mix; smear it on.

140. If a nail comes off a hand.

Take some wheat kernels; pound; mix with honey. Apply to the finger. Boil blackthorn bark; wash with that same liquid.

141. For a cough.

Boil knapweed roots; make into a powder. Give it to him to drink in wine. The coughing will straightway cease.

142. Wið magan wærce, ond gif he bið toblawen se innoð.
Wringc pollegian in ceald wæter oððe in win. Syle drincan; him bið sel.

143. Wið ðon ðe wif færunga adumbige.
Genim pollegian, ond gnid to duste, ond in wulle bewind.
Alege under þæt wif; hyre bið sona sel.

144. Wið þeor.
Rose ond rude, elene ond feferfuge, rædic ond bisceopwyrt, salvie ond savine, eferðrote.

145. Eft oþer.
Fanu ond feferfuge, garleac ond rædic, ellenrind inneweard ond cyrse, netele, pipor, minte þe wyxð be þære ea. Nim mealt-eala. Ofgeot ða wyrta nygon niht ond syle drincan nyxtnig.

146. Gif þu wille wyrcean godne drænc wið ælc in-yfel, sy hit on heafde sy þær hit sy, þonne genim þu salvian leaf ond rudan leaf ond heldan leaf ond finoles, ond cerfillan leaf ond hegeclifan leaf ond persoces leaf ond reades seales leaf, ealra efenfela. Cnoca hy tosomne ond lege on wine oððe on hluttran ealað; ond wring þonne of þa wyrta, ond nim þonne hunig be dæle ond swet þone drænc. Drinc hine þonne anre tide ær þu þe wille blod lætan. Beþa þe þonne þa hwile to hatum fyre, ond læt yrnan þone drænc into ælcan lime. Gif þu him ænige hwile befylgest, þu ongitst þæt he is frymful to beganne.

142. For stomachache, and if the belly is distended.

Wring pennyroyal into cold water or into wine. Give to drink; the person will be better.

143. In case a woman suddenly loses her voice.

Take pennyroyal, and pound to a powder, and wind it around with wool. Place it underneath the woman. She will straightway be better.

144. For theor-disease.

Rose and rue, elecampane and common centaury, radish and marshmallow, sage and savine, carline thistle.

145. And another.

Iris and common centaury, garlic and radish, the inner part of elder-tree bark, cress, nettle, pepper, mint that grows by the streamside. Take malt ale. Soak the plants for nine days, and give to drink on an empty stomach.

146. If you wish to make a good drink for every internal affliction, whether in your head or wherever it may be, then take leaves of sage and leaves of rue and leaves of tansy and of fennel, and leaves of chervil and leaves of cleavers and leaves of a peach tree and leaves of red willow, all in an equal amount. Mash them together and place them in wine or in clear ale; and strain off the plants, and then take a little honey and sweeten the drink. Then drink it one hour before you wish to let blood. Warm yourself then at a hot fire for that while, and let the drink penetrate into every limb. If you follow this course for any while, you will understand that it is a beneficial practice.

147. Wið mete-cweorran.

Genime eorðgeallan. Drig to duste; scad on eala oððe on swa hwæt swa þu drincan wille. Þe bið sel.

148. Wið þæt man ne mage slapan.

Genim hænnebellan sæd ond tunmintan seaw. Hrer togædere ond smyre þæt heafod mid. Him bið sel.

149. Þonne þe mon ærest secge þæt þin ceap sy losod, þonne cweð þu ærest, ær þu elles hwæt cweþe:

2    Bæðleem hatte seo buruh   þe Crist on acænned wæs.
Seo is gemærsod   geond ealne middan-geard;
swa þyos dæd for monnum   mære gewurþe.
Þurh þa haligan Cristes rode, amen.

3    Gebide þe þonne þriwa east ond cweþ þonne þriwa: "Crux Christi ab oriente reducat." Gebide þe þonne þriwa west ond cweð þonne þriwa: "Crux Christi ab occidente reducat." Gebide þe þonne þriwa suð ond cweþ þriwa: "Crux Christi ab austro reducat." Gebide þonne þriwa norð ond cweð þriwa:

4    Crux Christi ab aquilone reducat. Crux Christi abscondita est et inventa est. Iudeas Crist ahengon; dydon dæda þa wyrrestan, hælon þæt hy forhelan ne mihtan. Swa þeos dæd nænige þinga forholen ne wurþe. Þurh þa haligan Cristes rode, amen.

150. Contra oculorum dolorem.

2    Domine, sancte Pater, omnipotens æterne Deus, sana oculos hominis istius NOMEN, sicut sanasti oculos filii Tobi

147. For indigestion.

Take yellowwort. Dry it to a powder; sprinkle into ale or into whatever you wish to drink. You will be better.

148. For insomnia.

Take henbane seed and spearmint juice. Stir together, and daub the head with it. The person will be better.

149. As soon as someone tells you that your cattle are lost, first say this, before you say anything else:

> Bethlehem is the name of the town where Christ was 2
> born.
> It is famed throughout the whole earth;
> so may this deed become notorious among humankind.
> Through Christ's holy rood, amen.

Then pray three times to the east and say three times: 3 "May Christ's cross lead them back from the east." Then pray three times to the west and say three times: "May Christ's cross lead them back from the west." Then pray three times to the south and say three times: "May Christ's cross lead them back from the south." Then pray three times to the north and say three times:

May Christ's cross lead them back from the north. 4 Christ's cross was stolen away and was found. The Jews hung Christ; they did the worst of deeds, they concealed what they could not hide away. So may this deed be in no way hidden. Through Christ's holy cross, amen.

150. For eye pain.

Lord, holy Father, omnipotent eternal God, heal the eyes 2 of this man NAME, just as you healed the eyes of the son of

et multorum cecorum quos [. . .]. Domine, tu es oculus cae-
corum, manus aridorum, pes claudorum, sanitas egrorum,
resurrectio mortuorum, felicitas martyrum et omnium
sanctorum.

3    Oro, Domine, ut eregas et inluminas oculos famuli tui
NOMEN; in quacumque valitudine constitutum medelis celestibus sanare digneris, tribuere famulo tuo NOMEN ut,
armis iustitiae munitus, diabolo resistat et regnum consequatur aeternum. Per.

151. Domum tuam quesumus, Domine, clementer ingredere
et in tuorum tibi cordibus fidelium perpetuam constitue
manstionem, ut cuius edificatione subsistit huius fiat habi-
tatio preclara.

152. Gif hors bið gewræht, þonne scealt þu cweþan þas word:
"*Naborrede* unde venisti" tribus vicibus. "Credidi propter"
tribus vicibus. "Alpha et O, initium et finis." "Crux mihi vita
est et tibi mors, inimici." "Pater noster."

153. Wið cyrnel.

2    Neogone wæran Noðþæs sweoster. Þa wurdon þa nygone
to eahte, ond þa eahte to seofone, ond þa seofone to siexe,
ond þa siexe to fife, ond þa fife to feowere, ond þa feowere
to þriowe, ond þa þriowe to tweowe, ond þa tweowe to ane,
ond þa ane to nanum.

3    Þis þe libbe cyrneles ond scrofelles ond weormeþ ond
æghwylces yfeles. Sing "Benedicite" nygon siþum.

Tobit and of the many blind ones who. . . . Lord, you are the eye of the blind, the hand of the crippled, the foot of the lame, the health of the sick, the resurrection of the dead, the joy of the martyrs and of all the saints.

I pray, Lord, that you raise up and illumine the eyes of your servant NAME; may you deign to heal him with celestial remedies in whatever state of health he may be, to grant to your servant NAME that, fortified with the arms of justice, he may resist the devil and may reach the eternal kingdom. By the Father and the Son and the Holy Spirit, amen. 3

151. We beseech you, Lord, that you mercifully enter your house and that you establish a perpetual mansion for yourself in the hearts of your faithful, so that it may be a glorious habitation for the one by whose construction it abides.

152. If a horse has sprained a leg, then you should intone these words:

"*Naborrede* whence you came," three times. "I have believed, therefore," three times. "Alpha and Omega, the beginning and the end." "The cross is life to me and death to you, O enemy." The "Our Father."

153. For a hard swelling.

Nine were they, the sisters of Noth. Then the nine became eight, and the eight became seven, and the seven became six, and the six became five, and the five became four, and the four became three, and the three became two, and the two became one, and the one became none. 2

This is your remedy for a hard swelling and for scrofula and infection and anything else that is noxious. Sing the "Benedicite" nine times. 3

154. Þis mæg horse wiðþonþe him bið corn on þa fet.

*Geneon genetron genitul catalon care trist pan bist et mic for-*
*rune naht ic forrune nequis annua maris sanctana nequetando.*

155. Gif hors bið gesceoten.

"Sanentur animalia in orbe terre" et "Valitudine vexan-
tur." In nomine Dei Patris et Filii et Spiritus Sancti, extin-
guatur diabolus per inpositionem manum nostrarum. "Quas
nos separavimus a caritate Christi?" Per invocationem
omnium sanctorum tuorum, per eum qui vivit et regnat in
secula seculorum. Amen. "Domine quid multiplicati sunt,"
þriwa.

156. Gif wif ne mæge bearn beran.

"Solve iube Deus," ter, "catenis."

157. Ad articulorum dolorem constantem malignantem, medicina.

Diabolus ligavit, angelus curavit, Dominus salvavit. In
nomine, amen.

158. Contra dolorem dentium.

2      Christus super marmoreum sedebat. Petrus tristis ante
eum stabat; manum ad maxillum tenebat. Et interrogebat
eum Dominus, dicens: "Quare tristis es, Petre"? Respondit
Petrus et dixit: "Domine, dentes mei dolent." Et Dominus
dixit:

3      "Adiuro te, migranea vel gutta maligna, per Patrem et Filium et Spiritum Sanctum, et per celum et terram et per
viginti ordines angelorum et per sexaginta prophetas et per
duodecim apostolos et per quattuor evangelistas et per

154. This is for a horse that is suffering from corns on its feet.

*Geneon genetron genitul catalon care trist pan bist et mic forrune naht ic forrune nequis annua maris sanctana nequetando.*

155. If a horse is shot.

"May the animals in the world be healed" and "They are troubled in health." In the name of God the Father and of the Son and of the Holy Spirit, may the devil be destroyed by the laying on of our hands. "Which ones have we separated from the charity of Christ?" By the invocation of all your saints, through the one who lives and reigns forever and ever, amen. "Lord, how are they increased," three times.

156. If a woman cannot give birth to a child.

"God, command release from chains," three times.

157. For constant malignant pain of the joints, a remedy.

The devil has bound you, an angel has cured you, the Lord has saved you. In the name of the Father and the Son and the Holy Spirit, amen.

158. For toothache.

Christ was sitting on a marble stone. Peter was standing before him, looking miserable; he was holding his hand to his jaw. And the Lord questioned him, saying: "Why are you miserable, Peter?" Peter answered and said: "Lord, my teeth hurt." And the Lord said:

"I adjure you, whether migraine or malignant drop, by the Father and the Son and the Holy Spirit, and by heaven and earth and by the twenty ranks of angels and the sixty prophets and the twelve apostles and the four evangelists

omnes sanctos qui Deo placuerunt ab origine mundi, ut non possit diabolus nocere ei, nec in dentes, nec in aures, nec in palato—famulo Dei illi—non ossa frangere nec carnem manducare, ut non habeatis potestatem nocere illi, non dormiendo nec vigilando, nec tangatis eum usque sexaginta annos et unum diem."

4  *Rex pax nax* in Christo Filio, amen. "Pater noster."

159. . . . Deus, qui dixisti, "Venite ad me, omnes qui laboratis et onerati estis, et ego reficiam vos": hos famulos tuos laborum suorum premio refice sempiterno. Per Dominum.

160. Wið utsihte.

2  Þysne pistol se ængel brohte to Rome þa hy wæran mid utsihte micclum geswæncte. Writ þis on swa langum bocfelle þæt hit mæge befon utan þæt heafod, ond hoh on þæs mannes sweoran þe him þearf sy. Him bið sona sel.

3  *Ranmigan adonai eltheos mur O ineffabile O miginan midan-mian misane dimas mode mida memagartem orta min sigmone Beronice irritas venas quasi dulaþ fervor fruxantis sanguinis sic-catur fla fracta frigula mir gui etsihdon segulta frautantur in arno midomnis abar vetho sydone multo saccula pp pppp sother sother miserere mei Deus Deus mini Deus mei. Amen, alleluia alleluia.*

161. Se wifman se hire cild afedan ne mæg: gange to ge-witenes mannes birgenne and stæppe þonne þriwa ofer þa byrgenne, and cweþe þonne þriwa þas word:

and by all the saints who have pleased God from the beginning of the world, that the devil may have no power to harm
this man, whether in the teeth, nor in the ears, nor in the
palate—this man, the servant of God—nor to break his
bones nor to chew his flesh, so that you will have no power
to harm him, whether sleeping or waking, nor may you
touch him for sixty years and a day."

*Rex pax nax* in Christ the Son, amen. The "Our Father."  4

159. . . . God, who said, "Come to me, all you that labor and
are heavy-laden, and I will refresh you": refresh these ser-
vants of yours with an eternal reward for their labors. By the
Lord.

160. For dysentery.

The angel brought this epistle to Rome when they were  2
badly afflicted with dysentery. Write this on a piece of
parchment long enough that it can be wrapped around the
head, and hang it on the person's neck when he has need. He
will straightway be better.

*Ranmigan adonai eltheos mur O ineffabile O miginan midan-*  3
*mian misane dimas mode mida memagartem orta min sigmone*
*Beronice irritas venas quasi dulap fervor fruxantis sanguinis sic-*
*catur fla fracta frigula mir gui etsihdon segulta frautantur in arno*
*midomnis abar vetho sydone multo saccula pp pppp sother sother*
*miserere mei Deus Deus mini Deus mei. Amen, alleluia alleluia.*

161. A woman who cannot nourish her child: she should go
to a deceased person's grave and step over the grave three
times, and then say these words three times:

2  Þis me to bote þære laþan læt-byrde,
  þis me to bote þære swæran swært-byrde,
  þis me to bote þære laðan lam-byrde.

3 And þonne þæt wif seo mid bearne and heo to hyre hlaforde on reste ga, þonne cweþe heo:

4  Up ic gonge, ofer þe stæppe
  mid cwican cilde, nalæs mid cwelendum,
  mid fulborenum, nalæs mid fægan.

5 And þonne seo modor gefele þæt þæt bearn si cwic, ga þonne to cyrican, and þonne heo toforan þan weofode cume, cweþe þonne: "Criste ic sæde, þis gecyþed!"

162. Se wifmon se hyre bearn afedan ne mæge: genime heo sylf hyre agenes cildes gebyrgenne dæl. Wry æfter þonne on blace wulle ond bebicge to cepemannum. Ond cweþe þonne:

2  Ic hit bebicge; ge hit bebicgan,
  þas sweartan wulle ond þysse sorge-corn.

163. Se man se ne mæge bearn afedan: nime þonne anes bleos cu-meoluc on hyre handæ, ond gesupe þonne mid hyre muþe. Ond gange þonne to yrnendum wætere ond spiwe þærin þa meolc, ond hlade þonne mid þære ylcan hand þæs wæteres muðfulne, ond forswelge. Cweþe þonne þas word:

> This is my cure for hateful delayed birth,  2
> this is my cure for grievous black birth,
> this is my cure for hateful lame birth.

And when the woman is with child and she goes to her hus-  3
band in bed, she should say:

> Up I go, over you I step  4
> with a living child, not with a dying one,
> with a child brought to full term, not with one destined
>     to die.

And when the mother feels that the child is alive, she should  5
go to church, and when she comes in front of the altar she
should say: "To Christ I have declared it, I have made this
known!"

162. A woman who cannot nourish her child: she herself
should take up a portion of her own child's grave. After that
she should wrap it up in black wool and sell it to peddlers.
And then she should say:

> I sell it; now *you* sell it,  2
> this black wool and this harvest of grief.

163. A woman who cannot nourish her child: she should take
in her hand some milk from a cow of a single color, then take
a sip of it in her mouth. And then she should go to running
water and spit the milk into it, and she should take up a
mouthful of the water, using the same hand, and swallow it
up. Then she should say these words:

2    Gehwer ferde ic me þone mæran    maga-þihtan
mid þysse mæran    mete-þihtan:
þonne ic me wille habban    ond ham gan.

3    Þonne heo to þan broce ga, þonne ne beseo heo no, ne eft
þonne heo þanan ga. Ond þonne ga heo in oþer hus oþer heo
ut ofeode ond þær gebyrge metes.

164. *Ecce dolgula medit dudum beðegunda breðegunda elecunda
elevachia mottem mee renum orþa fueþa letaves noeves terre dolge
drore uhic.* Alleluia.

Singe man þis gebed on þæt se man drincan wille nygan
siþan, ond "Pater noster" nigan siþan. Wið cyrnla.

165. *Arcus supeð assedit; virgo canabið; lux et ure canabið.*

Sing ðis nigon siþan, ond "Pater noster" nigon, on anum
berenan hlafe, ond syle þan horse etan.

166. Wyrc lungen-sealfe.

Nim cost ond suðernewuda, hylwyrt, garclife, bete þe bið
ansteallet.

167. Wið gedrif.

Nim snægl ond afeorma hine, ond nim þæt clæne fam.
Mengc wið wifes meolc. Syle þicgan; him bið sel.

168. Wið hors-oman ond mannes.

2    Sing þis þriwa nygan siðan on æfen ond on morgen, on
þæs mannes heafod ufan, ond horse on þæt wynstre eare, on
yrnendum wætere, ond wend þæt heafod ongean stream.

Everywhere I have carried this splendid belly-strong one    2
along with this splendid food-strong one:
I will keep him as mine and go home.

When she goes to the brook, she should not look round    3
about her, nor likewise when she goes from there. And she
should then go into a different house than the one she set
out from, and there she may partake of food.

164. *Ecce dolgula medit dudum beðegunda breðegunda elecunda*
*elevachia mottem mee renum orþa fueþa letaves noeves terre dolge*
*drore uhic.* Alleluia.

This prayer should be sung nine times onto whatever the
person wishes to drink, and the "Our Father" nine times.
For hard swellings.

165. *Arcus supeð assedit; virgo canabið; lux et ure canabið.*

Sing this nine times and the "Our Father" nine times on a
barley loaf, and give it to the horse to eat.

166. Make a salve for the lungs.

Take costmary and southernwood, pennyroyal, agrimony,
and beets that have a single stalk.

167. For diarrhea.

Take a snail and wash it, and take the clean foam. Blend it
with a woman's breast milk. Give to eat; the person will be
better.

168. For inflammation in a horse or in a person.

Sing this three times nine times in the evening and in the    2
morning, above the person's head and into the horse's left
ear, in running water, and turn the horse's head against the
current:

3    *Indomo mamosin inchorna meoti otimimeoti quoddealde otuuo-*
*tiva el marethin.* Crux mihi vita est, tibi mors, inimici. "Alfa
et O, initium et finis," dicit Dominus.

169. Wið oman.

2    Genim ane grene gyrde ond læt sittan þone man on middan huses flore, ond bestric hine ymbutan, ond cweð:

3    "O pars, et O *rillia pars,* et pars inopia est. 'Alfa et O,
initium.'"

170. Arestolobius wæs haten an cing. He wæs wis ond
læcecræftig. He þa gesette forþon godne morgendrænc wið
eallum untrumnessum þe mannes lichoman iondstyriað innan oððe utan.

2    Se drænc is god wið heafod-ecce, ond wið brægenes
hwyrfnesse ond weallunge, wið seondre exe, wið lungen-adle
ond lifer-werce, wið seondum geallan ond þære geolwan
adle, wið eagena dimnessa, ond wið earena swinsunge ond
ungehyrnesse, ond wið breosta hefignesse ond hrifes aþun-
dennesse, wið miltan wærce ond smæl-þearma, ond wið
ornum utgange, ond wið þon þe mon gemigan ne mæge,
wið þeor-ece ond sina getoge, wið cneow-wærce ond fot-
geswelle, wið ðam micclan lice ond wið oþrum giccendum
blece, ond þeor-geride ond æghwylcum attre, wið ælcre untrumnesse ond ælcre feondes costunge.

3    Gewyrc þe dust genoh on hærfeste, ond nytta þonne þe
þearf sy. Wyrc þonne drænc of þyssum wyrtum. Nim merces

*Indomo mamosin inchorna meoti otimimeoti quoddealde otuuo-* 3
*tiva el marethin.* The cross is life to me, death to you, O enemy. "Alpha and Omega, the beginning and the end," says
the Lord.

169. For acute inflammation.

Take a freshly cut rod and have the person sit in the mid- 2
dle of the floor of the house, and draw a circle around him,
and say:

"O part, and O *rillia pars,* and the part is useless. 'Alpha 3
and Omega, the beginning and the end,' says the Lord."

170. There was a king named Arestolobius. He was wise and
was skilled in medicine. For that reason he made up a good
morning drink for all infirmities that disturb a person's body,
whether within or without.

The drink is good for headache, and for dizziness and un- 2
steadiness of the brain, for discharge from the brain, for
lung disease and liver pain, for oozing from the gallbladder
and for jaundice, for dimness of vision, and for ringing in the
ears and for loss of hearing, and for heaviness in the chest
and swelling of the belly, for pain of the spleen and of the
small intestines, and for excessive excretion, and for a person's inability to urinate, for pain from theor-disease and
from contraction of the sinews, for knee pain and for a
swollen foot, for a body that is all swollen up and for other
itchy skin conditions, and for the inflammation from theor-
disease and every kind of poison, and for every weakness
and for every temptation sent by the devil.

Make up plenty of powder in the fall, and use it when you 3
need. Make up the concoction from the following plants.

sæd drige, ond finoles sæd ond petersylian sæd ond feldmoran sæd ond felterran sæd (þæt is, eorðgeallan) diles sæd ond
rudan sæd, cawel sæd ond cyllendran sæd ond feferfuigan
sæd, ond twa mintan (þæt is, tunminte ond horsminte)
ond betonican sæd ond luvestices sæd ond alexandrian sæd
ond salvian sæd ond slarian sæd ond wermodes sæd ond
sæþerian sæd ond biscopwyrte sæd ond horselenan sæd ond
beolonan sæd (þæt is, hænnebelle) acrimonian sæd (þæt is,
garclive) ond stancroppes sæd, marubian sæd (þæt is, harehune) ond neptan sæd ond wuduhrofan sæd ond wudu-
merces sæd ond eoforþrotan sæd. Do ealra þyssa wyrta
efenfela.

4      Nim þonne þyssa wyrta ælcre anre swa micel swa þara
oþra twa: þæt is, cymen ond cost ond piper ond gingifra ond
hwit cudu.

5      Wyrc þas wyrta ealle to swiþan smalan duste, ond do þæs
dustes godne cucelere fulne on ane scænce-cuppan fulle
cealdes wines, ond syle drincan on niht-nyhstig. Nytta þys
drænces þonne þe þearf sy.

171. Gif man scyle mugcwyrt to læcedome habban, þonne
nime man þa readan wæpnedmen ond þa grenan wifmen to
læcecræfte.

172. Þis deah wið fot-ece.
      Genim elenan moran ond eferþrotan moran ond doccan
moran; wyll swiðe well on buteran. Dreahna ut þurh wyllene
clað; læt colian æfter. Smyre syþþan þæt geswel. Him bið
sona sel.

Take dry celery seed and fennel seed and rock-parsley seed and wild carrot seed and seed of felterre (that is, of yellow-wort) and dill seed and rue seed, cabbage seed and coriander seed and seed of common centaury, and two kinds of mint (that is, spearmint and horse mint) and betony seed and lovage seed and alexanders seed and sage seed and seed of clary sage and wormwood seed and seed of wild basil and marshmallow seed and seed of horse elecampane and henbane seed (that is, of hennebelle) and agrimony seed (that is, of *garclife*) and stonecrop seed and seed of *marrubium* (that is, white horehound) and catmint seed and woodruff seed and wild celery seed and seed of carline thistle.

Use an equal amount of all these plants. Then take twice   4
as much of each one of the following plants as of the others:
that is, cumin and costmary and pepper and ginger and gum
mastic.

Work all these plants into a very fine powder, and put a   5
good cupful into a drinking cup full of cold wine, and give to
drink on an empty stomach. Use this drink when you need
to.

171. If mugwort is required for a remedy, then take the red
type for a man and the green type for a woman as your medicine.

172. This is good for pain in the foot.

Take roots of elecampane and roots of carline thistle and
roots of the dock plant; boil them very well in butter. Drain
them out through a woolen cloth; then let them cool. Afterward apply to the swelling. He will straightway be better.

173. Wið hwostan, hu he missenlice on man becymð ond hu his man tilian sceal.

2  Se hwosta hæfð mænigfealdne tocyme, swa ða swat beoð missenlicu. Hwilum he cymð of ungemætfæstre hæto, hwilum of ungemetfæstum cyle, hwilum of ungemetlicre wætan, hwilum of ungemætlicre drignesse.

3  Wyrc drænc wið hwostan. Genim mascwyrt; seoð on cyperenan cytele, ond wyll oððæt heo sy swiþe þicce, ond heo sy of hwætenum mealte geworht. Genim þonne eofor-fearnes (mæst), biscopwyrt, hindhæleþan, dweorgedwost-lan, singrenan; do eall on fæt. Syle drincan middeldagum, ond forga sur ond sealtes gehwæt.

174. Wið hwostan eft.

Genim hunan; seoð on wætere; syle swa wearme drincan.

175. Eft.

Genim clifwyrt—sume men hatað "foxesclife," sume "eawyrt"—ond heo sy geworht ofer midne sumor. Seoð ða on wætere oððæt. . . .

176. Gif wænnas eglian mæn æt þære heortan.

Gange mædenman to wylle þe rihte east yrne, ond gehlade ane cuppan fulle forð mid ðam streame, ond singe þæron Credan ond "Pater noster." Ond geote þonne on oþer fæt, ond hlade eft oþre ond singe eft Credan ond "Pater nos-ter," ond do swa þæt þu hæbbe þreo. Do swa nygon dagas. Sona him bið sel.

173. For the cough, how it comes upon a person in various ways and how one should treat it.

The cough has a manifold onset, just as its discharges are 2 diverse. Sometimes it comes from immoderate heat, sometimes from immoderate cold, sometimes from immoderate moisture, sometimes from immoderate dryness.

Make up a drink for cough. Take mashwort; cook it in a 3 copper kettle, and boil it until it is very thick; and it should be made of wheaten malt. Then take common polypody (the biggest part), marshmallow, hindheal, pennyroyal, and houseleek; put them all into the vessel. Give to drink at midday, and abstain from sour food and anything salty.

174. Another cure for cough.

Take black horehound; cook it in water; give to drink while warm.

175. Again.

Take cleavers — some people call it "fox's cleavers," others "eyebright" — and it should be prepared over midsummer. Then cook it in water until. . . .

176. If wens are hurting someone at the heart.

Have a virgin go to a stream that flows directly east, and collect a cupful of water running down the stream, and sing the Creed and the "Our Father" over it. And then pour it into another vessel, and gather another cupful in turn, and again sing the Creed and the "Our Father," and do this until you have three cupfuls. Do this for nine days. The person will straightway be better.

177. Wið heort-wærce.

Rudan gelm. Seoð on ele, ond do alwan ane ynsan to. Smyre mid þy; þæt stilð þæm sare.

178. Wið heort-ece.

Gif him on innan heard heortwærce sy, þonne him wyxt wind on þære heortan ond hine þegeð þurst, ond bið unmihtiglic. Wyrc him þonne stanbæð, ond on þæm ete suþerne rædic mid sealte. Þy mæg seo wund wesan gehæled.

179. Wið heort-ece eft.

Genim giðrifan; seoð on meolce; syle drincan syx dagas.

180. Eft.

Neoþeweard eoforfearn, giðrifan, wegbrædan. Wyl to-somne; syle drincan.

181. Wið breost-nyrwette.

Þus sceal beon se læcecræft geworht, þæt man nime ane cuppan gemeredes huniges ond healfe cuppan clænes ge-myltes spices, ond mængc on gemang þæt hunig ond þæt spic togædere, ond wylle hit oððæt hit beo wel briwþicce, forþan hit wile hluttrian for þan spice. Ond drige mon beana ond grinde hy syðþan, ond do þærto be þæs huniges mæþe, ond pipra hit syþþan swa swa man wille.

182. Þry dagas syndon on geare þe we "Egiptiaci" hatað— þæt is on ure geþeode "plihtlice dagas." On þam natoþæs-hwon for nanre neode ne mannes ne neates blod sy to wani-enne. Þæt is þonne utganggendum þam monþe þe we Aprelis hatað, se nyhsta monandæg an; þonne is oþer in-gangendum þam monþe þe we Agustus hatað, se æresta

177. For heart pain.

A handful of rue. Cook it in oil, and add an ounce of aloe. Rub this in; it will calm the pain.

178. For heart pain.

If a person is feeling persistent internal pain, then wind is accumulating at the heart and he will be suffering from thirst and will be weak. Fashion a stone bath for him, and while he is in it, have him eat southern radish with salt. This is how the hurt can be healed.

179. Again, for heart pain.

Take cockle; cook it in milk; give to drink for six days.

180. Again.

The lower part of common polypody, cockle, common plantain. Boil together; give to drink.

181. For tightness of the chest.

Here is how this medicine shall be prepared. Take one cup of purified honey and half a cup of clean melted lard, and mix the honey and the lard together, and boil it until it reaches the consistency of porridge, for it will want to clarify on account of the lard. And dry some beans and afterward grind them, and add to them a measure of honey, and afterward pepper it as you like.

182. There are three days in the year that we call "Egyptian"—that is, in our language, "perilous days." During those days, for no cause whatsoever is the blood of either man or beast to be diminished through bloodletting. The first day is the last Monday at the end of the month that we call April; the second is the first Monday at the start of the month that

monandæg; þonne is se þridda se æresta monandæg æfter utgange þæs monþes Decembris.

2 Se-þe on þysum þrim dagum his blod gewanige, sy hit man, sy hit nyten—þæs þe we secgan gehyrdan—þæt sona on þam forman dæge oþþe þam feorþan dæge his lif geændað; oþþe gif his lif længre bið, þæt he to þam seofoþan dæge ne becymð. Oððe gif he hwilcne drænc drincð þam þrim dagum his lif he geændað binnan fiftienum dagum. Gif hwa on þis dagum acænned bið, yfelum deaðe he his lif geændað; ond se-þe on þys ylcum þrim dagum gose flæsces onbyrigeð, binnan feowortiges daga fyrste he his lif geændað.

183. ✚ In nomine Patris et Filii et Spiritus Sancti, amen. Nomen. In adiutorium sit salvator. Nomen.

2 Deo celi, regi regum, nos debemus reddere
gratiarum actionem adque se petere
ut a nobis lues iste huius pestis careat,
et in nobis quam donavit salus vera maneat.

5 Ihesu Christe, me Nomen defende per tuam potentiam
adque nobis nunc extende benignam clementiam,
quia solus ipse potes prestare auxilium
te petentibus ex toto corde donare presidium.

Patrem pium, dignum, verum, summum adque optimum
10 ter rogamus, audi preces famulorum famularumque
tuarum.
Domine Ihesu Christe, vite altor, subveni auxilio,
et salutis tue pelta defende presidio summo.

we call August; the third is the first Monday after the end of the month of December.

Whoever diminishes his blood on these three days, 2 whether human or animal—as we have heard say—straightway he will die on the first day or on the fourth day; or if his life lasts longer, he will not live as long as the seventh day. Or if he drinks any beverage during those three days, he will die within fifteen days. If anyone is born on these days, he will die an evil death; and he who eats goose meat during these same three days, he will die within forty days.

183. ✚ In the name of the Father and the Son and the Holy Spirit, amen. NAME. May the savior be our aid. NAME.

> To the God of heaven, king of kings, it is our duty 2
> to pay an act of thanks and to ask of him
> that the curse of this pestilence be removed from us
> and that the health that he gave us remain sound within
> us.
>
> Jesus Christ, defend me, NAME, through your power 5
> and extend your beneficent mercy to us now
> for you yourself alone can offer help,
> can give protection to those who entreat you from their
> whole hearts.
>
> Holy, worthy, true, highest, and best Father, we ask you
> three times: hear the prayers of your servants, whether 10
> male or female.
> Lord Jesus Christ, nurturer of life, come to our aid
> and with your shield of salvation defend us with your
> supreme protection.

Et digne te obsecro intende ad ilia
mei cordis; adque peto angelorum milia
15 aut me NOMEN salvent ac defendant doloris igniculo
et potestate variole ac protegant mortis a periculo.

Tuas Ihesu Christe aures nobis inclina clementie,
in salute ac virtute intende potentie;
ne dimittas nos intrare in hanc pestilentiam,
20 sed salvare nos dignare per potentiam tuam.

Filii Dei vivi Ihesu Christe, qui es vite dominator,
miserere adque nos huius mundi salvator.

3 Deus libera illam, Domine, de languoribus pessimis et de
periculis huius anni, quia tu es salvator omnium, Christe qui
regnas in secula. Fiat sanitas Domini super me NOMEN.
Amen.

4 *Brigitarum dricillarum tuarum malint voarline deamabda*
*murde murrunice domur brio rubebroht.*

5 Sancte Rehhoc et sancte Rehwalde et sancte Cassiane et
sancte Germane et sancte Sigismundi regis, gescyldað me
wið ða laþan poccas ond wið ealle yfelu. Amen.

184. Benedictio herbarum.

Omnipotens sempiternae Deus, qui ab initio mundi
omnia instituisti et creasti, tam arborum generibus quam
erbarum seminibus, quibus etiam benedictione tua benedi-
cendo sanxisti eadem: nunc benedictione holera aliosque

And in fit fashion I beseech you to turn your attention to
    the inner workings
of my heart; and I ask that a thousand angels
may save me, NAME, and may defend me from the fire    15
    of pain
and, through their diverse powers, may protect me from
    the threat of death.

Incline to us, Jesus Christ, your ears of mercy;
harken to us in the safety and strength of your power;
do not dismiss us to enter into this pestilence
but deign to save us through your might.    20

Son of the living God, Jesus Christ, who is the Lord of
    life,
have mercy upon us, savior of this world.

My God, my Lord, deliver this woman from the worst ill-    3
nesses and from the perils of this year, for you are the savior
of all, Christ who reigns forever. May the health of the Lord
be upon me, NAME. Amen.

    *Brigitarum dricillarum tuarum malint voarline deamabda*    4
*murde murrunice domur brio rubebroht.*

    Saint Rehhoc and Saint Rehwald and Saint Cassian and    5
Saint Germanus and Saint Sigismund the king, shield me
against the hateful pocks and against all evils. Amen.

184. A blessing of the plants.

    Omnipotent, everlasting God, who from the beginning
of the world has established and created all things, both the
species of trees and the seeds of plants, which by your blessing you have sanctified with a benediction: now may you
deign to sanctify and to bless by your benediction the herbs

fructus sanctificare ac benedicere digneris, ut summentibus ex eis sanitatem conferant mentis et corporis ac tutelam defensionis eternamque vitam. Per salvatorem animarum Dominum nostrum Ihesum Christum, qui vivit et regnat Deus in secula seculorum. Amen.

### 185. Alia.

Deus qui hec holera que tua iussione et providentia crescere et germinare fecisti, etiam ea benedicere et sanctificare digneris, et precamur ut quicumque ex eis gustaverint incolumes permaneant. Per.

### 186. Benedictio unguentum.

Deus Pater omnipotens et Christe Ihesu Filii Dei, rogo ut mittere digneris benedictionem tuam et medicinam celæstem et divinam protectionem super hoc unguentum, ut perficiat ad salutem et ad perfectionem contra omnes egritudines corporum vel omnium membrorum intus vel foris, omnibus istud unguentum sumentibus.

### 187. Alia.

In nomine Patris et Filii et Spiritus Sancti, et per virtutem Dominice passionis et resurrectionis a mortuis, ut sanctificentur tuo verbo sancto et benedicentur omnes fideles cum gustu huius unguenti adversus omnes nequitias inmundorum spirituum, et contra valitudines et infirmitates que corpus affligunt.

### 188. ... arbor ... sint sanctificati, per.

### 189. ... manducare ... pane ordeaceo.

and other fruits, so that to those who consume them they may confer health of mind and of body and a safeguard of protection and eternal life. By the savior of souls our Lord Jesus Christ, God who lives and reigns forever and ever. Amen.

185. Another.

God who by your command and providence has made these herbs to grow and to germinate, deign as well to bless and sanctify them, and we pray that whoever tastes of them may remain unharmed. By the Father, the Son, and the Holy Spirit.

186. A blessing of ointments.

God the omnipotent Father and Jesus Christ the Son of God, I pray that you will deign to send your blessing and heavenly medicine and divine protection upon this ointment, so that it may bring about health and a state of perfection against all sicknesses of the body and of all its parts, whether within or without, for all those who use this ointment.

187. Another.

In the name of the Father and the Son and the Holy Spirit, and by the power of the Lord's passion and resurrection from the dead, may all the faithful be sanctified and blessed by your holy word, with the taste of this ointment, against all wickedness of unclean spirits and against the sicknesses and infirmities that afflict the body.

188. ... tree ... may they be sanctified, by your holy name.

189. ... to eat ... barley bread.

190. Medicina ad cancrum: accipe . . . farina de sancti . . . cancri.

191. A os freint en teste. Amerusche et herbe terestre. Boilli en miel, et puis melle od birre e fet oignement. En gete le os, et garist la teste.

190. A remedy for a tumor: take . . . wheat of holy . . . for a tumor.

191. For a fractured skull. Stinking chamomile and garden herbs. Boil in honey, and then mix with butter and make a salve. Apply it to the bone, and the head will heal.

# PERI DIDAXEON

# Peri Didaxeon

1. Her onginþ seo boc *Peri Didaxeon*—þæt ys, seo swytelung, hu fela gera wæs behuded se læcecræft, and be his gewitnesse þa gelæredustan læce gewislice smeadon. Þæt was se ærusta Apollo, and his suna Esculafius and Asclepius; and Asclepius was Ypocrates yem. Þeos feower gemetun ærest þa getymbrunga þare læcecrafte. After Noes flode ymba wintra a þusund and fif hund wintra, on Artaxis dæge, se was Persa cingc, hy aluste þa leoht þæra læcecræfte. Giwislica se Apollon ærest he gemetta *meþodicam,* þæt syndon, sa ysene þa mann mid cnifun hæle menn; and Scolafius *empiricam,* þæt is, ilæcnunga of læcecrafta; and Asclepius *loicam,* þæt ys, seo gehealdenysse þære æ and þæs lifæs; and Ypocras *theoricam,* þæt ys, foresceawunga þara seocnesse.

2     Þannun Plato and Aristotiles, þa gelæredusþan uþwytyna, æfterfyligdun þas forecwedenan læcun. And hi gesæddun þæt feower wætun syndon on þan manniscen lichama, forþam byð wylyd ealswa middangeardes boga: þat ys, þa wæte on þan heafode, and þæt blod on þara breosta, and se ruwa gealla on þan innoþe, and se swerta gealle innan ðare blædran. And hyra an gehwylce rixaþ ðra monþas: þæt ys, fram 18 kalendas Januari usque in 8 kalendas Aprilis, þæt on ðan heafde se wæte byð wexende; and fram 18 kalendas

# Peri Didaxeon

1. Here begins the book called *Peri Didaxeon*—that is, the account of how for many years the science of medicine lay hidden, and how the most learned physicians thought deeply about its proper written expression. The first of these was Apollo, and his sons Esculafius and Asclepius; and Asclepius was Hippocrates's uncle. These four were the first to establish the true basis of the science of medicine. It was fifteen hundred years after Noah's flood, in the time of Artaxis, king of the Persians, that they brought forth the light of medicine. In truth, Apollo was the first to establish *methodica,* that is, the means by which people are healed with surgical implements; and Scolafius founded *empirica,* that is, pharmaceutically based cures; and Asclepius founded *loica,* that is, the preservation of law and life; and Hippocrates founded *theorica,* that is, the diagnosis of illnesses.

Then Plato and Aristotle, the most learned philosophers, 2 succeeded the aforesaid physicians. And they declared that there are four humors in the human body, for which reason the body is furnished with wellsprings of fluid just as is the circuit of earth; namely, the fluid in the head, the blood in the chest, the rough gall in the intestines, and the black gall in the gallbladder. And each of these fluids is dominant for three months: that is, from the eighteenth calends of January until the eighth calends of April, when the fluid in the head is ascendant; and from the eighteenth calends of April

Aprilis usque in 8 kalendas Julii, þæt ðæt blod biþ wexinde on þan breosten; ab 18 kalendas Julii usque in 8 kalendas Octobris, þæt sa ruwa gealle byð wexenda on þan innoþe. Forþan synd þa dæges genemnede *cinotici*—þæt sindan, þa dæges caniculares. And þara byd fif and feowertig dæga, and on þan dægen ne mæg nan læce wel don fultum ænigen seoce manne. And þe feorðan gescornesse ys ab 18 kalendas Octobris usque in 8 kalendas Januari, þat se blace gealle wixt on þara bladre.

3      Þis gescead ys hæfter þam feorwer heorren heofenes and eorðan and þara lyfte and þare dupnesse. Þa wæs ealswa Drihtne licede: ealswa was se man geset þæt, þur þara smeagunga and þarre endbirnesse. Utan nu nymen æryst gewislice þane, æt fruman of þan heafde.

2. Þus man sceal wyrcen þa sealfe wið oman, and þus he sceal beon gehæled. Nim litargio twentige scillinga gewyht, and niwes limes twentiga scillinga gewihte, and anne healfne sester ecedes, and feorwer scillinga gewiht de oleo mirtino, and meng togadere and gnid swiþe ætsomne mid þan ecede. And þanne nima man oðder ele and meng þarto, and smyre þæt sare mid.

3. Wið þæt heafod þe byð toswollen þæt Grecas *ulcerosus* hatað—þæt is, heafod-sar, þa bula þe betwyx felle and flæsce arisað and on mannes anwlytan ut-bersteþ swa grete swa beane. Þus he scel beon gehaled. Nim wingeardes sæt and gnid on wæte, and lege uppan þat sar, and he byð sona hæl.

until the eighth calends of July, when blood in the chest is ascendant; from the eighteenth calends of July until the eighth calends of October, when the rough gall in the intestines is ascendant. For that reason those days are called *cinotici*—that is, the dog days. And there are forty-five of those days, and during them no physician can be of much help to anyone who is ill. And the fourth division is from the eighteenth calends of October until the eighth calends of January, when the black gall in the gallbladder is ascendant.

This distinction is in accord with the four cardinal points 3 of heaven and earth and of the sky and of the nether realm. These things came about just as was pleasing to the Lord: it is thus that humankind was fashioned, through careful deliberation and in accord with the divine order of the universe. Let us now take up this matter judiciously, starting from the head.

2. This is how one should prepare a salve for erysipelas, and this is how the person shall be healed. Take twenty shilling-weights of litharge, and twenty shilling-weights of fresh lime, and one half sester of vinegar, and four shilling-weights of oil of myrtle, and blend these together, and whip them vigorously together with the vinegar. And then take a different oil and blend it in, and apply this ointment to the sore.

3. For swellings on the head that Greeks call *ulcerosus*—that is, head sores whereby boils arise between the skin and the flesh, breaking out as big as beans on a person's face. This is how the person is healed. Take grape seeds and work them into a liquid, and apply this to the sore, and he will soon be well.

4. Eftsona wid þat ylca: nim swearte beanen and cnuca hy swiðe smale, and bynd hy to þare wunda, and selest heo hit gehaleð.

5. Eftsona, nim mintan and cnuca hy smale, and lege uppan þa wunda. And ealle þa wæten ðe þar-ut gað of þan sare, eall heo hit adrigh, and gehælð þæt sare.

6. Eftsona wið, gif þeo ylca adle cilde egelie on geogeþe. Nim garluces heafud—swa gehæl, mid felle and mid ealle—and bærne hit to axan. And nim þanne þa axan and ele, meng togadere, and smire þæt sar mid. And þæt byd selysþe wið þa adle.

7. And eftsona, gif þa wunda to-ðindaþ. Nim fyrs and cnuca hine, and lege uppa þat geswollene, and hyt sceal sana settan.

8. Wið tobrocenum heafod, oððer gewundedun þe of þan wætan byð acenned of þan heafode: nim betonica and cnuca hi, and lege to þare wunda, and eal þat sar heo forswylhþ.

9. *Cefalaponia*—ðæt ys, heafod-sar, and þat sar fylgþ lange þan heafode. And þis synda þa tacnu þæs sares: þæt is, ærest þa ðunewenga clæppaþ, and eal þat heafod byð hefi, and swagoð þa earan, and þa sinan on þan hneccan særgiað.

2    Þis sceal to botan þan sare. Do þane mann innan to ana huse þe be no to leoht no þustre. And begyte man hym rudan, swa mycel swa he mæge mid hys han byfon, and eorð-ivi ealswa micel, and laur-treowes leaf em-mycel oððer þæra

4. Alternatively, for the same ailment: take black beans and pound them very fine, and bind them to the afflicted part, and this will provide an effective cure.

5. Alternatively, take some mint and pound it finely, and apply it to the afflicted part. And this will dry out all the liquid discharge from the sore, and it will cure the sore.

6. Another remedy, if the same ailment should affect a child in his youth. Take a head of garlic—the whole head with its skin and all—and burn it to ashes. And then take the ashes and some oil, mix them together, and apply this to the sore. And that is the best treatment for the ailment.

7. And alternatively, if the afflicted parts swell up. Take some furze and pound it, and apply this to the swollen flesh, and the swelling will soon subside.

8. If the head has been fractured, or if it has been injured on account of the fluid that originates there: take betony and pound it, and apply it to the afflicted part, and it will absorb all the pain.

9. For *cefalaponia*—that is, a head pain, one that persists in the head for a long time. And these are the signs of this ailment: that is, first of all the temples are pulsing, and the whole head feels heavy, and the ears ring, and the tendons at the back of the neck are sore.

This is the cure for that ailment. Take the person inside a house that is neither too brightly lit nor too dark. And get some rue, as much as one can grasp with one's hand, and the same amount of ground ivy, and an equal amount of bay 2

beriga nigon, and seoþ hit eall togadere on wætera. And do
þarto ele, and smere þæt heafod myd. Hyt byð sona hæl.

10. To þan mann þæt hys heafod æcþ, oððer wurmas on þan
heafedon rixisiad. Nim senep-sæd and næp-sæd, and meng
eced, and cned hyt mid þam ecede þæt hit si swa þicce
swa doh, and smyre þæt heafoð foreweard mid. And þis his
anredes læcecræft.

11. Eft, nim ladsar—þæt teafur—and galpani, oþþres healfes
panige whit, and gnid hyt togadere mid wlacan ecede. And
nim þanne þa sealfe and geot on þæs seocys mannes eare.
And læt hyne liggen swa lange fort þan eara hit habben eal
gedrucan, and he byð wundelice hraþe hal.

12. Eft: nim ellenes piþan and ecede, and wull eall togadere,
and geot þa sealfan in þat eare. Gif se wyrme ys þarinnan,
sona he sceal hutgan of þan earen, gif he þarinna hys.

13. Þis ys se lacecræft be þan manne þat hym þing þæt hyt
turnge abotan hys heafod and farþ furwendun brachenum.
Nim man rudan and cervillan and enneleac, and cnuca þa
wurtan togadere. Nim þanne eale and buteran and ecede
and hunig, and meng togadere þa sealfe mid þare wulle þe ne
com næfre awæxan. And do inna þa sealfen, and wæte þa
sealfen inne ane panne mid wulle and mid ell. Nim þane þa
wulle werme and beþece þæt heafod mid, and him byð sona
bet.

leaves or else nine bay berries, and simmer it all together in water. And add oil, and apply to the head. It will soon be well.

10. For a person whose head aches, or when insects have taken over the head. Take mustard seed and rapeseed, and add vinegar, and blend the seeds with the vinegar until it is as thick as dough, and rub the front part of the head with it. And this is an outstanding medication.

11. In turn, take some laserwort sap—the dye—and a penny-weight and a half of galbanum, and blend them together with some lukewarm vinegar. And take that salve and pour it into the sick person's ears. And have him lie flat until the ears have absorbed it all, and he will be well remarkably soon.

12. Alternatively: take the pith of an elder tree and some vinegar, and boil them up together, and pour the liquid into the ear. If the *wyrm* is within, it will soon come out of the ear, if it is in there.

13. This is the cure for a person who feels like his head is spinning and his brains are turned about. Take rue and chervil and onion, and pound the plants together. Then take oil and butter and vinegar and honey, and mix the salve together with wool that has never been washed. Work the salve into the wool, and soak the salve in a saucepan along with the wool and some oil. Then take the wool while still warm and cover the patient's head with it, and he will straightway be better.

14. Eftsane, nim renwæter, oððer wullewæter þa upwærd wyllð and clæne byð. Do hyt in an fæt. Nim þann anne lin-nenne cla ð, and do hine eal wate on þan wætere. And byn hine syðþan twyfeald uppe þan heafode of se claþ drige beon, and hym byð sone bet.

15. Eftsona, nim balsmeðan and ele, and cnuca þane bals-meþan, and menge sydðe wið hlutre ele, and nim þann ane panne and wyrme þa sealfe innan. Nim þann þa sealfe swa wearme and bebind þæt heafed mid. And nim eftsona platagine—þæt ys, webrædan—and cnuca þa wurt togadere, and meng hecede þarto. Wyrce syðan anne cliþan þarof. Nim þann þane clyðan and bynd to þan sare.

2      Þanne scealt þu wyrcen ðus þane dreng þarto. Nim savi-nam and ambrotena and cnuca hi, and do hi syþþan on win, and meng piper þarto and sum dal huniges. And þige þarof anne cuppan fulle on ærne morge and oþerne an niht þann he gad to bedde.

16. Wið þat þæ mannes heafod clæppitað, and to ealre þare clansunge þas heafedes. And hit ys nidþearf wið ælc yfel þæt man ærest hys heafod clænsige.

2      Þæt ys ærest twegen sestres sapan and twege hunies and þre sestres ecedes, and se sester sceal wegan twa pund be sylfyrgewyht. And nim hwytne stor and senep and gingiber, æl þissa twelf peniga gewihte. And nim rudan ane handfulle and organe ane handfulle and ane gelare pina-hnutte. And do eal þys innan anne niwne croccan, and amorgen þann seoð þu hyt swa swiðe þat se þriddan dæl beo besodan. Nim hit þanne and do in an glæs-fat, and man machiæ stufbæþ and baþege hine þaron, and smyrige þann þæt heafod mid þare scealfe.

14. Alternatively, take rainwater, or water from a spring that wells up and is pure. Put it in a vessel. Then take a linen cloth and soak it thoroughly in the water. And then tie two folds of it onto the patient's head until the cloth is dry, and he will soon be better.

15. Alternatively, take balsamint and oil, and pound the balsamint, and then blend it with clarified oil, and then take a saucepan and warm the salve in it. Then take the salve, while still warm, and bind it onto the person's head. And moreover take some plantain—that is, common plantain—and pound the plants together, and mix in some vinegar. Then make a poultice from it. Then take the poultice and tie it to the sore area.

Then you should make up a healing drink for the same ailment. Take savine and southernwood and pound them, and then put them into wine, and add some pepper and a portion of honey. And have the person drink one cupful of it in the early morning and another at night when he goes to bed.

16. If a person's head is pulsing, and for all cleansing of the head. And whatever the illness may be, it is necessary that one first of all cleanse one's head.

First of all take two sesters of soap and two of honey and three sesters of vinegar, and the sester should weigh two pounds as measured in silver. And take white storax and mustard and ginger, twelve pennyweights of each. And take a handful of rue and a handful of marjoram and a yellow pine nut. And put all this into a new earthenware vessel, and then in the morning cook it long enough to reduce it by one third. Then take it and put it into a glass vessel, and prepare a steam bath and have him steam himself in it, and then apply the salve to the head.

17. Þis sceal to þan earan þe wind oþþer wæter forclyst. Þus man hy læcnian sceal. Gif þar sy sweg oþþer sar innan þan heafedan, on fruman do þas scealfe.

2    Nim twegen styccan fulle godes eles, and grene diles twa handfulle and rudan ealswa micel, and wyl on an niwen crocen—næs to swiðe, ðe læsse þe ele his mægn forleosen. Wryng þann þur linne clæþ and do hyt on an glæs-fæt. Wyrme þanne mann þæt heafod and smyre mid þare sealfe, and he binde þanne þæt heafod mid ane clæþe ane niht. Wring þanne garleyc inne þa earre alche dæg. After þat he byd hæl.

18. *Ad parotidas*—þæt ys, to ðan sare þe abutan sa earan wycst, þæt man nemneð on ure geðeode "healsgund." And þe healsgun ys twera cunna; and he becum oþerhwylum on man þur þa awergeda adle, and þam mannan swyðest, se on sara seocnesse cealdne wætan drincaþ. Þo oþer byd eaðe to halene, and þæge non dolh ne wyrces. And oþer syndun þe Grecas *cacote* hateð—þæt synde, "awyrgede," and þæge syndan to agytene ealswa hit her beforen segð, forþan þe færunga hy atyweþ and færinga aweg gewiteþ buta ælce læcecrafte. And swa þeah micele frecnysse getacnæð, forþan þe hi beoð acennede of þan swertan wætan, and hy reade atywþ.

2    Þus hy man sceal hælen. Nim webrade-leaf ar sunne up-gange. Nym þanne hlaf and sealt and swamm, and cnuca hyt eal togadere, and wyrce to clyðan, and lege to þan sare. Þanne sceal hit bersten, and hælige sona after.

17. This is for ears that have been stopped up through the action of wind or water. This is how one should treat them. If the ears are ringing, or if there is pain within the head, use this salve first of all.

Take two portions of good oil, and two handfuls of fresh dill and the same amount of rue, and boil them in a new earthenware vessel—not too briskly, lest the oil lose its potency. Then strain it through a linen cloth and put it into a glass vessel. Then have the head warmed and apply the salve to it, and have him bind his head with a cloth for one night. Then wring garlic into the ear every day. After that he will be well.

18. For *parotidas*—that is, for a sore that develops near the ears, the ones that in our language are called "neck swellings." And the neck swellings are of two kinds; and one kind affects a person through the accursed illness, particularly if he drinks cold water while he is severely ill. The other kind is easy to heal, and that type causes no open wounds. The swellings of the former type the Greeks call *kakote*—that is, "accursed," and they are contracted just as has just been said, for they appear suddenly and they go away suddenly without any medical treatment. And all the same they signify a major attack of illness, for they are produced by the black humor, and they have a red appearance.

This is how one should treat them. Pick some leaves of common plantain before sunrise. Take bread and salt and mushrooms and pound them all together, and work this into a poultice, and apply it to the sore. Then the swelling will burst, and soon afterward the sore will be healed.

19. Þis scal wyð þare eagene tyddernesse, heallswa Hypcras
þe læce hyt cydde. Þæt ys ærest þæt ðæt sar becymþ on ða
eagen mid mycelre hætan. Hwilum hit cymð on mid wæten
þæt hi beoð toþundene, and hwilum buton ælce sore þat
hi ablindiað, and hwilum of þan flewsan þe of þan eagean
yrnaþ.

2      Þanne sceal hy man þus lacnian. Gif seo unhælþe cymþ of
þare drigan hætan, þanne niman man ane clæþ and waxen þa
eagan mid þan claðe. Dyppe hine on watere and gnide þa
eagean mid. And gif hi beoþ toswollene oððer blodes fulle,
ðanne scel mann settan horn aþ þunwangan. And gif hy
ablindiaþ butan ælcon sare, sylle hym drincan *catarcum,* and
he byð gehaled. And eftsona, gif ani þing innan þa eagen
byfulþ, þanne sceal man nime mede oððer wyfes meolc and
do innan þa eagen, and him byð sona bet.

20. Þis sceal to þan eagen se geslegen byð oððer toregan.
Nim berbene leaf and cnuca hy swyþe. Wyrc anne cliðan
swylc an litel cicel, and lege uppan þæt eagan anne dæge and
ana niht. Eftsona, nim attrumu and hunig and þæt hwita of
æge. Meng togadere; lage to þan eagean. Him byð sona sel.

21. Eftsona wið þan ylcan: nim niwne cysan and screda hyne
on weallendan wætere. And nim þanne cyse and maca elswa
litles cicles, and bynd to þan eagean ane niht.

19. This remedy is for impaired vision of the eyes, just as the physician Hippocrates identified that ailment. This comes about first of all when soreness affects the eyes with great heat. Sometimes it is the result of fluid that causes the eyes to be swollen, and sometimes the eyes go sightless without any soreness, and sometimes it comes about from discharge that runs from the eyes.

This, then, is how one should treat them. If the disorder 2 comes about from dry heat, then one should take a cloth and wash the eyes with the cloth. Dip it in water and rub the eyes with it. And if the eyes are swollen or bloodshot, then one should place a cupping horn to the temples. And if they go sightless without any soreness, give the patient a purgative drink, and he will be healed. And again, if there is any corruption inside the eyes, then one should take mead or a woman's breast milk and put it into the eyes, and he will soon be better.

20. This is a remedy for an eye that has suffered a blow or is bleared. Take gypsywort leaves and pound them well. Make up a poultice in the shape of a little cake, and apply it to the eye for a day and a night. Alternatively, take black vitriol and honey and egg white. Mix them together; apply this to the eye. He will soon be better.

21. Alternatively, for the same ailment: take fresh cheese and shred it into boiling water. And take the cheese and make it into cakes of the same small size, and tie these to the eyes overnight.

22. Þis sceal wyð eagena dymnysse þæt Grecas nemniað *glau-comata*—þæt ys, eagena dymnesse. Þus me hyne sceal læc-nige. Nim wifes meolce þry sticcæs fulla, and cyleþena, *id est* celidonia wos, anne sticce fulne, and alewan and croh and safran gallice, and meng æl þas togadere and wring ðurh lin-nenne claþ, and do þanne þa sealfan inna þa eagen.

23. Þis sceal wyð eagen tyddernyssa þe beoþ on þan ægmo-ran sara. Nim myrta and lege hy on hunige, and nym þanne ða myrta and lege to ðan eagean þæt þa eagen toðinden. And nim þanne rudan and cnuca hy and meng axan to, and lege sydþan to þan eagen. Þanne ærest byt heo swyle þa brewas, and after þan heo hyt glewlyce gehælð.

24. *Ad nectalopas;* þæt ys on ure þeodum, þe man þe ne mæge nenge geseo after sunna upgange ær sunna eft on setl ga. Þanne is þis ðe læcecræft þe þærto gebyreþ. Nim buccan hwurfban anð bræde hit, and þanne þeo bræde geswate, nim þanne ðæt swot and smyre mid þa eagen; and after þan, ete þa ylcan braden. And nim þanne niwe assan tord and wrynge hit. Nime ðanne þæt wos and smyrege þa eagen mid, and hym byð sone bet.

25. Þis sceal wyð þat þe on eagen beoþ þæt Grecas hatað *or-diolum.* Þæt ys þe læcecræft ðe þarto gebyreð. Nim beremele and cned hyt mid hunige; lege to þan eagen. Þes læcecræf hys fram manigum mannum afanded.

26. Eftsona: nim beana melu and sapan, meng togadere, and lege to þan eagen.

22. This remedy is for the ocular impairment that the Greeks call *glaucomata*—that is, dimness in the eyes. This is how one should treat it. Take three spoonfuls of a woman's breast milk, and one spoonful of greater celandine—that is, of celidonia juice—and aloes and saffron and French saffron, and mix these all together and strain through a linen cloth, and then put the salve into the eyes.

23. This remedy is for the impaired vision of eyes that are sore at the roots. Take myrtle berries and lay them in honey, and then take the berries and apply them to the eyes so that the eyes become swollen. And take some rue and pound it and mix it with ashes, and then apply this to the eyes. Then at first the eyelids will swell up, and after that, this will cunningly cure it.

24. For *nectalopas;* that is, in our language, when a person can see nothing from sunrise until the sun sets again. This is the remedy for that condition. Take a buck's thigh joint and grill it, and when the grilled meat sweats, then take the drippings and apply this to the eyes; and after that, have the patient eat the same roast. And then take some fresh ass's dung and wring the juice from it. Then take the juice and apply it to the eyes, and he will straightway be better.

25. This remedy is for the eye ailment that the Greeks call *ordiolum.* This is the remedy that pertains to it. Take barley meal and knead it with honey; apply to the eyes. This remedy has been tested by many people.

26. Alternatively: take bean meal and soap, mix them together, and apply to the eyes.

27. Þis man sceal don þan mane þe ne mæg slapan. Nim wermod and gnid on wine odðer on wearme wætere and drinca, and hym byð sona bet.

28. Þis ys þa tylung to þan manne, þe wel gefnesan ne mæge and micel nearnesse on þa heafedan habbaþ. Þis ys þe læcecraft þe þarto gebyreð. Nim *castorium* odðer elleborum and wyrc to duste, and do hyt innan þa nosan, and hyt bring forð þane fnæst.

29. Eftsona, þes læcedom sceal þan manne ða hyra lippa beoð sare, odðer hyra tunga and seo ceola swa sær byþ þæt he hearfoðlice hys spatel forswelgan mæg. Þus hym man sceal tiligan. Nim fifleafan and drige to duste, and meng hunige þanne þarto. Nim ðanne se sealfe and smire mid þa lippa and ða geaglas innan, and hym byð sona bet.

30. Þisne læcecræft man sceal don þan manne ða færinga adumbiaþ. Nim dworge-dwostlan, *hoc est pollegia,* and do hi on ecede. And nim þanne anne linnenne clæð and do þa dworgedwostlan on innan, and do þanne benyþan his nosu, and he mæg specan sona.

31. Þisne læcecræft mæn sceal don þan mannum þe se streng under þare tunge toswollen byð; and þurh þanne streng ærest ælc untrumnesse on þane man becumð. Þanne nim þu ærest þane cyrnel þe byð innan þan persogge, and cyrfetan cyrnel, and cawel-stelan; bærne togædere. And ceorf þane streng under þara tunga and do þat dust on innan, and hym byd sona bet.

27. This is what should be done for a person who cannot sleep. Take wormwood and grate it into wine or warm water and have him drink it, and he will straightway be better.

28. This is the treatment for someone who has trouble sneezing and who suffers from severe head congestion. This is the medicine that pertains to that condition. Take castoreum or white hellebore and work it into a powder, and put it into his nose, and it will induce sneezing.

29. In turn, this is a remedy for a person whose lips are sore, or whose tongue and throat are so sore that he has trouble swallowing his own saliva. This is how to treat him. Take cinquefoil and dry it to a powder, and then mix it with honey. Then take this salve and apply it to the lips and the jowls, and he will soon be better.

30. This is the remedy for someone who is suddenly struck dumb. Take pennyroyal—that is, *pollegia*—and put it into vinegar. And then take a linen cloth and wrap the pennyroyal in it, and hold this beneath his nose, and he will soon be able to speak.

31. This is the remedy for someone who suffers from swelling of the ligament beneath the tongue; and it is through that ligament that every kind of infirmity first enters a person. First of all take a peach pit, and the seeds of a gourd, and cabbage stalks; burn them together. And make an incision in the ligament beneath the tongue and put this powder into it, and he will soon be better.

32. *Ad gincivas* þe Grecas hæteð—þæt ys on ure þeodum, þæt flæsc ðe abute þa teþ wuxt and þa teþ awegð and astyreþ. Nim forcorfen leac and cnuca hyt and wring þæt wos of, anne sticcan fulne, and ecede anne sticcan fulne and huniges þry sticcan fulne, and do þæt hyt welle þrywa. Nim þanne swa hætte swa he hattest forbere mæge, and habban an dæl on hys muþe forte acoled beo. Þanne eftsona oðer dæl ealla swa, þane þæt þriddan dæl eallswa.

33. Þes lacecraft ys to ðan menniscan toþan ðat Grecas nemneþ *organum*. Þæt ys on ure geþeddan "bysse" genemned, forþan þurh þa teþ seo blissa sceal uppspringan and manna arwyrþnys. And ealle nydþearfnys on þan toþan ys, æl man wyte mæg, and þan toþa þa tunga to spæce gesteal ys.

2     Þanne þæt Greccas nemnes ys ærest *tritumes*. Þæt synden þa fyrst teþ, þe ærest þane mete underfoð. Oþre Greccas nemneð *eumotici*. Þæt sindon þe teþ þe þane mete brecaþ syþþe þa forme hyne underfangene habbæt. Þann Grecces nemneþ sume *molides* þæt we hæteð "grindig-teþ," fore hy grindeþ æl þæt man bygleofaþ. And oft mann smeaþ, hwæþer teþ bænene beon, forþan þe ælc ban mearh hæfþ and hy nan mearh nabbaþ. And oþre bæn, þeah hi beon tobrocene, mid suman læcecræfte hy man mai hælen, and nafre þane toþ, gif he tobrocen beoþ. Oft of þan hevede se wyrsta wate cumþ to þan toþan on þare gelicnesse, þe hyt of huse dropað on stan þan hyt vinð, and þane stan þurhþurleþ

32. For what the Greeks call *gingivas*—that is, in our language, the flesh that grows at the base of the teeth and that lifts up the teeth and loosens them. Take a leek that has been chopped up and pound it and wring the juice from it, a single spoonful, and take one spoonful of vinegar and three spoonfuls of honey, and bring it to a boil three times. Then take some of it, as hot as the person can stand, and have him keep some of it in his mouth until it has cooled. Then do the same thing in turn with the second portion, then with the third portion just the same.

33. This medicine is for the human teeth that the Greeks call *organum*. In our language these are called "bysse," because it is on those teeth that bliss depends, as does a person's dignity. And those teeth are useful for all manner of purposes, as all people know, and it is through them that the tongue is the foundation of speech.

The ones that the Greeks name first of all are *tritumes*. These are the most prominent teeth, the ones that take in food first of all. There are others that the Greeks call *eumotici*. These are the teeth that break the food apart once the first teeth have taken it in. Then there are others—ones that the Greeks name *molides*—that we call "grinding teeth," for they grind up whatever one eats. And people often debate whether teeth are made of bone, for every bone has marrow and they have no marrow. And if other bones are broken, they can be healed by some kind of medical procedure, but if teeth are broken they can never be healed. Often the worst fluid comes into the teeth from the head, in the same manner that liquid drops from a house onto a stone pavement when it makes its way through, and it wears through

and þurhþreawþ. Ealswa þa yfel wæte of þan heafod fylþ uppan þa teþ and hy þane þurhþreaþþ and deþ þæt hy rotigeþ
and toþinddaþ, þat þa teþ þoligean ne mæge ne hæte ne
ceald—and swyþest þa grindig-teþ, þe alc mid feower wyrtrume gefæstned byð; and þane hy hero wurtruma forleataþ,
þanne sweratiged hy and fealled.

3      Þanne ys þe læcecræft þarto. Nim sumne dæl of heortes
hyde and anne niwne croccan, and do wæter on, and seoþ
swa swyþe þæt hit þriwa wylle, swa swyðe swa wæter-flæsc.
Nim þanne þat wæter, and habbe on hys muþe, swa wearm
swa he forbere mæge, fort hyt acoled beon. And þanne hyt si
col, wyrpe hyt ut of hys muþe. And nime eft wearmre and do
yt eft col ut, and byð sona bet.

34. Eftsona, nim piper and alewen and sealt and leaces sæd
and hunig, and meng eal togadere. Nim þanne se sealfe and
gnid þa teþ mid, and þa sealfe aflymþ fram þa toþa eall þæt
yfel.

35. Eftsona: hwitne stor and laurberigie and ecede; meng
heal togadere. Nym þanne ane panne and wlece hyt eall
togadere þæt hyt wlæc beo, and habbe on hys muþe swa
wlac.

36. Þes lacecræft deah wyð þone huf. Nim piper and cumyn
and rudan, þreora scyllinga gewyht, and do þarto anne sticcan fulne huniges. Nimm þanne ane clæne panne and seoð
þa scealfe þæt heo wel wealle, and styre hy swyþe gemang
þan þe heo welle. Nim anne clæne fæt and do hy on. Etan
þanne twegen sticcan fulle a æfen, twegen a morgen, and
byþ sona hæl.

the stone and eats it away. And in like manner the noxious fluid from the head drips upon the teeth and twists through them and causes them to decay and erode, so that the teeth cannot tolerate either heat or cold—and especially the molars, each of which is held in place by four roots; and when their roots cease to hold firm, then the teeth turn black and fall out.

This is the remedy for that condition. Take a piece of deer hide and a new earthenware vessel, and put water into it, and cook it so briskly so that it boils three times, as briskly as for boiled beef. Then take that water and have the person hold it in his mouth, as hot as he can stand, until it has cooled. And when it is cool, have him spit it out of his mouth. And take hotter broth again and spit it out again once it has cooled, and he will soon be better. 3

34. Alternatively, take pepper and aloes and salt and some leek seeds and honey, and mix them all together. Then take this salve and rub it on the teeth, and the salve will drive everything noxious out of the teeth.

35. Alternatively: white storax and laurel berries and vinegar; mix everything together. Then take a saucepan and warm it all together until it is lukewarm, and have the person keep it in his mouth while it is lukewarm.

36. This remedy is effective for the uvula. Take pepper and cumin and rue, three shilling-weights, and add a spoonful of honey. Then take a clean saucepan and cook the salve until it is boiling briskly, and stir it around thoroughly while it is on the boil. Take a clean vessel to put it in. And have the person eat two spoonfuls in the evening, two in the morning, and he will soon be well.

37. Þes lacecræft deah gif þæs mannes þrota toswellon byð, and þa ceola þæt Greccas *brahmas* hataþ. Þis ys þe læcecræft. Sule hym supan gebræddan hrere ægeran, and hunig to. And do hym bryð of meolce gemaced, and syle hym cervillan etan, and fæt flæsc þæt beo wel gesoden eta. And he byd sona hal.

38. Þes lacecraft sceal þan manne þe nerwnysse byð æt þare heortan and æt ðare þrotu, þæt he uneþe specan mægan. Þæt scealt þu hym þus don. Nim leac and cnuca hit and wring þat wos of. Syle hym supan, and hym byð sona bet.

39. Eft: nim beana and ele, and seoð þa beana on eala and syle hym etan, and hy doþ þa nearwnysse aweg.

40. Þisne læcecraft man sceal don þan manne þe hura stemna of-fylþ. Ðæt Greccas nemneþ *catulemsis*. Þus þu hine scealt lacnian. Do hym forhæfædnysse on mete, and læt hine beo on stille stowe. Nim þanne godre butere, twegen sticcan fulle, and anne sticcan fulne huniges, and wyll togadere. And læt hine swelgan þa sealfe leohtlice, and sile hym þanne leohtne mete, and drica win, and hym cymþ bote.

41. Þisne læcecræft man sceal don manne þe byð þe ceola sar, þæt Greccas hæteþ *gargarisis*. Nim niwe beane and puna. Nim þanne eced oþþer win and seoð se beanna, and nim ele and meng þarto, oþþer spic gif man ele nabbe, and do þarto. Wille on ana panna. Nim þanne wylle and dype on þare scealfe, and bind þa wulle to þare ceolan.

37. This remedy is effective if a person's throat is swollen, along with the soft tissues at the back of the mouth that the Greeks call *brahmas*. This is the remedy. Give him half-cooked eggs to eat, along with honey. And make him a milk-based broth, and give him chervil to eat, and some fatty meat that has been well boiled. And he will straightway be well.

38. This remedy is for a person who is suffering from constriction at the heart and at the throat, so that he is barely able to speak. This is what you must do for him. Take a leek and pound it and wring the juice from it. Give it to him to sip, and he will soon be better.

39. Again: take beans and some oil, and cook the beans in oil and give them to him to eat, and they will put an end to the constriction.

40. This is the medicine to give to a person who has lost his voice, a condition the Greeks call *catulemsis*. This is how you should treat him. Have him refrain from food, and let him be in a quiet place. Then take good butter, two spoonfuls of it, and one spoonful of honey, and boil them together. And have him swallow the salve gently, and then give him a light meal and let him drink wine, and this will cure him.

41. This is the medicine one should give to a person suffering from sore tissues at the back of the mouth, a condition the Greeks call *gargarisis*. Take fresh beans and pound them. Then take some vinegar or wine and cook the beans, and take some oil and mix it in, or lard if one does not have oil, and add it in. Boil in a saucepan. Then take some wool and dip it into the salve, and tie the wool to the gullet.

42. Þes læcedon is god manne þe hyra hnecca sær byd, and eal se swyra sargiað swa swiðe þæt he þane muþ uneaþe todon mæg. Þæt sar Greccas nemneþ *spasmus;* þæt ys on ure leodene "hnecca-sar."

2     Þis ys se læcedom þarto. Nim ane handfulle mintan and cnuca hy, and nim þanne anne sester fulne wines and ane pundes gewyht eles. Meng þane eall togadere, and seoð hit swa swyðe þæt þæs wines and þæs eles ne sy na mære þane ær wæs þæs eles, þa hit drige wæs. Wring þanne þurh claþ and wurp aweg þa mintan, and nim wulle and wyrcean twegen cliðan of þare wulle. Duppe þanne ðonne cliþan on þare sealfe and lege to þan hneccan. Þanne eftsona þane oþþerne, and do þanne oþþerne aweg. Do þus fiftine syþun. Nim þanne oþþre wulle and wyrm to heorþe þæt heo beo swyþe wearm, and bynd to þan hneccan; þanne byn twan tide do þa wylle aweg and nim þa ylcan clyþan þe þar ær wæran. Do þarto on þa ylcan wisan þe þu ær dydest.

43. Þisne læcecræft man sceal don mannum þe hyra swyran mid þan sinum fortogen beoþ þæt he hys næn geweald nah. Þæt Greccas hætað *tetanicus.* Þys adle ys þreora cynna. Þæt an cynn Greccas hætað *tetanicas.* Þat syndan þa menn þa rihte gað uppaþenedan swyran and ne magan abugan fora untrumnesse. And þa oþer adle sit þus on þan swuran þæt sa syna teoð fram þan cynne to þan breostan, þæt he þane muþ atyne ne mæg fore syna getoge. And þæge adle Greccas nemneð *brostenus.* And þe þrydde adle sitt þo on þa swyran

42. This remedy is good for a person whose neck is sore, and the whole of his neck hurts so badly that he can scarcely open his mouth. The Greeks call this pain *spasmus;* in our language it is "neck pain."

This is the cure for this condition. Take a handful of mint 2 and pound it, and then take a sester-full of wine and one pound-weight of oil. Then mix it all together, and cook it enough that there is no more of the wine and oil than there was previously of oil, when it was unmixed. Then wring it through a cloth and throw the mint away, and take some wool and make two poultices of the wool. Then dip the poultice into the salve and apply to the back of the neck. Then do the same thing in turn with the other one, and put the first one aside. Do this fifteen times. Then take some more wool and warm it at the hearth until it is nice and warm, and tie it to the neck; then within two hours put that wool aside and take the same poultices as were used before. Apply them in the same way as you did before.

43. This is the medicine that should be given to someone whose neck is pulled down by his sinews so that he has lost control of it. The Greeks call this condition *tetanicus.* This disorder is of three kinds. The first of these the Greeks call *tetanicus.* This refers to people who walk upright with the neck extended upward, and who cannot bend it because of this infirmity. And the second ailment settles on the neck such that the sinews contract from the chin to the chest, so that the person can scarcely open his mouth because the sinews pull it down. The Greeks call this condition *brostenus.* And the third disorder affects the neck such that the sinews

þæt sa syna teoþ fram þan cynnbane to þan sculdre and þane muþ awoh breddad.

2    Do hym ærest þanne þisne læcecræft. Wyrce hym arest hnesce bedd and macian wearm fyr. Þanne sceal hym man læten blod on þan earme, on þan middemyste ædra. And gif þan gehæled ne byd, þanne teo hym man blod ut betweoxan þan sculdran mid horne. Nim þanne eald wyn and ealde rusel. Nim þanne ane panne, and seoð þane rusel and þat wyn swa swyþe fort se rusel habbe bedruncan þat wyn. Nim þanne wulle and tæs hy and maca hy swylce anne clyþa, and lege þa scealfe on uppan, and bynd þanne to þan sare myd ane clæþe.

44. Eftsona: nym buteran and ele and meng togædere. Nim þanne winberian coddes and galpanan and cnuca eall togadere. And wyl in ðare buteran and on þan ele and do to þan sare, ealswa hyr beforan seið. Do hym þanne hnesce mettas and godne drincan, healswa hit beforen seið, swylce hwile swa hym he beþurfe.

45. Þes læcecræft is god wyð sare handum and þara fingra sare þæt Greccas hataþ *pormones,* and on leden *perniciam* man hyt hæt. Nim hwitne stor and seolferun syndrun and swefel, and meng togadere. Nim þanne ele and meng þarto. Wurm þanna sa handa and smyra þarmid. Bewynd þanne þa handan mid linnen claþe.

46. Þis hys to þan handan þæt þat fel of gæþ and þan flæsc tospringad. Nym winberian þe beoþ acende æfter oþþre berigian, and cnuca hy swyþe smale, and do hy on buteran

pull down from the jaw to the shoulder and stretch the mouth awry.

Make this medicine for him first of all. First prepare a soft bed for him and kindle a warm fire. Then he should be bled from the arm, from the midmost vein. And if he is still unwell, then one should draw his blood out from between his shoulders, using a cupping horn. Then take some old wine and some old fat. Then take a saucepan, and cook the fat and the wine vigorously enough for the fat to absorb the wine. Then take some wool and tease it out and shape it like a poultice, and apply the salve to it, and then tie this to the sore with a cloth.

44. Alternatively: take butter and oil and mix them together. Then take raisins and galbanum and pound them all together. And boil them in the butter and oil and apply it to the sore spot, just as has been said here before. Then give him soft food and good drink, as has just been said, for as long as he requires.

45. This remedy is good for sore hands and for the soreness in the fingers that the Greeks call *pormones,* and in Latin people call it *perniciam.* Take white storax and bits of the dross left over from silver extraction, and sulfur, and mix them together. Then take oil and blend it in. Then warm the hands and apply the salve to them. Then wrap the hands in a linen cloth.

46. This is for hands that have lost their skin so that the flesh is exposed. Take grapes that have matured later in the year than other berries, and pound them very finely, and put

swyþe, and smure þæt sar gelomelice mid. Bærne þanne streuw, and nime þa axan and strewe þaruppe.

47. Eft: nim dracentan wyrtrume and puna hy smale, and wyll hy on hunige, and lege þanne uppan hændan.

48. Þis lacecræft sceal to þan handan þe þæt fell of pyleþ. Nim betan ane handfulle and lactucan ane handfulle and coliandrone ane handfulle, and cnuca eall togadere. Nim þanne cruman and do on wæter, and þa wyrt mid, and wurme þanne wel þa wurtan on þan wæter, and þa cruman mid. Wyrc þanne clyþan þarof, and bind uppan þa handan ane niht. And do þus þa lange þe hit beþurfe.

49. Þis sceal to scurfedan næglum. Nim plum-sewes anes scyllinges gewyht, and swegles æpples twegean scyllenges gewyht, and cnuca hy togadere. Smyre þa næglas mid, and læt hy beon swa gesmyrede.

50. Þis hys god ta þan mann þe hura metes ne lyst. Þæt Greccas hataþ *blaffesis*. And Ypocras seggeþ þæt seo untrumnyss cymþ of þrim þingum: oþþer of cyle, oþþer of miclum æte and drince oþþer of lytte æte and drince, oþþer of miclum wernesse.

2     Gif hyt cumeþ of þan cyle, þanne scealt þu hym helpan mid baþe. Gif hyt cymet of mycele drence, þanne scel he habba forhæfdnysse. Gif hyt cymeþ of mycle swynce oþþer of earfodnysse, þanne scealt þu hym don eced wyd hunige gemengded, oþþer drinccan ecede wyd leac gemengded. Gif þa untrumnysse cumþ of þan cyle, þanne nim þu beferes herþan and barne to duste, and grind piper and meng piper

them into some butter, and apply this frequently to the sore. Then burn some straw, and take the ashes and sprinkle them on top.

47. Alternatively: take dragonwort roots and pound them small, and boil them in honey, and then apply to the hands.

48. This remedy is for hands whose skin is peeling off. Take a handful of beets and a handful of lettuce and a handful of coriander, and pound them all together. Then take some breadcrumbs and put them in water, and the plants too, and then warm the plants well in the water, and the breadcrumbs too. Then make a poultice of it, and bind this to the hands overnight. And do this for as long as is needed.

49. This is for scabrous fingernails. Take a shilling-weight of plum juice, and two shilling-weights of swail's apples, and pound them together. Apply this to the fingernails, and leave the ointment on them.

50. This is good for a person who has no appetite for food. The Greeks call this condition *blaffesis*. And Hippocrates states that this infirmity arises from three things: either from chills, or from too much food and drink or too little food and drink, or from extreme fatigue.

If it comes from chills, then you should help him by making a bath. If it comes from overdrinking, then he should maintain abstinence. If it comes from hard work or from exhaustion, then you should give him vinegar mixed with honey to drink, or else vinegar mixed with leeks. If the ailment comes from chills, then take castoreum and burn it to a powder, and grind some pepper, and mix the pepper and

and þæt dust togader, and nim sticcan fulne þas gemeng-
dede dustes and do in ane cuppefulle wynes, and wlece
þanne þæt win mid þæt duste, and sile hym drinca. Oþþer
nim peretrum wyð mede gemengded, swa micel swa gemenged wæs þæs oþþres, and sile hym drince.

51. Þisne læcedon do þan manne þa hym beoð on hyra
brosten nearuwe. Þat Greccas hæteð *asmaticos*—þæt ys,
"nearunyss." And uneaþe mæg þane fnæst to do and ut-
abringan, and hæfd hæte breost and byd innen mid micle
nearnysse; and hwilan he blod hræcþ and hwylum mid blode
gemenged, and hwile he wriþaþ swylce he on dueorge sy;
and micel spatel on ceola wyxeþ and syhþ adun on þara lungane. And þus byð þat yfel acenned: ærest þur mycele æteþ
and drincas þæt yfel hym on innan wyxt and rixað, swa
swyþe þæt hym næþer ne meteþ ne ealaþ ne lyst.

2    Þus þu scealt hine halan. Do hyne into þan huse þe beo
næþer ne to hæt ne to ceald, and læt hym læce blod on þan
wynstran earme, gef he þare ylde hafeþ. Gif þu þanne on þan
earme ne mæge, þanne sceal þu hym læten blod mid cyr-
fetum betwex þan scoldrum on þa ylcan wysa þe mann mid
horne deð. Gyf wyntra sy, þanne scealt þu niman pollegian
and seoð hy on watere. Nim þanne þa wyrta and wyrce togædere swa micel swa celras. Þacc yt þanne gelomelice mid þan
wermum wætere betwex þan sculdrun, oþþer mid harehunan, gif þu dueorgeduostle næbbe.

3    And gif þur þis hæl ne beon, nim *ventuosam* and lege under þa earmes and anbutan þane mægen, and nim þane fele

the powder together, and take a spoonful of this composite powder and add a cupful of wine, and then warm the wine with the powder, and give him that to drink. Or else take common centaury mixed with mead, the same quantity as with the other medicine, and give him that to drink.

51. Give this medicine to a person whose chest is constricted. The Greeks call this condition *asmaticos*—that is, "constriction." And it is hard for the person to breathe in and to expel his breath, and his chest is hot and he experiences great constriction within; and sometimes he coughs up blood and sometimes his phlegm is bloodshot, and sometimes he twists and turns as if he were delirious; and a great deal of spittle accumulates in his throat and percolates down into the lungs. And this is how this illness comes about: it is first of all through excessive eating and drinking that the illness arises within him and comes to dominate, to the point that he has no desire for either food or ale.

This is how you shall heal him. Bring him into a house 2 that is neither too hot nor too cold, and also have him be bled from the left arm, if he is of age. If you cannot bleed him from the arm, then you should draw his blood using a cupping glass between the shoulders in the same way as one docs with a cupping horn. If he is of age, then you should take pennyroyal and cook it in water. Then take that plant and work it up to the size of curds. Then pat it frequently between the shoulders along with the warm water, or else do this with horehound, if you have no pennyroyal.

And if he is not well despite all this, take *ventuosa* and 3 apply it to the underside of his arms and round about his

cyne wyrta and wyrc to sealfe and smeri abotan þan mæge mid sare selfe. Nim þane hnesce wulle and dupe on ele þe beo of cypressan, and smyre anne clæþ mid þan ele, and wrið þane clæþ abutan þan mægan, and smyre abutan þane swyran mid þan ele and abutan þa hrig-bræde geloemelice.

4     Wyrc þanne clydan of eorþan þa mann nemneþ "nitro," þa byþ fundan on Ytalia, and do þar piper to, and lege to þan sare fort þe man wearmie. Nym þanne narð and pin-treowes sæp and panic and wyrc þær drenc, and syle hym drince. Nim þanne eft cicenamete ane handfulle and þry æpple of celidonia; nim þanne ane healfne sester wynes, and seoþ hi fort hy beon wel gesodene. Syle hym þanne drincan þry dæges, ælce dæg ane cuppan fulne.

52. Þes læcedom sceal to þan mann þe byð yfele on þan breo-stam. Þur þa breost fela freccenysse synden þe on þe manne becumeþ. And soþ hys þæt ælc wætt cymd ærest ut of þan magan, and þur þane wæten þa breorst beoþ geheafugede and þa heorte ge sydu byð gefullede mid yfele blode. And æfter þan ealle þa ædran slapad, and þa sina fortogiað, and eal se lichama byþ fah, and þa eaxle særgeaþ, and sa sculdraþ teoþ togadere, and hyt pricaþ innan þan sculdru and on þan hrigge swilce þar þornas on sy, and hys andwlita byð eall awend.

2     Þanne þu þas tacnunge seo on þan manna, þanne scealt þu hym blod lætan. And gif þu ne dest, hit cym hym to mucele and stranga adle, forþan þa æddra and þa lime beoþ gefullede myd mucellere fulnesse.

stomach, and then take many kinds of plants and work them into a salve and rub the salve all around the stomach. Then take soft wool and dip it in oil of cypress, and apply the oil to a cloth, and wrap the cloth around the stomach, and rub the oil frequently around the neck and around the flesh on either side of the spine.

Then make a poultice of the earth that is called "niter," 4 which is found in Italy, and add pepper to it, and apply it to the sore area until he becomes warm. Then take spikenard and pine tree sap and panax and work up a drink from them, and give it to him to drink. Then take a handful of chickweed and three quinces; then take a half sester of wine, and cook them until they are well boiled. Give him this to drink for three days, one cupful a day.

52. This medicine is for someone suffering from a chest ailment. There are many infirmities that come to a person through the chest. And it is the case that each humor originates first of all in the stomach, and it is through that humor that the chest is weighed down and the heart and sides are filled with noxious blood. And after that all the veins go lax, and the sinews contract, and the whole body becomes discolored, and the shoulders are sore and the shoulder blades draw together, and there is a stinging sensation between the shoulders and in the upper spine as if thorns were lodged there, and a person's facial appearance is wholly altered.

When you see these signs in a person, then you should let 2 his blood. And if you fail to do so, he will contract a severe and racking illness, for his veins and limbs will be filled with much foul matter.

3    Forþan we byddaþ æræst þæt mann hym wyrce speau-
drenc, forþan eal þæt yfel þe byð on þare heorta and on þan
breoste, eall yt sceal þanne ut, and beo þa heorta and þa
breost and þæt heafod swa wel geclansæd. And gif he þanne
þa spatl swyþe utspæte, þanne hys þat þe hyfela wæte, þe on
þan heafoded rixaþ; and eall se lichama geswæred byþ and
gehefeguþ eal swylc he of mycele swynce come, and ealle he
byþ geswenced. And bute he þe hraþur gehæled beo, hyt
cum hym to mycele yfele.

4    Þus man hine sceal læcnie. He hine forhabban sceal wyð
feala cunna metas and drencas and wyð gebræd flæsc and
wið ælces orffes flæsc þe cudu ceowe, and drince leoht wyn
þæt hym ne þyrste. Ac ceowe hwytes cuduwys sæd and fif-
fingran ælce dæg ær he etan; and wite þu gewyslice gif
he mid earfodnysse hwest and yt ut-hræcþ. Þanne ys þæt
clænsunga þara breosta. Þanne sceal he etan drigne hlaf and
cyse. And ne cume he on nane cyle þe hwile þe he seoc beo,
ac beo hym on wermum huse, and hæte hym man bæþ swa
hraþa swa hys wisa godige.

5    Nim þanne earixena wyrtrumma and glædene more and
swearte mintan and mucgwurt, and drige to duste, and do
þær æcern to, oþþer hwætena flysma. Menge togædera.
Meng þar þanne hunig to and wynberigera coddes and pices
sum dæl and hwyttre gosu smere. Seoð þanne eall togadera
on anu niwe croccan. Nim þanne wulle þe ne com næfre
awaxen and wyrc cliþan þærof. Lege þæruppa þa sealfe, wel
þicce. Wryð þanne to þan breostan, swa hæt swa he hattest
forberan mæge. Þanne þeo beo acoled, lege oþerne wearme
þarto, and do þus ðe hwyle hym þearf sy.

Therefore we ask first of all that he be prepared a drink 3
that will induce vomiting, for everything that is noxious in
the heart and in the chest will then be expelled, and the
heart and the chest and the head will thus be thoroughly
cleansed. And if he then spits up a lot of spittle, that is the
evil humor that is ascendant in the head; and his whole body
will be weighed down and will become heavy just as if he had
come from hard labor, and he will be utterly exhausted. And
unless he is cured straightway, he will be in a very bad way.

This is how he should be treated. He must refrain from 4
many kinds of food and drink, and from grilled meat and
from meat of any kind of beast that chews the cud, and he
should drink diluted wine to slake his thirst. But he should
chew gum-mastic seeds and stavesacre seeds every day before he partakes of food, and you should take care to see if
he has difficulty coughing and coughs up phlegm. If so, that
represents a purging of the chest. Then he should eat dry
bread and cheese. And he should not be allowed to suffer
any chills as long as his illness persists, but have him stay in a
warm house, and have a bath heated for him as soon as his
condition improves.

Then take the roots of rushes and the edible root of iris 5
and take dark mint and mugwort, and dry them to a powder,
and add some mast to it or some wheat meal. Blend them.
Then mix in honey and some raisins and a bit of pitch and
some grease from a white goose. Cook them all together in
a newly made earthenware vessel. Then take some wool
that has never been washed and fashion a poultice from it.
Spread the salve on it, a thick amount. Then tie this to the
chest, as hot as he can stand. When this one has cooled,
then apply another warm one, and do this as long as he requires.

6    Wyrce hym drenc gode, þe ægþer clænsige ge þa breost ge þane innoþ. And bace hym man þanne wearmen hlaf be heorþe, and ete þane manige dæges þane hlaf þe wyrm. Nim eft cicenemete and wermod and lauberigan, and do hwyt cudu oþer gerusodne ele to, and gnid eall togadere mid ele mid eall. Wyrme þane þa breost to heorþan, and smyre hy þanne mid þare scealfe.

53. Eft: nim cicenemete and seoþ on wine. Do þanne ele to þe beo of frencissen hnutu, and drince þæt.

54. Þus man sceal wyrcean þane cliþan to þann scearpan bane þe betweox þan breostum byð, gif hyt sar sig. Nim ealdne swynes risel twegea punda gewiht, and wexas syx scyllinga gewyht, and eles swa mycel, and þæt sæpp of cypresso swa micel, and fearres smere fif scillinga wyht, and panecis fif scillinga gewyht, and ysopa feorwer scillinga wyht, and galpanan feorfer scellinga wyht, and beferes herþan feorfer scillingaþ wiht, and hwitere gose smere anes sceallinges wyht, and euforbeo swa micel, and pyne æl togadere, and do in ane boxs, and nime syþþan swa oft swa he beþurfe.

55. Eftsona to þan ylcan: nim niwe butera twegen dæles, and þane þriddan dæl niwes huniges, and ane gode cuppan fulle wines, and hæt þat wyn on ane clæne panne. And þanne hyt wel hæt byð, do þæt hunig and þa butera þærto, and syle hym þanne drinca fæstende ane cuppan fulle.

56. Þisne læcedon man sceal do þan manne se his nafulsceaft intyhþ. Nim eormeleaf and seoþ, and wryð þanne swa hæt uppan þane nafelon.

Then make up a good drink for him, one that will purge 6
both the chest and the inner organs. And bake him a warm
loaf at the hearth, and for many days afterward have him eat
the bread that you have warmed. Then take chickweed and
wormwood and laurel berries, and add gum mastic to it or
extract of roses, and knead it all thoroughly together with
the oil. Then warm the person's chest at the fireside, and
then rub the salve into it.

53. Alternatively: take chickweed and cook it in wine. Add
walnut oil and have him drink it.

54. This is how one should make up a poultice for the breastbone, if it is sore. Take two pound-weights of aged pig's fat,
and six shilling-weights of wax, and the same amount of
oil, and the same amount of cypress sap, and five shilling-
weights of the fat of a young pig, and five shilling-weights of
panax, and four shilling-weights of hyssop, and four shilling-
weights of galbanum, and four shilling-weights of beaver's
testicles, and one shilling-weight of the fat of a white goose,
and the same amount of spurge, and pound them all together, and put them in an ointment box, and take from it
afterward it as often as he requires.

55. Alternatively, for the same ailment: take two parts of
fresh butter, and a third part of new honey, and a good cupful of wine, and heat the wine in a clean saucepan. And when
it is good and hot, add the honey and the butter, and then
give him a cupful to drink on an empty stomach.

56. This is the remedy for someone with an ingrown navel.
Take common mallow and simmer it, and then apply this to
the navel while hot.

57. Eftsona to þan ylcan: nim hwit cudu and weremod and cicenamete and wyll eall togadere. Nim þanne þa wyrta and streuwa uppa ane clæþe, and bynd swa hate uppa þane nafelan.

58. Þisne læcedom mann sceal do þan mann þeo beo on heora heortan ge sidu unhale. Þus þu scealt þat yfel ongyta on þan manne. Hym byd hynnene eall swylce he si eall to-brocen, and he hwest swyþe hefelice and micelne hefe gefret æt hys heortan, and þat he ut-hræcþ byþ swyþe þicce and hæfet hwyt hyw.

2       Þan scealt þu hine þus læcnigean. Nim grene helda and cnuca hy swyþe smale, and nim ane æg and þa wurt and swyng togadere. Nim þanne swynes smere and ana clæna panne. Wylle þanne þa wurt mid þan æge on þan swunes smere innan þare panne fort hyt genoh beo, and sile him fæstenda eta. And æfter þan he sceal fæsten seofan tide ær he ænigne oþerne mete etan. And gif nabbe grene helda, nime þat dust and mæcige mid þan æge, and bruce þysses læcecræft fort he byð hæl.

59. Þis sceal þan manna to læcecræfte þe swyþe hyra spatl ut-spiwaþ and hy habbaþ swyþe heve magan. Þanne ys god þæt mann foresceawie hwanne seo seocnysse sig, forþan þeos ædle ne eglad ælce manne gelice. Sume men hyt eglas of þas heafedes wæten, and sume men hyt eagles þanne hi fæstende beoþ, and hy swyþust hyre spatl ut-spiwaþ gyf hy fulle beoþ, and næfre hy ne swycaþ ac þanne hi hungrie beoþ. Þu miht þa adle gecnawa forþan of þara hæten byþ þæt spatl tolysed, and þa micele spatl of þara mycele hæte, ealswa þæt treow þæt man on heorþe leges. For þare mycele hæten þe þæt treow barned beoþ, þare wylþ ut of þan ende water.

57. Alternatively, for the same ailment: take gum mastic and wormwood and chickweed and boil them all up together. Then take the plants and spread them on a cloth, and tie this to the navel while hot.

58. This is the remedy for a person whose heart or whose sides are unwell. This is how you shall diagnose this disorder. His insides will seem to be all broken up, and he will cough severely, and he will feel a great weight at his heart, and what he coughs up will be very thick and will have a white color.

This is how you should treat him. Take fresh tansies and pound them very fine, and take an egg and those same plants and whip them together. Then take pig fat and a clean saucepan. Then simmer the plants along with the egg in the pig fat in the saucepan until you have produced enough of it, and give it to him to eat on an empty stomach. And after that he should fast for seven hours before he eats any other food. And if you do not have fresh tansies, take tansy powder and stir it up with the egg. And have him partake of this medicine until he is well.

2

59. This is the remedy for people who cough up a great deal of phlegm and have a very heavy stomach. It is advisable, then, to diagnose the source of this sickness, for the disease does not affect everyone in the same way. It affects some people through the humor in the head, and it affects some people when they are fasting, and they cough up phlegm excessively when they have a full stomach, and this never stops except when they are hungry. You can recognize this illness because the phlegm is released because of heat, just as with a log that is put on the hearth. On account of the great heat by which the log is burned, water oozes out of its end.

2     Þus þu hyne scealt læcgnie. Nim gingyfran twelf penega wyht, and piperes feower and twentiga penega gewhyt, and hunige heahta and feorwertig penega gewyht. Meng þanne eal þas togadere, and sille hym fæstende etan þarof, twege sticca fulle oþþer þru.

60. *Ad acidiva;* þæt hys, þæt hæte wæter þe scet upp of þan breostan, and hwylan of þa mæge. Þanne sceal he drinca fif handfulle scealtes wæteres. And nim eftsona wermodes sæd and seoþ hyt on watere, and menge þærto wyn, and drince hyt þanne. Eallswa: nim þro pipercorn, oþþer fif, and hete hyt. Eft, nim bettonica anes scyllingas gewyht and seoþ on wætere, and sile him drinca fæstinda. Nim eft rudan and cnuca, and lege hy þanne on eced, and sile hym fæstende drinca. Eftsona, nim lufestices sæd ane handfulle and ete hyt.

61. Þes lacecræft sceal þan mann þæt spiwan wyllan. Wyte þu gewyslice þæt se speau-drenc deaþ hym mycel god and fultum, ge on þa breostan and on heort ge sidu, and on þarra lungane and on þare milta and on þan innoþ and on þan mæga, ge on ealle þa yfele wæta þe wyþinna þe mægen beoþ and abeotan þa heortan. Eall þe drenc afyrsaþ and aclænsaþ þa-hwylc þing swa þar weaxan þe byð to yfele in þan mann. Þur þane drenc he sceal beon gelyþegod and alysed. Þe spæu-drenc ys god ær mete and betra æfter mete, forþan þe ealde læces hyt þus wrytan, þat seo fastnysse þæs yfeles wætan on þan heafede, and þæt oferflowende yfel on þan breostan, byð astired æfter þan mete, and se yfela wæta on þan gellan byð eac astired. Þanne þur þane dreng he byd

This is how you should treat the person. Take twelve    2
pennyweights of ginger, and twenty-four pennyweights of
pepper, and forty-eight pennyweights of honey. Then mix all
these together, and give him two or three spoonfuls to eat
on an empty stomach.

60. For heartburn; that is, the hot liquid that shoots up from
the chest, and sometimes from the stomach. For this he
should drink five handfuls of salt water. And likewise he
should take some wormwood seeds and cook them in water,
and mix in some wine, and then drink it. Likewise, take
three peppercorns, or five, and have him eat them. Again,
take a shilling weight of betony and cook it in water, and
give him this to drink on an empty stomach. Or take rue and
pound it, and then steep it in vinegar, and give it to him to
drink on an empty stomach. Alternatively, take a handful of
lovage seed and have him eat it.

61. This remedy is for a person who wants to vomit. You
should have a clear understanding that a drink that induces
vomiting does a person much good and much benefit,
whether in the chest or in the heart or the sides, and in the
lungs, the spleen, the intestines, and the stomach, or with
regard to all the noxious humors that are inside the stomach
and close by the heart. This drink expels and purges whatever accumulates there that has a harmful effect on one.
Through this drink one is soothed and relieved. The purgative drink is good to take before a meal and better to take
afterward, for physicians of old have written that the firm
hold of the noxious fluid in the head, along with the over-
flowing of noxious fluid in the chest, is stimulated after a
meal, as is the noxious humor in the gallbladder. Through

afeormud, and ne geþafaþ þæt þær ænig yfel wæta beo ge-
samnud innan þan mægen.

2   Þus þu scealt þane spæaw-drenc wyrcean. Nim smale
napes and lege hy on eced, and do þar hunig to, and læt hy
licgean ane niht þæron ofgotene. Ete þanne a morgen fort
he full sy. Drince þanne æfter wearm wæter. Nim þanne an
feðere and dyppe on ele and stynge on hys muþe, oþþer his
fingers do on hys muþ, þæt he þane spæu-drenc astyrie.

3   And eftsona: nim cuppan fulle wæteres and sealt and
meng swyþe togadere of þæt sealt moltan sy. Do hyt þanne
on ane croccan an nyht. Nim hyt a morgen and dreahne hit
þurh linnen clæþ and syle hym drinca. Þanne se drenc hyne
styrge, þanne sile him drince gelomlice wearm wæter þæt he
þa bet spiwe.

62. And eft, gyf þu wylle sile hym leohtran dreng: nim þanne
wearm wæter and syle hym drincan. Duppe þanne a feþer on
ele and do on hys muþ, oþþer hys fingres, and he spiþ sona.

2   Eftsona: endlufun leaf of bulgagine. Ofgeot hy ane niht
mid wyne; þanne on morgen nim þa leaf and cnuca hy on
treowenum fæte, and ofgeot hy mid þan ylcan wyne þe hy ær
ofgotene wæran, and sile hym drincan.

3   Nim eft eallanwyrte wos swa wearm twegea dæles and
huniges þan þriddan dæl and meng togadere, and sile hym
drincan fastende. And nim eftsana grene cyrfætan an handfulle and do hy on wyn, and do þarto hunige, and do hy on

this drink, then, he will be purged, and it will prevent any noxious humor from accumulating in his stomach.

This is how to work up the purgative drink. Take small rape plants and place them in vinegar, and add some honey, and let them sit overnight steeped in that liquid. Have the person eat them in the morning until he is full. After that have him drink some warm water. Then take a feather and dip it in oil and poke it into his mouth, or have him insert his fingers into his mouth, so that he activates the purgative drink. 2

As an alternative: take a cupful of water and some salt and stir them vigorously together until the salt is dissolved. Then let this stand in an earthenware vessel overnight. In the morning, take it and strain it through a linen cloth and give it to him to drink. Once the potion begins to have an effect on him, give him warm water to drink frequently so that he may vomit more easily. 3

62. And as an alternative, if you wish to give him a lighter drink, then take warm water and give it to him to drink. Then dip a feather in oil and put it into his mouth, or else his fingers, and he will straightway vomit.

Alternatively: eleven asarabacca leaves. Steep them in wine overnight; then in the morning take the leaves and pound them in a wooden vessel, and pour the same wine over them in which they were previously steeped, and give this to him to drink. 2

In turn, take two portions of warm dwarf-elder juice and one portion of honey and blend them, and give him this to drink on an empty stomach. Alternatively, take a handful of fresh bryony and soak it in wine, and add honey, and put it 3

ealu, and sile drinca. And eftsona: nim curfettan wyrtruman and cnuca hy, and wring þærof anes æges-sculle fulle þæs woses, and eles æne æges-sculle fulle; and ellanwyrte wurtrumen nim þanne and cnuca hy and wring þærof ane sculla fulle, and twegra ægersculle fulle wynes, and meng eall togadere, and sile hym drincan on stufbaþe.

63. Þisne lacecræft mann sceal don manne þæt swyþe spywaþ, gif wullaþ þat hit astonden. Þæt Greccas hateþ *apoxerrisis*. Þæt sinden þa menn þa after þan þe hy hure mete habbaþ geþiged þæt hine sceollan aspywan, and hwylan ær hy etan hy spiwaþ, and þe mæga sargað, and þe innoþ toswylþ, and he byþ on ælce lime werig, and singalice hym þurst, and se ansine and þa fet beoþ toswollen, and his anwlita byþ blac, and his migga byþ hwit, and he sceal gelomelice migan.

2    Þus þu scealt hine hrædlice læcnige. Gif þa ylda habbe, læt hym blod of bam þa foten byneoþan ancleowe. Swa si þæt blod forlæte þæt eallunga se seocca ne getorige, and þa þing þe þane magen healdeþ þæt hy næfre forþan forwyrþan. And þeo oþru blodlæse ys þe þu þane seoccan læcnige scealt. Þæt ys, þæt þu hym scealt lætan blod under þare tuncgan, þæt þeo blodlæse þane mann alihte. And æfter þæt seo blodlæse si gefylled, þu hine scealt scearpigean. Nim þann sealt and gnid þa wunda mid. Nim þanne cicenamete and wyllecærsen and eormeleafes sæd and seoþ hy on watere. Hwonlice meng þarto ele and hunige, and wyrc þanne clyþan þerof, and lege þarto þru dæges and þre niht.

3    Eftsona: nim gladenan and hlutter pic and meng togadere, and do to ele and wex and beferes herþþan and

into some ale, and give to drink. And alternatively: take bryony roots and pound them, and wring out an eggshell-full of
the juice, and an eggshell-full of oil; and then take dwarf-
elder roots and pound them and wring an eggshell-full from
them, and take two eggshell-fulls of wine, and mix them all
together, and give him this to drink in a steam bath.

63. This is the cure for people who are vomiting excessively,
if they want it to stop. The Greeks call this condition *apoxer-*
*risis*. These are people who have to vomit after they have had
a meal, and sometimes they vomit before they eat, and the
stomach is sore, and the intestines swell up, and the person
is exhausted in each limb, and he is constantly thirsty, and
his face and feet are swollen, and he has a pale complexion,
and his urine is colorless, and he has to urinate frequently.

This is how to heal such a person quickly. If he is of suffi- 2
cient age, have him bled from both feet beneath the ankles.
The blood should be drawn in such a way that the patient
suffers no discomfort, and the tissues that support the
stomach must not suffer damage on this account. And a
second bloodletting is to be done when it is time to treat the
patient. That is when you should bleed him from under the
tongue, so that the bloodletting provides relief for the person. And once you have finished bleeding him, you should
make an incision in him. Then take salt and rub it on the
cut. Then take chickweed and watercress and seeds of common mallow and cook them in water. Stir in oil and honey
from time to time, and then make a poultice from it, and
apply this to the sore for three days and three nights.

Alternatively: take iris and resin and mix them together, 3
and add oil and wax and castoreum and galbanum and panax

galpanan and panic and hwyt cudu. Cnuca þanne eall þas togadere and magce togadere. Meng þarto þanne ecede, and wyrce clyþan of þissum and lege þarto.

4    Eftsona: nim alewen and myrra and hwit cudu and ægra hwit. Meng eall togadere. Nim þona acuma and do wylle þaron, and lege angen þane mæge. And after þyssun nim weremod and dyle; cnuca togadere. Nim þanne ele; seoð þa wyrta. Wyrma þanne þa fet and þa handa. Wyrce þanne clyþan of þisse wyrta and bynd swyþe to þan handan and to þan fotum, and myd swyþe drigeon handum straca geornlice þane innoþ. And æfter þissum unbynd þa fet and þa handa and smyre hy lange hwile mid þare sealfe. And forhabban hyne wyð micele gangas. And nim getemsud melu and bac hym anne cicel of, and nim cumin and merces sæd and cnede to þan hlafe, and syle hym etan hnesce ægere mid þan hlafe and hetan pin-hnutena cyrnles and amigdalas and oþera hnutena cyrnlu, and wyrce hym blacne briuþ, and forhabbe hyne wyð ælc þweal. And gif he after spiwe, sile him drincan hlutur ecede ær he eta and after hys mete.

5    Wyð þan ylcan: nym betonican swa grene and gnid hy on wætera, and do þonn sum dæl huniges to, and sile drincan fæstende ane cuppan fulle. Nim eft bettonican þreora scyllange gewyht and seoð hy on hunige sweþe, and stire hy gelomlice. Wyrc þann swa greate clympran, feowur swa litle æceran, and sile hym þann fæstende etan on wearmum wæteran feowur dages, alc dæ ane clyne.

6    Eft: nim salvian ane handfulle and cnuca hy swyþe smale, and nim twelf pipercorn and grind hy smæle, and nim þann ægru, and swing ho togædere mid þam wyrtum and mid þan

and gum mastic. Then pound them all together and mash them together. Then add some vinegar to this, and make a poultice from it, and apply.

Alternatively: take aloes and myrrh and gum mastic and egg white. Mix them all together. Right after that take some tow and put wool on it, and lay it against the stomach. And after this take wormwood and dill; pound them together. Then take some oil; cook the plants. Then warm the patient's feet and hands. Then make a poultice from these plants and bind it tightly to his hands and feet, and massage his inner organs firmly with very dry hands. And after this untie his feet and hands and rub them for a long time with that ointment. And keep him from taking long walks. And take finely sifted meal and bake him a small cake from it, and take cumin and wild celery seed and knead it into the bread; and give him soft eggs to eat along with the bread, and some pine nuts and almonds and some other nuts to eat, and prepare a clear broth for him, and do not let him wash himself at all. And if he is still vomiting, give him clarified vinegar to drink before he eats and after his meal. 4

For the same condition: take betony, as fresh as you can get, and grate it into water, and add some honey, and give him a cupful of this to drink on an empty stomach. Alternatively, take three shilling-weights of betony and cook them thoroughly in honey, and stir them frequently. Then make up clumps of this of equal size, four little heaps the size of acorns, and then give him this to eat on an empty stomach for four days in warm water, a single lump each day. 5

As an alternative: take a handful of sage and pound it very finely, and take twelve peppercorns and grind them finely, and then take eggs, and whip them up together with the 6

pipore. Nim þane ane clæne panne and hyrste hy mid ele. And þann hy beon cole, ete hy þann fæstinde. Nim eft dyles sædes twelf penega gewiht, and piperes ælswa fela, and cimenes swa fela, and gnid hit to duste. Nim þann mintan and seoð hi on wætera, and do þærto gehwæde wyn. Drinca þane he wylle to hys bedde.

7    Eftsona, gif se man spiwan and he ne mage etan, syle him drincan elenam wyrtrumann oþþer valerianam leaf oþþer myllefolyam wyð wyne gemengged. Eftsona, gif man sy ge-wanulic þæt hyne þyrete: nym lubestican nyþewearde and gnid on wine and on watera, and sile hym drincan. Eftsona: nim elenam and spelter and seoþ on wine, and sile hym drin-can. Þis ys seo selesta drenc wyð þæt broc.

8    Wyð þan ylcam: genym hwit cude and alewan and mirra and gingiferan and cymen, and grind hy eal togadere and do hunig to, swa fela swa þærf sy. Nim þann linnenne clæð and lege þa sealfe uppan. Bynd þann ofer þane mægen. Þann clansaþ þa scealfe þane innoþ and þa werinysse aweg gedeð and þann magan gewyrmþ.

9    Wyþ þan ylcan: nim swefles ehta penega gewyhta and cnuca hine smale. Nim þann an hrere bræd æg and do hyt on innan, and sile hym etan. Eftsona, gif þu wylt þe werinyssa aweg don of þan mann þann þat yfel hyne geþreadne hæfð ofðe þurft aweg adon: nim hwyt cudu and gyngyfere and re-cels and lauwurberigean and cost, ælces þissa emfela. Nim þann of oþþrum pyhmentum ane sticcan fulne, and gnid hy eal togadere. Nim þann wateres twegen daleles and wines þann þriddan dal. Meng þann eall togadere and syle him drican.

plants and the pepper. Then take a clean pan and fry them in oil. And once they have cooled, have him eat them on an empty stomach. In turn, take twelve pennyweights of dill seed and an equal amount of pepper and an equal amount of cumin, and pound it to a powder. Then take some mint and cook it in water, and add a little wine. Have him drink it when he means to go to bed.

Alternatively, if the person is vomiting and he cannot eat, 7 give him elecampane roots to drink, or valerian leaves, or yarrow mixed with wine. Alternatively, if a person is accustomed that . . . , take the lower part of lovage and grate it into wine and into water, and give him this to drink. Alternatively: take elecampane and spelt and cook them in wine, and give him this to drink. This is the best drink for that affliction.

And for the same disorder: take gum mastic and aloes and 8 myrrh and ginger and cumin, and grind them all together and add honey, as much as is needed. Then take a linen cloth and spread the salve on it. Then tie this over the stomach. The salve will then purge the inner organs, put an end to the person's fatigue, and warm the belly.

For the same condition: take eight pennyweights of sul- 9 fur and pound it finely. Then take a half-cooked egg and blend it in, and give it to him to eat. Alternatively, if you wish to put an end to the person's fatigue when this malady has affected him, or if you need to do so: take gum mastic and ginger and incense and laurel berries and costmary, an equal amount of each. Then take a spoonful of other medicinal plants and grate them all together. Then take two parts of water and a third part of wine. Then mix everything together and give him this to drink.

64. *Ad emoptoycos,* þæt Greccas hateð *amatostax.* Þæt ys
on ledene ure genemned *reiectatio,* and on englisc ys haten
"blodrine." Þus him egleþ se blodrine. Hwilum þurh þa nosa
hym yrnþ þæt blod; hwylum þane on arsganga sitt hyt hym
fram yrnaþ. Ac þa ealde læces sædan þæt þeos þrowung ys
geset of feofer þingum; þæt ys, of þan breoste and of þan
magan and of ædran and of þan þearman. Galwenus se læce
hyt of hys snotornysse þus wrat. Gif hyt on þan breoste byd
oþþer on þan magan, þanne þurh þann spiþan þu hyt miht
gecnawen. Gif hyt byþ on þan ædran oþþer of þare bladre,
þu miht þurh þane miggan hyt gecnawan.

65. Gif hit byð of þan þerman, þanne myht þu þurh þane ars-
gang hyt gecnawan. Hit byþ ongyton on sume manne þæt
þæt blod hym ut of þan heafode ut-wylþ, and on suma
hwilum þæt hyt ut-sprinþ þur þa twa litlan þurlu þa innan
þara ceolan beoþ forþan þa ædran beoþ tobrocone, þa inna
þa þurlu beoþ; and hwilun of þare ceolan þæt blod ut-wylþ,
hwilum of goman, hwylum of þan scearpan banun þe
bytweox þan breostan byþ, and hwylum of þare lungone,
hwylum of þan magen, hwylum of þan innoþe, hwilum of
þan lendune.

2      Þis ys þæt gescead þara lacnunge. Gif þat blod of þan heafode wyll, þus þu scealt hyt agytan: he hwest hefelice, and
sindrig blod he ut-racþ. Þane gif þa adra byþ tobrocen innan
þan þurlu, of þan uve dropaþ uppan þa tunga, and of þara
tungan hyt ingehwyrfþ and he agynþ to brecanne, þane to
spiwanne. Þann gif hyt cumþ of þare þrotan, þus þu hyt
scealt agitan: þann he hwest, þann smyit hys tunge, and he

64. For *emoptoycos,* which the Greeks call *amatostax.* In our Latin this is called *reiectatio,* and in English it is called "hemorrhage." Hemorrhage affects people in the following way. Sometimes the blood flows out through the nose; sometimes it runs out from the anus. But the physicians of ancient times have stated that this affliction arises from four causes; that is, from the chest, from the stomach, from the veins, and from the intestines. Galen the physician wrote about it in this manner in his wisdom. If it pertains to the chest or the stomach, you can recognize it from the person's vomit. If it pertains to the veins or the gallbladder, you can recognize it from the urine.

65. If it pertains to the intestines, then you can recognize it from the stool. It can be observed that with some people the blood flows out of them from the head, while with other people it flows out sometimes through the two small tubes that are located inside the gullet because the veins are ruptured that are inside those tubes; and sometimes the blood wells up from the throat, sometimes from the gullet, sometimes from the breastbone, and sometimes from the lungs, sometimes from the stomach, sometimes from the intestines, and sometimes from the loins.

This is how the modes of treatment differ from one another. If the blood flows from the head, this is how you shall know it: he will cough severely, and he will spit up blood on its own. Then if the veins are ruptured inside the bronchial tubes, blood will drip down from the uvula onto the tongue, and from the tongue it will be taken in and he will start to retch, then to vomit. Then if it comes from the throat, this is how you shall recognize it; when he coughs, his tongue

ut-hræcþ wurmsig blod, and þeo þrutu byþ mid sare gemenged swa swiþe þæt he hyt utan gefret. Gif hyt of þan
goman byþ butan blode and swiðe ut-hreæcþ, þann todo þu
hys muþ, and hawa hwæþer hys ceaflas sin toswollene, and
he eaþelic nan þing forswoligon ne mæg.

3      Þann gif hyt of þan scearpe bane byþ þæt he sarlice hwest,
and blod ut-spiuwþ and micel blod astyreþ, and gif æt hys
breost beoð gesargude, þann wite þu gewyslice þæt þa adran
synd tobrocene þe on þa þurlun synd gesette. Þann gif þat
blod of þa lungune cymþ, þæt agyt þu hyt þus: gif þat blod
beo swyþe read and clane ut to spiwanne, and he mid
hwostan hyt ut-hræcþ butun alcum sare.

4      Gif þan blod of þan innoþe flowe, þæt wyte þu þæt sindon wunda on þan þearmum, and þann he to arsganga gæþ
þann þæt hym fram gæþ byþ swyþe wyþ blode gemenged.
And þann, gif hyt byþ of renys oþþer þan lendene, þane
cumþ þæt blod of þara blæddran; and þæt he myhþ byþ
sweart oþþer hwyt oþþer read, forþan of yfelre adle becymþ
þis þing on þan mann.

5      Þus þu hyne scealt lacnige. Do hyne on wearme huse and
on beorht, and bedde hys bed myd morsecge oppan þara
eorþa. And he hyne sceal forhabban wyþ fela þingas: þis ys
ærest wyþ micele spæce, and wyþ yrsunga, and wyþ hamed
þing, and fram alce furwerfetum flæsce, and fram smyce,
and fram alce ungeþilde, forþan þa addran berstað hwyla for
þan miceles blodes þinge þe on þan lichama and on addra
byþ.

will be discolored, and he will spit up blood mingled with pus, and his throat will be so stricken with pain that he will be feeling the physical effect from it outside. If it comes from the gullet without blood and he is spitting up violently, then open his mouth and ascertain if his gullet is swollen and if he has trouble swallowing anything.

Then if it is from the breastbone that he is coughing 3 badly, and he vomits blood and this provokes much bleeding, and if he has been suffering pain in his chest, then you will know with certainty that the veins are ruptured that are located inside the bronchial tubes. Then if the blood comes from the lungs, you will recognize it in this manner: if the blood is very red and it is vomited up without impurities, and if he spits it up without any pain while coughing.

If the blood flows from the inner organs, you will know 4 that the intestines are suffering internal damage, and when he goes to defecate, what comes out will be shot through with blood. And then, if the blood comes from his kidneys or loins, then it is issuing from the gallbladder; and what he urinates will be black or white or red, because it is from a noxious disease that this condition has come upon him.

This is how you should treat him. Put him in a warm 5 house and a well-lit one, and make up his bed with moor sedge on the ground. And he should refrain from many things: that is, first of all, from much speech, and from irascibility, and from sexual intercourse, and from the meat of all quadrupeds, and from smoke, and from all impatience, because the veins sometimes rupture on account of the excessive flow of blood in the body and in the veins.

66. Ipocras se læce atwuwde þæt on sumum lichama beoþ ma addra þane on sume, and þe lichama byþ wearmra þann se þe smalran addran and þa swa feawa ann beoþ. Þane þe lichama and þa addran beoþ þæs yfelan blodes fulle, þann scealt þu hym læten blod on þan earme, gif he þara hulde habban. And wyrc hym siþþan twegen firesce clyþan, and bind oþerne betwex þa sculdru, oþerne betweox þa breoste, and syle hym ealra ærest etan gebrædne swam. And gif þæt blod ut-wealle of þan heafode, þann cnuca þu swam, and nim wæter and hunig and meng togadere, and sile hym drincan. Nym þann ecede and hunig an meng togadere. Nim þane an feþere and dyppe þaron and smyra þann þa stowe mid.

2    Loca hwær þæt blod ut-wealle, gif þu þa stowe geracen mægen. Gif þat blod of þara ceolan ut-wealle, nym cole spongiam and swam and sealt and cnuca eall togadere, and bynd þann þane clyþan uppa þa þrotan. And sile hym ærest drincan finul on hluttrun wine, and sile hym etan nywe beoblæd, and hym byd sona bet. And gif þat blod on þara lungane si, þane nim wegbrædan and cnuca hig and wring þarof þæt wos, and drinc. Gif hyt byþ of þan scearpan bane þa betwex þa broesta byþ, þann nym þu cealde swam and scealt and cnuca togadere. Nym þane spongyam and lege þa scealfe on uppan, and bynd to þan breostan. Cnuca þann swam and do hine on watere and drinca hyne butan sealt. And gif he þare ylde habban, þann læt þu hym blod and bynd þa scealfe to þan breostan þane.

66. Hippocrates the physician demonstrated that there are more veins in some bodies than in others, and that one body is warmer than another one in which the veins are smaller and correspondingly fewer. When the body and the veins are full of noxious blood, then you should bleed the person from the arm, if he is of age. And then prepare two fresh poultices for him, and tie one between the shoulders and the other between the breasts, and give him first of all grilled mushrooms to eat. And if the blood wells up from the head, then pound some mushrooms, and take water and honey and mix these together, and give this to him to drink. Then take vinegar and honey and blend them. Then take a feather and dip it in and apply it to the spot.

Observe where the blood wells up, if you are able to mark the place. If the blood is welling up from the gullet, take a cool sponge and some mushrooms and salt and pound them all together, and then tie the poultice to the throat. And the first drink you should give him is fennel in clear wine, and give him fresh honeycomb with honey to eat, and he will straightway be better. And if the blood is issuing from the lungs, then take common plantain and pound it and wring the juice from it, and drink. If it is issuing from the breast-bone, then take cold mushrooms and salt and pound together. Then take a sponge and apply the salve to it, and tie this to the chest. Then pound some mushrooms and put them into water and have him drink this without salt. And if he is of age, then let his blood and then tie the ointment to his chest.

# MISCELLANEOUS REMEDIES

# Cambridge, Corpus Christi College, MS 41

1. Gif feoh sy undernumen.

2    Gif hit sy hors, sing þis on his fetera oððe on his bridel. Gif hit si oðer feoh, sing on þæt hofrec and ontend þreo candella; dryp ðriwa þæt weax. Ne mæg hit nan man forhelan. Gif hit sy oþer orf, þonne sing ðu hit on feower healfa ðin, and sing ærest uprihte hit.

3    Petur, Pol, Patric, Pilip, Marie, Brigit, Felic. In nomine Dei et Chiric. Qui querit invenit.

4    Crux Christi reducat, crux Christi periit et inventa est. Habracham tibi vias, montes, silvas, semitas, fluminas, andronas cludat. Isaac tibi tenebras inducat. Crux Jacob te ad iudicium ligatum perducat.

5    Iudei Christum crucifixerunt; pessimum sibimet ipsum perpetraverunt. Opus celaverunt quod non potuerunt celare. Sic nec hoc furtum celatum nec celari possit. Per Dominum nostrum.

2. Wið eah-wærce.

Genim læfre neoðowearde, cnuwa, ond wring ðurh hærenne cla ð, ond do sealt to. Wring þonne in þa eagan.

# Cambridge, Corpus Christi College, MS 41

1. If livestock is stolen.

   If it is a horse, sing this on its shackle or on its bridle. If it 2
is other livestock, sing it on the hoofprints and light three
candles; drip the wax three times. No one will be able to
conceal the theft. If it is a different kind of domestic animal,
then sing it in each of the four directions, and sing it first of
all straight upward.

   Peter, Paul, Patrick, Philip, Maria, Brigid, Felix. In the 3
name of God and Cyriacus. He who seeks will find.

   May the cross of Christ lead them back; the cross of 4
Christ was lost and was found. May Abraham shut off roads,
mountains, woods, paths, rivers, houses from you. May Isaac
bring darkness upon you. May the cross of Jacob bring you,
bound, to judgment.

   The Jews crucified Christ; they inflicted the very worst of 5
deeds on themselves. They concealed a deed that they were
unable to hide. In the same way, this theft, though hidden,
cannot be concealed. By our Lord.

2. For pain in the eye.

   Take the lower part of a rush, pound it, and strain it
through a hair cloth, and add salt. Then squeeze it into the
eye.

3. Wið ealra feonda grimnessum.

2    Dextera Domini fecit virtutem, dextera Domini exaltavit me; non moriar sed vivam et narrabo opera Domini. Dextera glorificata est in virtute; dextera manus tua confringit inimicos, et per multitudinem magestatis tue contrevisti adversarios meos: misisti iram tuam et comedit eos.

3    Sic per verba veritatis emendatio: sic eris inmundissime spiritus. Fletus oculorum tibi gehenna ignis.

4    Cedite a capite, a capillis, a labiis, a lingua, a collo, a pectoribus, ab universis conpaginibus membrorum eius, ut non habeant potestatem diabulus ab homine isto NOMEN.

5    De capite, de capillis, nec nocendi nec tangendi nec dormiendi nec tacendi nec insurgendi nec in meridiano nec in visu nec in risu nec in fulgendo, sed in nomine Domini nostri Iesu Christi, qui cum Patre et Spiritu Sancto vivis et regnas in unitate Spiritus Sancti per omnia secula seculorum.

4. Wið sarum eagum.

2    Domine, sancte Pater, omnipotens aeterne Deus, sana occulos hominis istius NOMEN, sicut sanasti occulos filii Tobi et multorum cecorum. Manus aridorum, pes claudorum, sanitas egrorum, resurrectio mortuorum, felicitas martirum et omnium sanctorum.

3. Against the ravages of all fiends.

The right hand of the Lord has conferred courage, the 2
right of the Lord has exalted me; I shall not die but
shall live and shall declare the works of the Lord. Your right
hand is glorified in its might; your right hand shatters your
enemies, and through the abundance of your majesty you
have confounded my adversaries: you have sent forth your
wrath and it has consumed them.

This therefore will be your punishment through words of 3
truth: you will be a most unclean spirit. The weeping of eyes
shall be your lot, the fires of hell.

Go out from his head, his hairs, his lips, his tongue, his 4
neck, his chest, and the framework of his limbs, so that devils may have no power over this man NAME.

Of his head, of his hairs, there shall be no harming or 5
touching either while sleeping or while remaining silent or
at his rising or at midday or in his sight or in his laughter or
while shining, but rather in the name of our Lord Jesus
Christ, you who live and reign with the Father and the Holy
Spirit in unity of the Holy Spirit forever and ever.

4. For sore eyes.

Lord, holy Father, omnipotent, eternal God, heal the 2
eyes of this person NAME, just as you healed the eyes of the
son of Tobit and of many of the blind. You are the hand of
the palsied, the foot of the lame, the health of the sick, the
resurrection of the dead, the joy of the martyrs and of all the
saints.

3    Oro, Domine, ut erigas et inluminas occulos famuli tui NOMEN. In quacumque valitudine constitutum medellis celestibus sanare digneris. Tribue famulo tuo NOMEN ut, armis iustitie munitus, diabolo resistat et regnum consequatur aeternum. Per.

5. Wið sarum earum.

Rex glorie, Christe, Raphaelem angelum, exclude Fondorahel auribus famulo Dei illi. Mox recede ab aurium torquenti, sed in Raphaelo angelo sanitatem auditui componas. Per.

6. Wið magan seocnesse.

Adiuva nos Deus, salutaris noster. Exclude angelum Lanielem malum, qui stomachum dolorem stomachi facit, sed in Dormielo, sancto angelo tuo, sanitatem servi tui in tuo sancto nomine tribuere. Per.

7. Creator et sanctificator, Pater et Filius et Spiritus Sanctus, qui es vera Trinitas et Unitas, precamur te, Domine, clementissime Pater, ut elemosina ista fiat misericordia tua, ut accepta sit tibi pro anime famuli tui, ut sit benedictio tua super omnia dona ista. Per.

2    ✝ SATOR · AREPO · TENET · OPERA · ROTAS ·

3    Dominus, qui ab initio fecisti hominem et dedisti ei in adiutorem similem sibi ut crescere et multiplicare, da super terram huic famulae tuae NOMEN ut prospere et sine dolore parturit.

I pray, Lord, that you raise up and illumine the eyes of 3
your servant NAME. May you deign to heal him with celestial
remedies in whatever state of health he may be. Grant to
your servant NAME that, fortified with the arms of justice,
he may resist the devil and may reach the eternal kingdom.
By the Father and the Son and the Holy Spirit, amen.

5. For sore ears.

King of Glory, Christ, through the angel Raphael, drive
out Fondorahel from the ears of this servant of God.
Straightway send him back from the torturing of his ears;
rather, restore health to his hearing, by the angel Raphael.
By the Father and the Son and the Holy Spirit, amen.

6. For a stomach ailment.

Help us, God our healer. Drive out the evil angel Laniel,
who brings stomachache to the belly, but by Dormiel, your
holy angel, grant health to your servant in your holy name.
By the Father and the Son and the Holy Spirit, amen.

7. Creator and sanctifier, Father and Son and Holy Spirit,
you who are the true Trinity and Unity, we beseech you,
Lord, most merciful Father, that this offering may become
your clemency, that it may be accepted by you for the soul of
your servant, so that your blessing may extend over all these
gifts. By the Father and the Son and the Holy Spirit, amen.

✚ SATOR · AREPO · TENET · OPERA · ROTAS · 2

Lord, who at the beginning made man and gave him one 3
like himself as his help so that they might grow and multiply,
grant that this woman, your servant on earth NAME, give
birth successfully and without pain.

# Cambridge, Corpus Christi College, MS 367, part II

8. Medicina contra febres.

2 ✚ In nomine Patris et Filii et Spiritus Sancti, amen.

3 ✚ *Ire* ✚ *arex* ✚ *Christe* ✚ *rauex* ✚ *filiax* ✚ *ara-fax* ✚ NOMEN.

# Cambridge, Gonville and Caius College, MS 379

9. Contra felon.

2 Super infirmum, dic mane et vespere, tres tribus vicibus.

3 *Thigat . thigat . calicet . archlo . cluel . tedes . achodes . arde . et herclenom . abaioth . arcocugria . arcu . arcua . fulgura . sophuinit . ni . cofuedi . necutes cuteri . nicuram . thefalnegal . uflem . archa . cunhunelaya.*

4 Querite et invenietis. Pulsante et aperietur. ✚ Crux Matheus ✚ crux Marcus ✚ crux Lucas ✚ crux Iohannes.

5 Adiuro te, pestiferum virus, per Patrem et Filium et Spiritum Sanctum, ut amplius homini huic non nosceas neque crescas sed arescas. ✚ In nomine Patris et Filii et Spiritus Sancti, amen. "Pater noster."

# Cambridge, Corpus Christi College, MS 367, part II

8. A cure for fevers.

✠ In the name of the Father and the Son and the Holy ₂
Spirit, amen.

✠ *Ire* ✠ *arex* ✠ *Christe* ✠ *rauex* ✠ *filiax* ✠ *ara-* ₃
*fax* ✠ NAME.

# Cambridge, Gonville and Caius College, MS 379

9. For a stomach ailment.

Say over the one who is ill, daily both morning and ₂
evening, thrice three times:

*Thigat . thigat . calicet . archlo . cluel . tedes . achodes . arde . et* ₃
*herclenom . abaioth . arcocugria . arcu . arcua . fulgura . sophuinit .*
*ni . cofuedi . necutes cuteri . nicuram . thefalnegal . uflem . archa .*
*cunhunelaya.*

Seek and you shall find. Knock and it shall be opened. ₄
✠ The cross of Matthew ✠ the cross of Mark ✠ the
cross of Luke ✠ the cross of John.

I adjure you, pestiferous venom, by the Father and the ₅
Son and the Holy Spirit, that you injure this man no more,
nor grow larger, but dry up. ✠ In the name of the Father
and the Son and the Holy Spirit, amen. "Our Father."

10. Aliud.

2    Super infirmum dicat ista orationem tribus vicibus in aqua, et da infirmo bibere tribus vicibus tribus diebus, et sanabitur.

✠ Et ecce crucis signum.
✠ Crux est reparatio rerum.

3    Per crucis hoc signum, fuge potestas inimici retrorsum. Iesus Christus cruce pugnat, vincit, regnat, imperat. Christiane tolle crucem sicque tutus ambula. "Pater noster" novem vicibus.

## Cambridge, Queen's College, MS 19 (formerly MS 7)

11. ✠ Contra febres.

2    In nomine Patris et Filii et Spiritus Sancti, amen. Coniuro vos, febres, per Patrem et Filium et Spiritum Sanctum et per sanctam Mariam genitricem Dei, ut non habeatis potestatem super hunc famulum Dei NOMEN.

3    Coniuro vos, febres, per Deum verum, per Deum sanctum, per septuaginta nomina Dei sancta et immaculata:

4    Elyon, Elyon, Elyon. Eloy, Eloy, Eloy. Us, Ne, Te. Adonay. Tetragramaton. Inimitabilis. Invisibilis. Eternus. Simplex. Summum bonum. Incorporeus. Creator. Perfectus. Christus. Messias. Sother. Emanuel. Dominus. Octogenitus homo. Ysyon. Principium. Finis. Immaculatus. Altissimus.

10. An alternative cure.

May the following words be said over the sick person ₂ three times on water, and have the sick person drink of it three times for three days, and he will be well.

✚ And behold the sign of the cross.

✚ The cross is the restorer of things.

Through this sign of the cross, flee the power of the enemy. ₃ Jesus Christ fights with the cross, wins, reigns, commands. Take up the cross, Christian, and thus walk safely. The "Our Father" nine times.

## Cambridge, Queen's College, MS 19 (formerly MS 7)

11. ✚ For fevers.

In the name of the Father, the Son, and the Holy Spirit, ₂ amen. I entreat you, fevers, by the Father and the Son and the Holy Spirit and by holy Mary the mother of God, that you will have no power over this servant of God NAME.

I entreat you, fevers, by the true God, by the holy God, ₃ by the seventy holy and immaculate names of God:

Elyon, Elyon, Elyon. Eloy, Eloy, Eloy. Us, Ne, Te. Adonay. ₄ Tetragrammaton. Inimitable. Invisible. Eternal. Uncompounded. Highest good. Incorporeal. Creator. Perfect one. Christ. Messiah. Savior. Emmanuel. Lord. Eighth-begotten man. Ysyon. Beginning. End. Immaculate. Most high.

Sapientia. Stella. Omnia leticia. Mercator. Sponsus. Othos. Sebes. Carus. Agathos. Primus et novissimus. Caritas. Gaudium. Sös. Splendor. Admirabilis. Paraclitus. On. Bonus. Nobilissimus. Aries. Leo. Vitulus. Serpens. Ovis. Agnus.

5   Per ista nomina et per omnia cetera Dei nomina, coniuro vos, febres, et per angelos ac per archangelos, thronos et dominationes, principates et potestates, per cherubin et seraphin et per omnes virtutes caelorum, ut non habeatis potestatem super hunc famulum Dei NOMEN.

6   Coniuro vos, febres, per omnes sanctos Dei qui in celo et in terra sunt, et per omnia que creavit Deus in septem diebus et in septem noctibus, ut non habeatis potestatem super hunc famulum Dei NOMEN.

7   Coniuro vos febres, sive cotidianas sive biduanas sive triduanas sive quatriduanas sive quintanas sive sextanas sive septanas sive octanas, et usque ad nonam generationem, ut non habeatis potestatem super hunc famulum Dei NOMEN.

8   Postea dicantur hii tres psalmi: "Ad te levavi oculos," "Deus misereatur," "Quicunque vult," cum "Gloria Patri." Et "*Kyrieleison, Christe eleison, kyrieleison,*" "Pater noster," et "Salvum fac servum tuum, Domine."

9   Memor esto congregationis, Domine Deus virtutum. Domine exaudi. Dominus vobiscum.

10   Oremus. Respice, Domine, super hunc famulum tuum NOMEN in infirmitate corporis sui laborantem et animae. Refove quem creasti ut dignis castigationibus emendetur et continuo se senciat esse salvatum. Per Dominum.

Wisdom. Star. All happiness. Merchant. Spouse. Othos. Sebes. Dear one. Noble one. First and newest. Love. Joy. Sös. Splendor. Admirable one. Paraclete. On. Good one. Most noble one. Aries. Leo. Vitulus. Serpent. Sheep. Lamb.

By these names and by all other names of God, I entreat 5 you, fevers, and by angels and by archangels, thrones and dominations, princes and powers, by cherubim and seraphim and by all the virtues of heaven, that you have no power over this servant of God NAME.

I entreat you, fevers, by all the saints of God who are in 6 heaven and on earth, and by all things that God created in seven days and seven nights, that you have no power over this servant of God NAME.

I entreat you, fevers, whether you are daily or come every 7 other day or third day or fourth day or fifth day or sixth day or seventh day or eighth day, and up to the ninth generation, that you have no power over this servant of God NAME.

Afterwards have these three psalms recited: "To you I 8 have raised my eyes," "God have mercy," and "Whosoever wishes," with "Glory to the Father." And "Lord have mercy, Christ have mercy, Lord have mercy," "Our Father," and "Make your servant well, O Lord."

Be mindful of your flock, Lord God of virtues. Hear us, O 9 Lord. The Lord be with you.

Let us pray. O Lord, look down upon this man, your ser- 10 vant NAME, laboring in the sickness of his body and of his spirit. Restore the one whom you created so that, by worthy chastisements, he may be put right and may always feel himself to be saved. By the Lord.

11    Istud carmen debet dici in primo die novies, in secundo octies, tertio die septies, quarto die sexies, quinto die quinquies, sexto die quater, septimo die ter, octavo die bis, nono die semel.

# London, British Library, MS Cotton Caligula A.xv

12. Wið gedrif.

2    ✚ In nomine Domini nostri Ihesu Christi.

3    *Tera . tera . tera . testis . contera . taberna . gise . ges . mande . leis . bois . eis . andies . mandies . moab . lib . lebes .*

4    Dominus Deus adiutor sit illi.

5    *Illi . eax . filiax . artifex .* Amen.

13. Wið poccas.

2    Sanctus Nicasius habuit minutam variolam, et rogavit Dominum ut quicumque nomen suum secum portaret scriptum. . . .

3    Sancte Nicasi presul et martir egregie, ora pro me Nomen peccatore, et ab hoc morbo tua intercessione me defende. Amen.

14. Wið geswell.

Domine Ihesu Christe, Deus noster, per orationem servi tui Blasii festina in adiutorium meum.

This prayer should be said nine times on the first day, 11
eight times on the second, seven times on the third day, six
times on the fourth day, five times on the fifth day, four
times on the sixth day, three times on the seventh day, twice
on the eighth day, and once on the ninth day.

# London, British Library, MS Cotton Caligula A.xv

12. For fever.

✚ In the name of our Lord Jesus Christ. 2

*Tera . tera . tera . testis . contera . taberna . gise . ges . mande . leis* 3
*. bois . eis . andies . mandies . moab . lib . lebes .*

Lord God, may you be his helper. 4

*Illi . eax . filiax . artifex .* Amen. 5

13. For pustules.

Saint Nicasius had a tiny pock, and he asked the Lord 2
that whoever carried his name written out on him. . . .

Saint Nicasius, leader and eminent martyr, pray for me 3
NAME, a sinner, and deliver me from this sickness by your
intercession. Amen.

14. For inflammation.

Lord Jesus Christ, our God, through the prayer of your
servant Saint Blasius, hasten to my aid.

15. Se engel brohte þis gewrit of heofonum and lede hit on uppan sanctus Petrus weofud on Rome. Se-þe þis gebed singð on cyrcean, þonne forstent hit him sealtera sealma. And se-þe hit singð æt his endedæge, þonne forstent hit him huselgang.

2    And hit mæg eac wið æghwilcum uncuþum yfele, ægðer ge fleogendes ge farendes. Gif hit innon bið, sing þis on wæter; syle him drincan. Sona him bið sel. Gif hit þonne utan si, sing hit on fersce buteran and smere mid þæt lic. Sona him kymð bot.

3    And sing þis ylce gebed on niht ær þu to þinum reste ga. Þonne gescylt þe God wið unswefnum þe nihternessum on menn becumað.

4    "Matheus, Marcus, Lucas, Iohannes. Bonus fuit et sobrius religiosus. *Me abdicamus, me parionus, me orgillus, me ossius ossi Dei fucanus susdispensator et pisticus.*

5    "Matheus, Marcus, Lucas, Iohannes, cum patriarchis fidelis, cum prophetis eternis, cum apostolis humilis. Ihesus Christus et Matheus cum sanctis ac fidelibus adiunctus est actibus.

6    "Matheus, Marcus, Lucas, Iohannes. Deum Patrem, Deum Filium, Deum Spiritum Sanctum, trinum et unum, et Iohannnem basileus fidelium Damasci, per suffragium Sancti Spiritus lucidum omnipotens virtutibus, sanctus est in sermonibus.

7    "Matheus, Marcus, Lucas, Iohannes. *Panpulo dimisit et addinetum.* Alpha et Omega. Per camellos *abiunctionibus* degestum sit pro omni dolore, cum *dubitu observatione* observator.

8    "Exultabunt sancti in gloria. Letabuntur. Exultationes Dei in faucibus eorum, et gladii.

9    "Laudate Deum in sanctis eius," oð ende.

15. An angel brought this inscription from heaven and laid it upon the altar of Saint Peter in Rome. Whoever sings this prayer in church, it is as good as singing psalms from the psalter. And whoever sings it when he is dying, it is as good as taking Communion.

And it has power to withstand each and every evil of an 2 unknown kind, whether flying or earth borne. If the illness is internal, sing this onto water; give it to him to drink. Soon he will be better. If the disease affects his outer body, sing it onto fresh butter and smear it on his body. Straightway he will be healed.

And sing this same prayer at night before you go to bed. 3 Then God will shield you from the bad dreams that people experience at night.

"Matthew, Mark, Luke, John. He was a good and sober 4 religious man. *Me abdicamus, me parionus, me orgillus, me ossius ossi Dei fucanus susdispensator et pisticus.*

"Matthew, Mark, Luke, John, with the faithful patriarchs, 5 with the immortal prophets, with the humble apostles. Jesus Christ and Matthew are joined with the saints in the acts of the faithful.

"Matthew, Mark, Luke, John. God the Father, God the 6 Son, God the Holy Spirit, triune and one, and John the king of the faithful of Damascus, by the shining approbation of the Holy Spirit, he is omnipotent in his virtues, holy in his speeches.

"Matthew, Mark, Luke, John. *Panpulo dimisit et addinetum.* 7 Alpha and Omega. By cups *abiunctionibus* let it be taken for every pain, with *dubitu observatione* observer.

"The saints rejoice in glory. Let us be joyful. Hymns of 8 praise to God in their throats, and swords.

"Praise God in his saints," to the end. 9

## London, British Library,
## MS Cotton Domitian A.i

16. Þas wyrta sceolon to wen-sealfe: elene, garleac, cearfille, rædic, næp, hremnes fot, hunig, ond pipur. Cnucige ealle ða wyrta ond wringe þurh claÞ, ond wylle þonne on þam hunige.

## London, British Library,
## MS Cotton Faustina A.x

17. Ðeos eah-sealf mæg wiþ ælces cynnes broc on eagon: wiþ flean on eagon, ond wiþ gewif, ond wiþ mist, ond wiþ ter, ond wiþ wyrmas, ond wiþ dead flæsc.

2    Eall niwne croccan sy asett on eorþan oþ brerd, ond þas wyrta sy swyþe smæl corflode ond gedon innan þam croccan. On uppan þam sy gedon . $\overline{GT}$ . oþþe wæta þæt hi þearle wel wese beon. Þæt is þonne twegra cynna bisceopwyrt ond glæppe ond ribbe ond gearwe ond fifleafe, dægesege, ond synnfulle, ond brune hofe. Sy syþþan æren fæt, læfel, oþþe cec nyþewerd abywed þæt he eall scine. Bysmyra eall þæt scinende mid hunigteare leohtlice. Sete þonne on uppan þone croccan þæt se æþem slea upp. Þonne binnan þrim dagum wæt þinne finger mid þinum spatle ond gledda þone læfel swyþe lytlum ond litlum, ond nim þær gode eah-sealfe.

## London, British Library, MS Cotton Domitian A.i

16. These plants make up a salve for wens: elecampane, garlic, chervil, radish, wild rape, crowfoot, honey, and pepper. Pound all the plants and strain them through a cloth, and then boil the mixture in the honey.

## London, British Library, MS Cotton Faustina A.x

17. This cyc salve is good for every kind of eye disease: for cataracts, and for web, and for dimness, and for laceration, and for mites, and for dead flesh.

Have a newly made ceramic vessel set into the ground up to the brim, and have the following plants cut up very fine and put into the vessel. On top of them put . $\overline{GT}$ . or some liquid until they are well saturated. Use two types of marshmallow, and buckbean, and plantain and yarrow, and cinquefoil, daisy, and houseleek, and brown ground ivy. Then have the lower inside part of a bronze vessel, bowl, or basin be scoured so that it is all shiny. Smear fresh honeydrops onto the scoured part. Then put this vessel on top of the crock so that the vapor that rises from the crock strikes it. Then within three days wet your finger with your spittle and scrape the bowl with it very little by very little, and take from it a good eye salve.

18. Sy gemenged togædere hunigtear ond win ond rudan seaw, ond efenfela gedon on cyperen fæt, oþþe mæstling, oþþe bræsen. Nim þær gode eah-sealfe.

19. Ðis man sceal singan nigon syþon wiþ utsiht, on an hrerenbræden æg, þry dagas:

2 &#10010; *Ecce dol gola ne dit dudum bethe cunda bræthe cunda . elecunda ele vahge macte me erenum . ortha fuetha la ta vis leti unda . noevis terrae dulge doþ.*

3 "Pater noster" oþ ende. And cweþ symle æt þam drore huic.

20. &#10010; Contra frigora.

2 Omnibus horis scribis in carta, et cum licio ligas ad collum egroti hora deficiente:

3 &#10010; In nomine Domini crucifixi sub Pontio Pilato, per signum crucis Christi, fugite, febres seu frigora cotidiana, seu tertiana, vel nocturna, a servo Dei NOMEN. Septuaginta quatuordecim milia angeli persequentur vos.

4 &#10010; Eugenius, Stephanus, Protacius, Sambucius, Dionisius, Chesilius, et Quiriacus.

5 Ista nomina scribe, et super se portat qui patitur.

21. &#10010; Contra febres.

In nomine sancte et individue Trinitatis. In Effeso civitate Chelde ibi requiescunt septem sancti dormientes Maximianus, Malchus, Martinianus, Iohannes, Seraphion, Dionisius, et Constantinus. Deus requiescet in illis. Ipse

18. Make up a mix of fresh honeydrops and wine and the juice of rue, and put an equal amount of each into a copper vessel, or one made of brass or electrum, or one made of copper or bronze. Take from it a good eye salve.

19. This should be sung nine times for diarrhea, over a half-cooked egg, for three days:

✝ *Ecce dol gola ne dit dudum bethe cunda brœthe cunda . ele-* 2 *cunda ele vahge macte me erenum . ortha fuetha la ta vis leti unda . noevis terrae dulge dop.*

Then "Our Father" to the end. And always speak this 3 when defecating.

20. ✝ For chills.

At every hour you should write the following on a piece of 2 parchment, and tie it to the neck of the sick person with a strap at the waning of the hour:

✝ In the name of the Lord crucified under Pontius Pi- 3 late, by the sign of the cross of Christ, flee, you fevers or daily chills, or tertiary chills, or nocturnal chills, from the servant of God NAME. Eighty-four thousand angels will pursue you.

✝ Eugenius, Stephanus, Protacius, Sambucius, Dioni- 4 sius, Chesilius, and Quiriacus.

Write these names, and the person who is ill should carry 5 it on his body.

21. ✝ For fevers.

In the name of the holy and indivisible Trinity. In the city of Ephesus, in Chaldia, repose the seven sleeping saints Maximianus, Malchus, Martinianus, John, Serapion, Diony-sius, and Constantinus. God will rest among them. May the

Dei Filius sit super me, famulum (vel famulam) tuum (vel tuam) Nomen, et liberet me de ista egritudine et de febre et de omni populo inimici. Amen.

22. Wið þa bleinna.

Þes se hoccas mora, gesodan. Puna, ond ald rusel smoru. Lea þerto.

23. Wið hefd-eca.

Þare clata mora: et raw, festende.

24. Item.

Cnuca betonicam, ond gnid þa þungana ond on ufan þæt hefd.

25. Wið hraca.

Nim atena gratan ond unflid ond ac-drenc god togedera, ond lege þerto heortes hornes ond etriman dust. Et wið eara pið.

26. Contra cotidianas febres.

Sume de urticis manipulum, et stans flexo contra orientem dic: "In nomine Patris quero te; in nomine Filii invenio te; in nomine Patris et Filii et Spiritus Sancti arripio medicinam contra febres pro ea." Dic "Pater noster" et "Credo" novem vicibus.

Son of God himself be present over me, your male (or female) servant NAME, and may he free me from this sickness and from fever and from every hostile people. Amen.

22. For boils.

For this, the edible root of common mallow, soaked. Pound, and add the grease of old lard. Lay it on.

23. For headache.

Burdock roots: eat raw, on an empty stomach.

24. Another cure.

Pound betony, and rub it into the temples and the top of the head.

25. For spitting up phlegm.

Take kernels of oats and some sour cream and a good oak drink and put them together, and add to it a powder made from deer antler and oat bran. Eat it along with the pith of the ears of grain.

26. For daily fevers.

Lay hold of a handful of nettles, and standing with bent knee toward the east, say: "In the name of the Father I seek you; in the name of the Son I find you; in the name of the Father and Son and Holy Spirit I seize medicine for fevers on their behalf." Say the "Our Father" and the Creed nine times.

# London, British Library,
# MS Cotton Galba A.xiv

27. To gehealdenne lichoman hælo, mid drihtnes gyfe, þis is æþele læcedom.

Genim mirran ond gegnid on win swylc sye tela micel steapful, ond þige on niht-nestig þonne restan wille. Þæt gehealdeð wundorlice lichoman hælo, ond hit eac deah wið feondes yflum costungum.

28. Þonne is eft se æðelesta læcedom wið þam ylcan.

2    Genim myrran ond hwit recels, savinan ond salvian ond wurman, ond þæs rycelses ond myrran sy mæst ond þa oþere syn awegene þæt þara efenfela. Ond ætsomne on mortere ond gegnide to duste. Sette under weofod þonne Cristes tid sy, ond gesinge man þreo mæssan ofer þa þry dagas oð midne winter, ond æt Sancte Stephanes tide ond Sancte Iohannes evangelista. Ond on þa þry dagas þige on wine on niht-nihstig. Ond þæt þær to lafe sie þæs dustes, hafa ond heald.

3    Hit mæg wiþ eallum færlicum untrymnessum, ge wið fefre ond wið lencten-adle ge . . . ge wið attre ge wið yfelre lyfte. Gewritu eac secgaþ: se þone læcedom begæð, þæt he hine mæge gehealdan tweolf monað wiþ ealre untrymnessa frecenissum.

# London, British Library,
# MS Cotton Galba A.xiv

27. To maintain your body in good health, by the grace of God, this is a splendid remedy.

Take myrrh and grate it into wine so as to make a generous cupful, and drink it on an empty stomach when you go to bed. This will keep your body wonderfully healthy, and it will also protect you from the devil's evil temptations.

28. This too is a most splendid remedy for the same condition.

Take myrrh and white frankincense, savine and sage and 2 woad, in such an amount that frankincense and myrrh make up the largest part while the other plants are weighed out to be equal to that. And put them in a mortar and pound them to a powder. Set them under the altar at Christmastide, and have three Masses sung over them three days before midwinter, and on Saint Stephen's feast day and on that of Saint John the Evangelist. And partake of them in wine on those three days while fasting. And whatever remains of the powder, keep good care of it.

This is effective for all illnesses that fall on one suddenly, 3 whether fever or Lenten illness or . . . , or for poison or for noxious vapors. Learned writings likewise state: if a person uses this remedy, he can maintain his health for twelve months against the ravages of all illnesses.

## London, British Library
## MS Cotton Otho A.xiii

29. ✚ In nomine Patris et Filii et Spiritus Sancti, amen. Sanctus Nichasius habuit minutam variolam in oculo, et oravit ad Dominum ut quicunque hoc malum habuerat et nomen suum super se scriptum portaret liberaretur ab hoc morbo.

2 ✚ Sanctus Nichasius, oret pro isto Rogero. ✚ Sanctus Nichasius, oret pro isto Rogero. ✚ Sanctus Nichasius, oret pro isto Rogero. Cristus vincit, Cristus regnat, Cristus imperat. Cristus, per intercessionem beati Nichasii, liberet hominem Rogerum ab hoc morbo et ab omni alio.

3 Tetragrammaton. ✚ Qui crediderit et baptizatus fuerit salvus erit. ✚

## London, British Library,
## MS Cotton Otho B.xi and MS Add. 43703

30. Wiþ utsyhte.

2 Nim galluc niþeweardne. Gewyl in þeorfum meolcum; supe gemet hat.

3 Nim swegles æppel, do in meolc, sele drincan.

4 Nim nioþewerde liðwyrt grene. Wel in beore oþþe in þeorfum meolcum; supe þæt ond drince swiþe.

# London, British Library
# MS Cotton Otho A.xiii

29. ✚ In the name of the Father and the Son and the Holy Spirit, amen. Saint Nicasius had a tiny spot in his eye, and he prayed to the Lord that whoever had this evil and should carry his name written upon him would be freed from this sickness.

✚ Saint Nicasius, pray for this man Roger. ✚ Saint 2 Nicasius, pray for this man Roger. ✚ Saint Nicasius, pray for this man Roger. Christ conquers, Christ reigns, Christ commands. May Christ, through the intercession of blessed Nicasius, free this man Roger from this sickness and from every other.

Tetragrammaton. ✚ He who has believed and was bap- 3 tized will be saved. ✚

# London, British Library,
# MS Cotton Otho B.xi and MS Add. 43703

30. For diarrhea.

Take the lower part of comfrey. Boil in fresh milk; take it 2 moderately hot.

Take swail's apple, put it into milk, give to drink. 3

Take the lower part of fresh limbwort. Boil in beer or in 4 fresh milk; take it and drink it quickly.

31. Wið utsyhte.
Nim gesodene haran. Ete wiþ þon.

32. Wið utsyhte.
Nim rudan, smeorowyrte. Wel in ealoð; sele drincan.

33. Gyf utgong forseten sy.
2     Nim henne-ægero—þone geolocan—ond lybcorna þryt-tig, ond hwites sealtes godne dæl, ond wad-meolo.
3     . . . Meng eal tosomne. Do in aeges scylle; sete on gleda.
Do ealoð ænne sopan to. Supe swa hat.

34. Gyf utgong forseten sie.
Genim grundeswelgian ond ellen-rinde; berynde utan.
Wel þonne swiþe in ealoð. Asioh þurh cla∂. Nim þonne siex
ond þritig lybcorna, gnid smæle, do wætes hwon in. Geot
innon þone drenc. Do butran on. Dringe swa hat swa he hat-
tost maege, ond þonne stande ond gange þær hine ne cæle.

# London, British Library,
# MS Cotton Titus D.xxvi

35. Wið þa blegene.
Genim nigon ægra ond seoð hig fæste, ond nim þa geol-
can ond do þæt hwite aweg. Ond smera ða geolcan on anre
pannan, ond wring þæt wos ut þurh ænne cla∂. Ond nim eall
swa fela dropena wines swa ðæra ægra beo, ond eall swa fela
dropena unhalgodes eles, ond eall swa fela huniges dropena,

31. For diarrhea.

Take a boiled hare. Eat it for that condition.

32. For diarrhea.

Take rue and birthwort. Boil in ale; give to drink.

33. In case of constipation.

Take hen eggs—the yolk—and thirty caper-spurge seeds, 2
and a good deal of white salt, and meal made of woad.

... Mix it all together. Add in some eggshells and set it on 3
some coals. Add a draft of ale. Sup it while hot.

34. In case of constipation.

Take groundsel and elder bark; peel off the outer part.
Boil vigorously in ale. Strain through a cloth. Then take
thirty-six caper-spurge seeds, grate them finely, and add a
little liquid. Pour into the drink. Add butter. Have the patient drink it, as hot as he can stand, and then have him
stand and go where he will not be chilled.

## London, British Library, MS Cotton Titus D.xxvi

35. For boils.

Take nine eggs and boil them hard, and take the yolks and
throw the whites away. And smear the yolks in a pan, and
wring the juice out through a cloth. And take just as many
drops of wine as there are eggs, and just as many drops of
unhallowed oil, and just as many drops of honey, and just as

ond of finoles more eall swa fela dropena. Genim þonne ond
gedo hit eall tosomne, ond wring ut þurh ænne cla𝔡, ond syle
þam menn etan. Him by𝔡 sona sel.

# London, British Library,
# MS Cotton Vespasian D.xx

36. Ad dentium dolorem.

2    Petrus sedebat super petram, et manus suas tenebat ad
maxillas suas. Et dixit Ihesus: "Petre, quid tristis sedes?"
"Domine, vermes . . . in me. Fac mihi benedictionem quam
fecisti Lazaro, quem resuscitasti de mortuis."

3    "In nomine Patris et Filii et Spiritus Sancti, amen. Adiuro
te, migranea, per Patrem et Filium et Spiritum Sanctum, ut
tu amplius non possis stare, nec in faucibus, nec in dentibus,
nec in capite tuo. In nomine Patris et Filii et Spiritus Sancti,
amen."

4    Accipe saxafriga—id est, grumin—et petrosino et ambrosiana et apia et tanesia, et mitte simul. Et de quo eas cum
vino in olla nova ita ut tribus vicibus suffundes eas de vino ut
ad medietatem perveniat, et postea da infirmum bibere.

many drops of the edible root of fennel. Take these then and mix them all together, and wring out through a cloth, and give to the person to eat. He will straightway be better.

# London, British Library,
# MS Cotton Vespasian D.xx

36. For toothache.

Peter was sitting on a stone, and he was holding his hands 2
to his jaws. And Jesus said, "Peter, why are you sitting so sad?" "Lord, worms . . . in me. Give me the blessing that you gave to Lazarus, whom you brought back from the dead."

"In the name of the Father and of the Son and of the Holy 3
Spirit, amen. I adjure you, migraine, by the Father and the Son and the Holy Spirit, that you no longer be able to remain, whether in the jaws or in the teeth or in your head. In the name of the Father and the Son and the Holy Spirit, amen."

Take saxifrage—that is, gromwell—and rock parsley and 4
ambrosiana and common parsley and tansy, and mix them together. And after this put them into a new pan with wine such that, suffusing them with wine three times, you arrive at a good mixture, and then give to the sick person to drink.

# London, British Library,
# MS Cotton Vitellius C.iii

37. Ad vertiginem.

2    Nim betonica and wæll swyðe on win oþþa on ald ealað, and wæsc þæt heafod mid þam wose. And leg siððen þæt wyrt swa wærm abutan þæt heafod, and wrið mid claðe, and læt swa beon ealla niht.

3    Eft, wið þæt ilce. Nim savina ond betonica and wermod ond merc, and seoð on win oððe on oðer wæt swyðe. And nim cal-stocces ond bærn to ascen, and nim þonne þæt wos of þa wyrtas ond ofergeot þa ascen mide, and mac swa to lega, and wæsc þa heafod þærmide. And nim siððon þa wyr-tas wærma, alla wiðutan savina, and bind to þam heafde alla niht.

38. Ad pectoris dolorem.

2    Nim hors-ellenes rota and eft gewæxen barc, and dry swyðe and mac to duste, and drif þurh claþ. And nim hunig and seoð swyðe. Nim siððen þæt dust and mencg þærto and styre swyðe togædera, and do on box, and nota þenna neod sig.

3    Eft, wið þæt ilce. Nim þa read-stalede harhuna and ysopo, and stemp, ond do on ænne neowna pott, an flering of ða harhuna and oðer of ysopo and ðridde of fersc buter, and eft þa wyrt and swa þa butra, forð þæt se pott beo full. And seoð hig swyðe togædra, and wring siððen þurh claþ, and nota

# London, British Library, MS Cotton Vitellius C.iii

**37. For giddiness.**

Take betony and boil thoroughly in wine or in old ale, and 2 wash the head with the infusion. And then apply that plant, still warm, to the head, and wrap around with a cloth, and leave it in place all night.

Again, for the same. Take savine and betony and worm- 3 wood and wild celery, and simmer thoroughly in wine or in another liquid. And take some cabbage stalks and burn them to ashes, and then take the juice of those plants and pour this over the ashes, and make this into a bed, and wash the person's head with it. And then take the plants, while still warm, all except savine, and bind them to the head all night.

**38. For pain in the chest.**

Take elecampane roots and the rind that has grown back 2 after having been cut, and dry thoroughly and make into a powder, and force through a cloth. And take some honey and cook thoroughly. Then take the powder and mix it in and stir them vigorously together, and store this in a container, and make use of it when needed.

Again, for the same. Take red-stalked horehound and hys- 3 sop, and pound them, and put them into a new pot, one layer of horehound and a second of hyssop and a third of fresh butter, and again layer the plants and then the butter, and so forth until the pot is full. And cook them thoroughly together, and then strain them through a cloth, and make

þonna þærf sig, fæstende cald and on niht on hat ala oððe broð oððe wæter.

39. Þis is seo seleste eah-salf wið eh-wærce, and wið miste, and wið wenne, and wið wyrmum, and wið gihðum, and wið teorendum eagum, and ælcum cuðum swile.

2   Genim feferfugean blostman and diles blostman and ðunorclafran blostman and hamorwyrte blostman, and twegra cynna wermod, and pollegian and neoðewarde lilian, and hawenhydele and lufestice and dolhrunan, and gepuna ða wyrte tosomne, and wæl tosomne in heortes mærige oððe on his smeoruwe, and meng ele to. Do þonne teala mycel in ða eagan, and smyra utewarde, and wyrm to fyre. And ðeos salf helpð wið æghwylcum geswelle, to þicganne and to smyrianne, in swa hwylcum lime swa hit on bið.

40. Þis mæg to eah-salfe. Genim geoluwne-stan and saltstan and pipor and weh on wæge, and drif þurh clað, and do ealra gelice micel, and do eal togædere, and drif eft þurh linene clað. Þis is afandan læcecræft.

41. Wið lungen-adle.
  Genim hwite harehunan and ysopo and rudan and galluc and brysewyrt and brunwyrt and wudemerce and grundeswylian, of ælcere þisre wyrte twentig penega wiht. And genim ænne sester fulne ealdes ealoð, and seoð þa wyrtan oððet se sester ealoð sy healf gesoden. And drinc ælce dæg fæstende neap fulne caldes, and on æfen wearmes, lætst. Hit is haluwende bote.

use of them when there is need, cold on an empty stomach and at night in hot ale or broth or water.

39. This is the best eye salve for eye pain, and for clouded vision, and for a stye, and for mites, and for itching, and for teary eyes, and for every kind of known inflammation. Take common centaury blossoms and dill blossoms and bugle blossoms and hammerwort blossoms, and two kinds of wormwood, and pennyroyal and the lower part of lily, and scurvy grass and lovage and pellitory-of-the-wall, and pound the plants together, and boil them together in deer marrow or deer grease, and stir in some oil. Then apply a generous portion to the eyes and daub it outside the eyes, and warm it at the fire. And this salve will relieve every kind of swelling, whether one consumes or applies it, no matter what part of the body is affected.

40. This is effective as an eye salve. Take yellow stone and rock salt and pepper and weigh them on a set of scales, and strain them through a cloth, using an equal amount of each, and mix them all together, and again strain them through a linen cloth. This is a proven remedy.

41. For lung disease.
Take white horehound and hyssop and rue and comfrey and daisy and figwort and wild celery and groundsel, twenty pennyweights of each of these plants. And take a pitcher full of old ale, and cook the plants until the pitcher of ale is reduced by half. And every day drink a cupful, cold, on an empty stomach, and drink a cupful, warm, the last thing in the evening. It is a wholesome remedy.

42. Wið fot-adle and wið þone dropan.

Nim *datulus* þa wyrt, oðer name *titulosa:* þæt is on ure geþeoda, þæt greata crauleac. Nim þes leaces heafda and dryg swiðe, and nim ðerof þriddan healves penincges gewihte, and peretreo and romanisce rinda and cymen and feorðan del lauwerberian, and þera oðera wyrta ælces healves penincges gewihta, and siex pipercorn unwegen. And grind ealle to duste, and do win, twa ægscille fulle. Þis is soð læcæcræft. Syle þan men drincan oþðæt he hal sy.

43. Ad corruptionem corporis.

Polleio, aneto, centauria minore, ruta, salvia, grana pionie; de his equaliter. Sume et tribula cum vino aut veteri cervisa, et da bibere ieiuno.

44. Ad vocem validificandam.

Peretro, cinamomo, sinapis semine, cumino asso, pipero. De his equaliter tere et confice cum melle despumato, et uteris cum opus habueris.

45. Ad fluxum sanguinis.

Accipe de confirma—hoc est, consolda—et fac idem ius, et da bibere femine patienti fluxum sanguinis, et sanabitur.

46. Ad recipiendam menstruam.

Warantie ius cum vino. Da ei bibere, aut de foliis fraxini.

47. Aliter.

Accipe satureia et bulli cum lacte, et da ei bibere.

42. For foot disease and for gout.

Take the plant named *datulus,* or by another name *titulosa:* in our language, that is "greater crow garlic." Take the head of this plant and dry it thoroughly, and take one-and-a-half pennyweights of it, and take common centaury and cinnamon bark and cumin and a fourth part of laurel berries, and of each of the other plants half a pennyweight, and six unweighed peppercorns. And grind all this to a powder, and add two eggshells full of wine. This is a genuine remedy. Give this to the person to drink until he is well.

43. For putrefaction of the body.

Pennyroyal, dill, lesser centaury, rue, sage, some peony seeds; an equal amount of each. Take and mix with wine or aged beer, and give to drink on an empty stomach.

44. For strengthening the voice.

Common centaury, cinnamon, mustard seed, dried cumin, pepper. Grate an equal amount of each and mix with purified honey, and use when you have need.

45. For a flow of blood.

Take some comfrey—that is, *consolda*—and make it into a broth, and give this to a woman to drink who suffers from a flow of blood, and she will be well.

46. To bring back menstruation.

Prepare a broth with wine. Give to her to drink, or else leaves from an ash tree.

47. An alternative cure.

Take wild basil and boil in milk and give to her to drink.

# London, British Library,
# MS Cotton Vitellius E.xviii

48. Gyf hryþeru beon on lungen coðon.

. . . ton hylle, ond bærn to axan on middansumeres mæssedæg. Do þærto hali wæter, ond geot on heora muð on middansumeres mæsse mergen. Ond sing þas þry sealmas þærofer: "Miserere nostri" ond "Exsurgat Dominus" ond "Quicumque vult."

49. Gif sceap si on ylon.

Genim lytel niwes ealoð, ond geot innon ælc þæra sceapa muð, ond do þæt hi hraðor swelgon þæt heom cymð to bote.

50. Þis is þinan yrfe to bote.

Sing ymb þin yrfe ælce æfen him to helpe: "Agios, agios, agios."

51. Genim twegen lange sticcan feðer-ecgede, ond writ on ægðerne sticcan on ælcere ecge an "Pater noster" oð ende. Ond lege þone forman þam berene on þa flore, ond þone oðerne on . . . ofer þam oðrum sticcan.

2      . . . þæt þær si rode tacen on. Ond nim of ðam gehalgedan hlafe þe man halgie on hlafmæssedæg feower snæda, ond gecryme on þa feower hyrnan þæs berenes.

# London, British Library, MS Cotton Vitellius E.xviii

48. If cattle are suffering from diseased lungs.

. . . and burn to ashes on Midsummer's Eve. To this add holy water, and pour it into their mouths the morning of Midsummer's Day. And sing these three psalms over them: "Have mercy on us" and "Let the Lord arise" and "Whosoever wishes."

49. If a sheep is diseased.

Take a little freshly brewed ale, and pour it into the mouth of each of the sheep, and have them swallow it quickly so that it will do them good.

50. This is a cure for your cattle.

Every evening sing this around your cattle so as to help them: "Holy, holy, holy."

51. Take two long, four-edged sticks, and write an "Our Father," to the end, on each edge of each stick. And lay the first one on the floor of the barn, and the other one . . . on top of the other stick.

. . . that the sign of the cross is on it there. And take four pieces from the consecrated bread that is hallowed on Lammas Day, and crumble them into the four corners of the barn.

# London, British Library,
# MS Harley 6258 B

52. Wið heafod-ece.

*Pollege*—þæt is, on englis, "dwyregedwosle." Wulle on ele oððer on clane butere, ond smyre þæt heafod mid.

53. Wið ealda ond singalum heafod-ece.

Cnuca þa wyrt þat "bete" hatað, ond gnid on þa þun-wunge ond ufan þæt heafod. Þu wundrast þas lacedomes.

54. Eft, wiþ þat ylce.

Cnuca cyleþene on ecede, ond smire mid þæt heafod bu-fan þa eagen. Sona byð hym sæl.

55. Wið slapende lice.

Wyrce bæð. Nim þæt mycele fearn niðeweard, ond ellan rinde. Cnuca tosomne ond mededrosna do þarto, ond beþweh hine wel wearme.

56. Gif syna scrincon.

2    Nim mucgwyrte, gebeatene ond wid ele gemenged. Gelo-gode, smyre mid.

3    Mucgwyrte seaw. Seoþ on ele, smira mid.

57. Wiþ heafod-ece.

Genim bettonican ond pipor. Gignid togadere. Læt ane niht hangie on claðe, ond smira mid þat heafod.

# London, British Library,
# MS Harley 6258 B

52. For headache.
*Pollege*—that is, in English, "pennyroyal." Boil in oil or in pure butter, and apply to the head.

53. For chronic and persistent headache.
Pound the plant that is called "beets," and rub onto the temples and the top of the head. You will be amazed at this cure.

54. Again, for the same.
Pound greater celandine in vinegar, and smear it on the head above the eyes. He will soon be well.

55. For a paralyzed body.
Draw a bath. Take the lower part of the great fern, along with elder-tree bark. Pound together and add the dregs of mead, and give him a good warm wash.

56. If the tendons in the neck have contracted.
Take mugwort, mashed and blended with oil. Once it has settled, smear it on. 2

The juice of mugwort. Simmer in oil, rub on. 3

57. For headache.
Take betony and pepper. Pound them together. Leave this hanging in a cloth overnight, and smear it on the head.

58. Wið sceancena sarnyssa ond fot-ece.

Bettonica ond geormaleaf ond finul ond ribban, ealra efenfela, ond gemeng wyd mylc ond wið wæter, ond beþa mid.

59. Ad tumorem nervorum.

Plantaginis folia contunde cum modico sale, et bibe ieiunus. Bete nigre succus, et radicis minus dimidio, melle admixto si naribus infundatur ita ut palatum transeat. Pituitas omnes defluunt, et naribus et dentibus dolentibus prodest.

60. Item.

Ysopi, satureie sicce, organi, fasciculos singulos, in sapone optime per triduum macerabis. Hoc per singulos menses. Non solum capite sanus, sed et pectore et stomacho eris.

61. Cui capud cum dolore findi videtur.

Succum edere cum oleo. Miscetur et accetum et unget nares, et statim sedabitur.

# London, British Library,
## MS Royal 2.A.xx

62. Rivos cruoris torridi
   contacta vestis obstruit.
   Fletu riganti supplicis,
   arent fluenta sanguinis.

58. For pain in the shins and for foot ache.

Betony and common mallow leaves and fennel and hound's-tongue, an equal amount of each, and blend with milk and with water, and bathe the person with it.

59. For swelling of the tendons.

Crush plantain leaves with a small amount of salt, and drink it while fasting. Juice of black beets, and less than half of a radish, mixed with honey and poured into the nostrils so that it passes to the palate. All the phlegm runs out, and this benefits the sore nostrils and teeth.

60. Another cure.

Soften individual sprigs of hyssop, dried basil, and marjoram in the best soap for three days. Do this every month. You will be well, not only in the head, but also in the chest and stomach.

61. For a person with a splitting headache.

Partake of juice with oil. Mix also with vinegar and apply to the nostrils, and straightway the pain will subside.

# London, British Library, MS Royal 2.A.xx

62. The garment, once touched,
     halts the streams of gushing blood.
     By the flowing tears of the suppliant
     the floods of blood dry up.

2 Per illorum quae siccata, dominica labante, coniuro sta. Per Dominum nostrum.

63. Ociani inter ea motus siderum motus vertat. Restringe trea flumina: flumen aridum vervens, flumen pallidum parens, flumen rubrum acriter. De corpore exiens restringe tria flumina: flumen cruorem restrigentem nervos, limitem cicatricis, concupiscente tumores fugante. Per Dominum nostrum Ihesum Christum.

64. Obsecro te, Ihesus Christus, Filius Dei vivi, per crucem tuam, ut demittas delicta mea pro beata cruce.

2 Custodi caput meum pro benedicta cruce. Custodi oculos meos pro veneranda cruce. Custodi manus meas pro sancta cruce. Custodi viscera mea pro gloriosa cruce. Custodi genua mea pro honorabili cruce. Custodi pedes meos et omnia membra mea ab omnibus insidiis inimici, pro dedicata cruce, in corpore Christi. Custodi animam meam, et libera me in novissimo die ab omnibus adversariis, pro clavibus sanctis quae in corpore Christi dedicata erant.

3 Tribue mihi vitam aeternam et misericordiam tuam, Ihesus Christus, et visitatio tua sancta custodiat spiritum meum.

65. *Eulogumen . patera . caeyo . caeagion . pneuma . caenym . caeia. caeiseonas . nenonamini.*

2 Adiuro te, satanae diabulus aelfae, per Deum vivum ac verum et per trementem diem iudicii, ut refugiatur ab homine illo qui habeat hunc a Cristo scriptum secum. In nomine Dei Patris et Filii et Spiritus Sancti.

By those in whom it has dried up, coming to a halt on the 2 Lord's Day, I entreat you to stand fast. By our Lord.

63. May the movement of the stars among themselves direct the movements of ocean. Bind fast three rivers: the dry river reverting, the pale river submitting, the red river surging. Bind fast three rivers exiting from the body: the river of blood binding the sinews, the opening of a wound, the inflammations that are put to flight at your will. By our Lord Jesus Christ.

64. I beseech you, Jesus Christ, Son of the living God, by your cross, that you dismiss my faults, by the holy cross.

Guard my head by the blessed cross. Guard my eyes by 2 the venerable cross. Guard my hands by the holy cross. Guard my inner organs by the glorious cross. Guard my knees by the cross that is to be honored. Guard my feet and all my limbs from all ambushes by the enemy, by the consecrated cross, in the body of Christ. Guard my soul, and free me from all adversaries in the present day, by the holy nails that were consecrated in the body of Christ.

Grant me eternal life and your mercy, Jesus Christ, and 3 may your holy visitation guard my spirit.

65. *Eulogumen . patera . caeyo . caeagion . pneuma . caenym . caeia . caeiseonas . nenonamini.*

I command you, devil of Satan's elf, by the living and true 2 God and by the trembling Day of Judgment, that you take flight from the person who has with him this inscription written by Christ. In the name of God the Father and of the Son and of the Holy Spirit.

66. Crux Christi Ihesu, Domini Dei nostri, ingeritur mihi.

2   ✚ Rivos cruoris torridi
contacta vestis obstruit.
Fletu rigantis supplicis
arent fluenta sanguinis.

3 Per illorum venas cui siccato, dominico labante, coniuro sta. Per Dominum nostrum Ihesum Christum Filium tuum, qui tecum vivit et regnat, Deus in unitate Spiritus Sancti, per omnia secula seculorum.

67. ✚ In nomine sanctae Trinitatis atque omnium sanctorum, ad sanguinem restringendum scribis hoc:

2   ϹΟΜΑΡΤΑ ΟϹΟΓΜΑ ϹΤΥ ΓΟΝΤΟΕΜΑ
ΕΚΤΥΤΟΠΟ.

3 ✚ Beronice, libera me de sanguinibus. Deus, Deus salutis meae.

4   ϹΑϹΙΝΑϹΟ ΥϹΑΡΤΕΤΕ.

5 Per Dominum Ihesum Christum. Christe adiuva ✚ Christe adiuva ✚ Christe adiuva.

6   ✚ Rivos cruoris torridi
contacta vestis obstruit.
Fletu rigante supplicis
arent fluenta sanguinis.

7 Beronice, libera me de sanguinibus. Deus, Deus salutis meae.

8   ΑΜΙϹΟ ϹΑΡΔΙΝΟΠΟ ΦΙΦΙΡΟΝ ΙΔΡΑϹΑϹΙΜΟ.

66. The cross of Jesus Christ, our Lord God, is borne to me.

✛ The garment, once touched,　　　　　　　　　　　2
halts the streams of gushing blood.
By the flowing tears of the suppliant
the floods of blood dry up.

By the veins of those in whom it dries up, ceasing on the　3
Lord's Day, I adjure you to stay in place. By our Lord Jesus
Christ your Son, who lives and rules with you, God in unity
of the Holy Spirit, forever and ever.

67. ✛ In the name of the holy Trinity and of all saints, write
the following so as to staunch the flow of blood:

ϹΟΜΑΡΤΑ ΟϹΟΓΜΑ ϹΤΥ ΓΟΝΤΟΕΜΑ　　　　2
ΕΚΤΥΤΟΠΟ.

✛ Veronica, free me from hemorrhages. God, God of my　3
salvation.

ϹΑϹΙΝΑϹΟ ΥϹΑΡΤΕΤΕ.　　　　　　　　　　4

By the Lord Jesus Christ. ✛ Christ be my help ✛ Christ　5
be my help ✛ Christ be my help.

✛ The garment, once touched,　　　　　　　　　　　6
halts the streams of gushing blood.
By the flowing tears of the suppliant,
the floods of blood dry up.

Veronica, free me from hemorrhages. God, God of my salva-　7
tion.

ΑΜΙϹΟ ϹΑΡΔΙΝΟΡΟ ΦΙΦΙΡΟΝ ΙΔΡΑϹΑϹΙΜΟ.　　8

9 Fodens magnifice contextu fundavit tumulum—*usugma*. Domine adiuva.

68. In Epheso civitate, in monte Celion, requiescunt sancti septem dormentes, quorum ista sunt nomina: Maximianus, Malchus, Martinianus, Dionisius, Iohannes, Serapion, Constantinus. Per eorum merita et piam intercessionem dignetur Dominus liberare famulum suum NOMEN de omni malo. Amen.

69. Domine Ihesu Christe, qui somno deditus in mare a discipulis tuis excitari voluisti, per intercessionem sanctorum septem dormientium, quorum corpora in monte Celion requiescunt: fac dormire hunc famulum tuum NOMEN, ut convalescat a somno quem amisit et tibi et sanctae genitrici tuae Mariae, sanctisque martyribus tuis et omnibus sanctis tuis grates referat. Qui vivis et regnas.

70. Domine Ihesu Christe, vere Deus noster, per intercessionem servi tui Blasii, succurre in adiutorium servi tui NOMEN. "Pater noster" þreowa.

71. In principio erat verbum, et verbum erat apud Deum et Deus erat verbum. Hoc erat in principio apud Deum. Omnia per ipsum facta sunt, et sine ipso factum est nichil. Dominus, propitius esto mihi peccatori famulo tuo (peccatrici famulae tuae) NOMEN, et de eius (vel meis) plaga (vel corpore) amplius gutta sanguinis non exeat. Sic placeat Filio Dei sanctaeque eius genitrici Mariae.

Digging out, he made a tomb with magnificent construc-  9
tion — *usugma.* The Lord be my help.

68. In the city of Ephesus, at Mount Celion, repose the
seven holy Sleepers, whose names are Maximianus, Malchus, Martinianus, Dionysius, John, Serapion, Constantinus. Through their merits and their pious intercession may
the Lord deign to free his servant NAME from every evil.
Amen.

69. Lord Jesus Christ, you who wished to be roused by your
disciples when you were given over to sleep on the sea,
through the intercession of the seven holy Sleepers, whose
bodies rest in Mount Celion: grant that this servant of yours
NAME sleeps, so that he may make up for the sleep that he
has lost and may give thanks to you and to your holy mother
Mary, and to your holy martyrs and to all your saints. By him
who lives and reigns.

70. Lord Jesus Christ, our true God, through the intercession of your servant Saint Blasius, hasten to the aid of your
servant NAME. Say the "Our Father" three times.

71. In the beginning was the Word, and the Word was with
God, and the Word was God. The same was in the beginning
with God. All things were made by him, and without him
nothing was made. God, be propitious to me, a sinner, your
servant (or maidservant) NAME, and from his wound (or
body, or from mine) let not a drop more of blood flow. Thus
may it be pleasing to the Son of God and to Saint Mary, his
mother.

2    In nomine ✚ Patris, cessa sanguis! In nomine ✚ Filii, resta sanguis! In nomine ✚ Spiritus Sancti, fugiat omnis dolor et effusio a famulo (famula) Dei Nomen. Amen. In nomine ✚ sancte Trinitatis. "Pater noster."

3    Hoc dic novies.

72. Sanctus Cassius minutam habuit, Dominumque deprecatus est ut quicunque nomen suum portaret secum hoc malum non haberet.

2    Dic "Pater noster" tribus vicibus.

# London, British Library, MS Royal 12.E.xx

73. Ad dentium dolorem.

In nomine Domini nostri Ihesu Cristi, Patris et Filii et Spiritus Sancti. *Cheilei cecce becce upservicce.* Sla mone wuerm. *Naco dicapron sed noli et coli.* Þanne hyt an yerþe hates byrnet, finit. Nostri famulum Dei Nomen. "Pater noster" novem.

74. ✚ In nomine Patris et Filii et Spiritus Sancti. "Quicunque vult."

2    Contra bonum malannum. Sancta crux, amen. Crux sancta, amen. *Obsorbis et opto cuni sorbis subalca edifa,* amen. Crux sancta, amen. Sancta crux, amen. ✚ "Pater noster."

In the name of the ✚ Father, cease, blood! In the name 2
of the ✚ Son, stop, blood! In the name of the ✚ Holy
Spirit, may all pain and shedding of blood take flight from
the servant (or maidservant) of God Name. Amen. In the
name of the ✚ Holy Trinity. "Our Father."
Say this nine times. 3

72. Saint Cassius had a speck, and he prayed to the Lord that
whoever carried his name with him would not experience
this evil.
Say the "Our Father" three times. 2

# London, British Library,
# MS Royal 12.E.xx

73. For toothache.
In the name of our Lord Jesus Christ, of the Father and
the Son and the Holy Spirit. *Cheilei cecce becce upservicce.*
Strike the *wyrm* of men. *Naco dicapron sed noli et coli.* When
it burns hottest on earth, it is ended. The servant of God
Name. "Our Father" nine times.

74. ✚ In the name of the Father and the Son and the Holy
Spirit. "Whosoever wishes."
For a good headache. The holy cross, amen. The holy 2
cross, amen. *Obsorbis et opto cuni sorbis subalca edifa,* amen.
The holy cross, amen. The holy cross, amen. ✚ "Our Fa-
ther."

3    Adiuro te per Dominum et Dominum nostrum, per Patrem et Filium et Spiritum Sanctum, gutta migranea, ut non habeas potestatem in capite isto stare, nec in dentibus morari, nec in manibus nec in pedibus nec in ullo loco dominari.

4    Libera famulum tuum NOMEN, Deus. ✚ AGYOS ✚ AGYOS ✚ AGYOS. Sanctus Deus, sanctus, fortis, sanctus et immortalis. *Criste eleyson.* "Pater noster."

75. Contra febrem.

Accipe septem oblationes. In singulis scribe singula nomina septem dormientium, et super illa nomina ista alia super singulas, et da infirmo comedere. Maximianus, Malcus, Martinianus, Dionisius, Constantinus, Serapion, Iohannes.

# London, Wellcome Historical Medical Library, MS 46

76. Wið heort-æce.

Genim brade bisceopwyrt ond feldbisceopwyrt ond greate wyrt ond galluc and gagul ond hindehæleþan ond organan ond æþelferþincwyrt ond harehunan ond salvian ond hofan ond garclivan ond fifleavan ond hamerwyrt ond feldwyrt ond mucgwyrt ond suðernewudu, ond cnuca ealle tosomne, ond do ealu. Drinc þonne ðe þearf si.

I adjure you—malignant drop, migraine—by the Lord 3 and by our Lord, by the Father and the Son and the Holy Spirit, that you have no power to stay in the head of this man, nor to linger in his teeth, nor to preside in his hands or feet or in any other place.

Free your servant NAME, O God. ✚ HOLY ✚ HOLY ✚ 4 HOLY. Holy God, holy, mighty, holy and immortal. Christ have mercy. "Our Father."

75. For fever.

Take seven Communion wafers. On each one write the individual names of the Seven Sleepers, and over them write these other individual names, and give them to the sick person to eat. Maximianus, Malcus, Martinianus, Dionysius, Constantinus, Serapion, John.

## London, Wellcome Historical Medical Library, MS 46

76. For chest pain.

Take broad-leaved mallow and field mallow and autumn crocus and comfrey and sweet gale and hindheal and marjoram and stitchwort and horehound and sage and ground ivy and agrimony and cinquefoil and hammerwort and gentian and mugwort and southernwood, and pound them all together, and add ale. Drink when you need to.

77. Wiþ lungen-adle.

Hennebelle moran ond harehunan; betonican. Wylle on ealoþ ond drince þa hwile þe him þearf si. Supe syþþan ane hennescille-fulle gemyltere buteran. Wreo hine syþþan wearme, ond beo him syþþan stille.

78. Hat wyrcean þe sylf wenn-sealfe.

Man sceal niman clæne hunig, swylc man to blacan briwe deþ, ond wyllan hit neah briwes þicnesse. Ond niman rædic ond elenan ond cyrfillan ond hrefnes fot. Cnocian swa man betst mæge, ond wringan þonne þa wyrta, ond geotan þæt wos þærto. Ond þonne hit beo forneah gewylled, cnucian godne dæl garleaces ond don þærto, ond piperian, swa-swa þe þince.

79. ✚ Wiþ wennas sealf.

2    Hwerhwettan moran ond ane handfulle sperewyrte ond wildne næp ond wuduwexan moran. Wylle on mealt-ealoþ. Wringe þurh linenne claþ. Wylle on hunigteare. Nime þonne clænne lengtenbere ond grinde on hand-cwyrna. Nime seþ-þan mæderan ond drige on hand-cwyrna, ond grinde reades caules sædes ane handfulle on pipor-cwyrna. Wylle hit eal togædere, na to hearde. Þyge on wucan þriwa, swa him betsþ to onhagige.

3    Þeos sealf deah wiþ wennas ond wiþ þone flowendan fic. Þeah heo styriende sy, ne onscunige he hi for þam.

80. Wiþ lifer-adle.

Nim liferwyrt ond bere hi man ham onder cneowe, ond wylle on anes hiwes cu-meolce. Mengge buteran to.

77. For lung disease.

Take the edible root of henbane and horehound; betony. Boil in ale and have him drink it for as long as he has need. Afterward have him eat an eggshell-full of melted butter. Afterward wrap him up warmly, and have him keep still.

78. Have a wen salve prepared for yourself.

One should take pure honey, such as is used for a light soup, and simmer it until it is nearly the consistency of soup. And take radish and elecampane and chervil and crowfoot. Pound it as well as you can, and then strain out the plants, and pour the juice into the salve. And when the salve is almost at the boiling point, pound a good deal of garlic and add that in, and pepper it, just as you like.

79. ✚ For a salve for wens.

The edible root of cucumber, and a handful of spearwort, 2 and wild rape, and the edible root of dyer's greenwood. Simmer in malt ale. Strain through a linen cloth. Simmer in honey fresh from the comb. Then take clean lenten-berries and grind them on a hand quern. After this take madder and dry it on a hand quern, and grind a handful of red cabbage seeds in a pepper grinder. Simmer it all together, not too hard. Make use of this salve three times a week, as suits him most conveniently.

This salve is good for wens and for hemorrhoids. Even 3 though it is stringent, one should not avoid it on that account.

80. For liver disease.

Pick some liverwort and have it brought home under the knee, and simmer it in the milk of a cow of a single color. Mix in some butter.

## Oxford, Bodleian Library, MS Auctarium F.3.6

81. ... ond *thebal guttatim aurum et thus de:*
2 ✚ *abra Ihesus* ✚ *alabra Ihesus* ✚ *galabra Ihesus* ✚
3 Wið þone dworh. On þreo oflætan writ: Thebal guttatim.

82. Gif men ierne blod of nebbe to swiðe, sume þis writað:
2 ✚ *Ær grin thonn struht fola · aer grenn tart strut onntria enn piathu morfona onnhel · ara carn leow gruth ueron ·* ℋℋ *· fil cron diw ·* ℳ *· inro cron aer crio ær mio aær leno.*
3 Ge horsse ge men blod seten.

## Oxford, Bodleian Library, MS Barlow 35

83. Wið blodryne. Writ ðis:
2 In nomine, breve, pro signo Domini. "Dum maritum cartum manet obrido arigetus deprete obrigate. Cristus adiuvat; Dominus dissolvat." Amen.
3 In nomine, in protectione bova femina bona. "Restet sanguis, anima mala!" Amen.

# Oxford, Bodleian Library, MS Auctarium F.3.6

81. . . . and *thebal guttatim aurum et thus de:*
✠ *abra Ihesus* ✠ *alabra Ihesus* ✠ *galabra Ihesus* ✠    2
For high fever. Write on three Communion wafers: THE-    3
BAL GUTTATIM.

82. If someone is suffering from severe nosebleed, some people write this:
✠ *Ær grin thonn struht fola · aer grenn tart strut onntria enn*    2
*piathu morfona onnhel · ara carn leow gruth ueron · ℍ · fil cron*
*diw · ⋈ · inro cron aer crio ær mio aær leno.*
To staunch the flow of blood in either horse or man.    3

# Oxford, Bodleian Library, MS Barlow 35

83. For the flow of blood. Write this:
In the name of the Lord, briefly, for the sign of the Lord.    2
"While . . . parchment remains . . . lose its strength, harden!
Christ helps; may the Lord put an end to it." Amen.
In the name of the Lord, for the protection of a good cow.    3
"May the blood stand still, O evil spirit!" Amen.

# Oxford, Bodleian Library,
# MSS Junius 85 and 86

84. Wið wif bearneacenu.

2    "Maria virgo peperit Christum; Elisabet sterelis peperit Iohannem baptistam. Adiuro te infans, si es masculus an femina, per Patrem et Filium et Spiritum Sanctum ut exeas et recedas, et ultra ei non noceas, neque insipientiam illi facias. Amen.

3    "Videns Dominus flentes sorores Lazari ad monumentum, lacrimatus est coram Iudeis et clamabat: 'Lazare, veni foras!' Et prodiit, ligatis manibus et pedibus, qui fuerat quatriduanus mortuus."

4    Writ ðis on wexe ðe næfre ne com to nanen wyrce, and bind under hire swiðran fot.

85. Wið gestice.

2    Writ "Cristes mæl," and sing ðrywe ðæran ðis and "Pater noster":

3    "Longinus miles lancea ponxit Dominum, et restitit sanguis et recessit dolor."

86. Wið uncuðum swyle.

Sing on ðine læce-finger þreo "Pater noster," and writ ymb þæt sare and cweð: "Fuge diabolus, Christus te sequitur. Quando natus est Christus, fugit dolor." And eft þreo "Pater noster" and þreo "Fuge diabolus."

87. Wið toð-ece.

Sanctus Petrus supra marmoream. . . .

# Oxford, Bodleian Library, MSS Junius 85 and 86

84. For a woman who is with child.

"The Virgin Mary gave birth to Christ; Elizabeth, who 2 was sterile, gave birth to John the Baptist. I adjure you, infant, whether you are male or female, by the Father and the Son and the Holy Spirit, that you come out and leave her body, and do not harm her anymore, nor do any foolish thing to her. Amen.

"When the Lord saw the sisters of Lazarus weeping at 3 the tomb, he burst into tears in the presence of the Jews and called out: 'Lazarus, come forth!' And he came forward, his hands and feet bound, he who had been dead for four days."

Write this on wax that has never previously been used, 4 and tie it under her right foot.

85. For a stabbing pain.

Write "Christ's cross," and sing this and "Our Father" 2 three times on it.

"The soldier Longinus pierced the Lord with his spear, 3 and the flow of blood was restored and his suffering diminished."

86. For inflammation of an unknown kind.

Sing three "Our Fathers" on your leech-finger, and inscribe around the sore and say: "Flee, devil! Christ follows you. When Christ is born, sorrow flees." And again three "Our Fathers" and three "Flee, devil" prayers.

87. For toothache.

Saint Peter over a marble stone. . . .

# Oxford, Bodleian Library,
## MS Bodley 163

88. *Tigað, tigað, tigað. Calicet ac locluel sedes adclocles arcre, en-crere erernem nonabaioth arcum cunat arcum arcua fligata soh wiþni necutes cuterii rafaf þegal uflen binchni. Arta, arta, arta. Tuxuncula, tuxuncula, tuxuncula.*

2    Quaerite et invenietis. Pulsate et aperietur vobis. Crux Matheus, crux Marcus, crux Lucas, crux Iohannes.

3    Adiuro te, pestiferum virus, per Patrem et Filium et Spiritum Sanctum, ut amplius non noceas, neque crescas, sed arescas. Amen.

# Oxford, Saint John's College, MS 17

89. Wið blodrine of nosu.

2    Writ on his forheafod on Cristes mel:

*stomen metafofu* . ✚ .
*stomen calcos* . ✚ .

# Oxford, Bodleian Library,
# MS Bodley 163

88. *Tigað, tigað, tigað. Calicet ac locluel sedes adclocles arcre encrere erernem nonabaioth arcum cunat arcum arcua fligata soh wiþni necutes cuterii rafaf þegal uflen binchni. Arta, arta, arta. Tuxuncula, tuxuncula, tuxuncula.* Seek and you shall find. Knock and it will be opened to 2 you. The cross of Matthew, the cross of Mark, the cross of Luke, the cross of John.

I adjure you, pestiferous venom, by the Father and the 3 Son and the Holy Spirit, that you do no more harm, nor grow great, but dry up. Amen.

# Oxford, Saint John's College, MS 17

89. For nosebleed.
Write on his forehead in the shape of Christ's cross: 2

*stomen metafofu* . ✚ .
*stomen calcos* . ✚ .

# Worcester, Cathedral Library, MS Q 5

90. Ðis mæg wið gedrif.

2 Genim nigon oflætan ond gewrit on ælcere on þas wisan: "Iesus Christus," ond sing þærofer nigon "Pater noster." Ond syle ætan, ænne dæg þreo, ond oðerne þreo, ond ðriddan þreo. And cweðe æt ælcon siðan þis ofer þone mann:

3 In nomine Domini nostri Iesu Christi, et in nomine sanctae et individuae Trinitatis, et in nomine sanctorum septem dormientium, quorum nomina hec sunt: Maximianus, Malchus, Martinianus, Iohannes, Seraphion, Constantinus, Dionisius. Ita sicut requievit Dominus super illos, sic requiescat super istum famulum Dei NOMEN.

4 Coniuro vos, frigora et febres, per Deum vivum, per Deum verum, per Deum sanctum, per Deum qui vos in potestate habet, per angelos et archangelos, per thronos et dominationes, per principatus et potestates, per totum plebem Dei, et per sanctam Mariam, per duodecim apostolos, per duodecim prophetas, per omnes martires, per sanctos confessores et sanctas virgines et per quatuor evvangelistas Matheum, Marcum, Lucam, Iohannem, et per viginti quattuor seniores et per CXLIIII[or] milia qui pro Christi nomine passi sunt, et per virtutem sancte crucis: adiuro et obtestor vos diaboli ut non habeatis ullam potestatem.

# Worcester, Cathedral Library, MS Q 5

90. This is effective against dysentery.

Take nine Communion wafers and write as follows on    2
each one: "Jesus Christ," and sing nine "Our Fathers" over
them. And give to eat, three on one day, and three on the
second, and three on the third. And each time say this over
the person who is ill:

In the name of our Lord Jesus Christ, and in the name of    3
the holy and indivisible Trinity, and in the name of the seven
sleeping saints, whose names are as follows: Maximianus,
Malchus, Martinianus, John, Serapion, Constantinus, Dionysius. As the Lord made his rest upon them, so may he
make his rest upon this servant of God NAME.

I entreat you, chills and fevers, by the living God, by the    4
true God, by the holy God, by the God who has you in his
power, by angels and archangels, by thrones and dominations, by principates and powers, by all the people of God,
and by holy Mary, by the twelve apostles, by the twelve
prophets, by all martyrs, by the holy confessors and the holy
virgins and by the four evangelists Matthew, Mark, Luke,
and John, and by the twenty-four elders, and by the 144,000
who have suffered in the name of Christ, and by the power
of the holy cross: I adjure you and call you to witness, devils,
that you have no power.

# Louvain-la-Neuve, Université Catholique, "Fragmenta H. Omont No. 3"

91. Wið yflum ond miclum foot-suilun.

Cnua beolonan moran suiðe; lege on ða foet. Hiæ geduinað ond alne ðone ece wundurlice afirrað.

92. Eft to ðon ilcan.

Genim ðære læssan mucgwyrte moran. Cnua wel. Gemæng suæ calde wið hunig. Ete ðæt gemong on marne, ond ðonne he restan welle. Ðæs ilcan lecedom mæg wið ðeh-wærce.

93. Gif foet oððo sconcan oððo cnio suellen.

Genim neodowarde beolonan oððo eolectran. Gecnua suiðe. Mæng wið smælum hwæte-meolwe; gecneed to dage. Clæm on ðæt asuoln limm. Wrið on næhterne; on mærne bið se suile geduinen. Sio duæning dæg ðaeh ðe hio on neate sie.

94. Wyrc god dust to rotigendum dolge.

Nim ac-rinde ond slahðornes wyrtruman rinde innewarde. Adryg; wyrc to duste. Ðæt bið god in dolg to scædenne. Ðis dust is good to scedenne gif men foet suellen ond ða tan cinen. Ne wæt ðu ða foet, ond hie sie stille ðendon ðu hine lecnige.

# Louvain-la-Neuve, Université Catholique, "Fragmenta H. Omont No. 3"

91. For severe and great inflammations of the feet.

Pound henbane root well; apply it to the feet. The inflammation will subside, and this will do marvelously to drive out all the pain.

92. Another cure for the same.

Take the root of little mugwort. Pound well. Blend it, cold, with honey. Have him eat that concoction in the morning and when he goes to bed. This same medicine is good for thigh pain.

93. If the feet or the legs or the knees should swell.

Take the lower part of henbane or lupin. Pound well. Mix with fine wheat meal; knead into a dough. Apply to the swollen limb. Tie it on for a night; in the morning the swelling will have subsided. The swelling will diminish even if it is livestock that are affected.

94. Make a good powder for a festering wound.

Take oak bark and the inner part of blackthorn roots. Dry them; work into a powder. This is good to sprinkle into the wound. The same powder is good to sprinkle in if a person's feet should swell and the skin of the toes is cracking open. Do not wet the feet, and see that he keeps his feet still while you treat him.

95. Wið foet-adle.

Wyrc scoas of seles felle. Do on ða foet, ond hiæ biað sona soel.

96. Wið lend-wærce.

Wyrc gyrdels of seles felle. Gyrd mið, ond he bið hal.

97. Wið ðeh-æci, wyrc drenc.

Nim eofordrotan ond walwyrt; do in win oððo in god waelc alo. Drince on naeht-nihstig bollan fulle. Wyrc him aec god bæð: eapul, ðorn, slahðorn, aesc, cuiicbeam, aac, eofordrote, æscðrote, eolone, biscopwyrt, ifign, betonica, ribbe, rædic. Waell ðis all in wætre, ond ðis is god baeð wið ðeh-æci.

98. Wið aslepnum lic.

2 Genim neoðowardne secg. Waell in wætre. Lege hatne stan in. Let rican on ðæt liic ðær hit aslepne sie.

3 Wyrc him aec gode salfe. Genim ða readan netlan ond hræfnes fot. Waell in suines smeorue ond scepes ond in butran ond scip-teorwe. Smire mið ða salfe ðæt lic ðær hit aslepen sie.

99. Aec swilce is oðer god salf to ðon ilcan: peopor, hwit cudu, swegles eapol, suefl, cost, ecid, oele.

95. For diseased feet.

Make shoes out of sealskin. Put them on his feet, and they will soon be better.

96. For pain in the loins.

Make a belt out of sealskin. Have him put it on, and he will be well.

97. For pain in the thigh, make up a drink.

Take carline thistle and dwarf elder. Put them in wine or in good Welsh ale. Have him drink a bowlful on an empty stomach. Also make him a good bath: apple, thorn wood, blackthorn, ash, rowan, oak, carline thistle, vervain, elecampane, marshmallow, ivy, betony, hound's-tongue, radish. Boil all this in water, and this will be a good bath for thigh ache.

98. For a paralyzed body.

Take the lower part of sedge. Boil in water. Add a hot 2 stone. Let it steam onto the body where it is numb.

Also make him a good salve. Take red nettle and crow- 3 foot. Simmer in pig's fat and sheep's fat and in butter and in ship's tar. Smear the body with the salve where it is numb.

99. Likewise there is another good salve for the same ailment: pepper, gum mastic, swail's apple, sulfur, costmary, vinegar, oil.

# Vatican City, MS Regina lat. 338

100. Þis man sceal wið þæt gedrif writan on þreom leac-bladan ond his naman þærmid: Eugenius, Stephanus, Por-tarius, Dyonisius, Sambucius, Cecilius, et Cyriacus.

101. ✚ Wið blodryne.

2     p · ꟿ · c · p · o · λ · o · x · λ · φ · ý · z · b ·

# Vatican City, MS Reg. lat. 338

100. For diarrhea, this should be written on three leek blades along with the person's name: Eugenius, Stephanus, Portarius, Dionysius, Sambucius, Cecilius, and Cyriacus.

101. ✚ For the flow of blood.

p · Ϻ · c · p · o · λ · o · x · λ · φ · ý · z · в ·            2

# Abbreviations

ASPR = The Anglo-Saxon Poetic Records, ed. Alexander Krapp and Elliott Van Kirk Dobbie, 6 vols. (New York, 1931–1942)

CCCC = Cambridge, Corpus Christi College

DOE = *Dictionary of Old English,* ed. Angus Cameron and others (Toronto, 1986–), letters *A* through *I* currently available

DOML = Dumbarton Oaks Medieval Library

EETS o.s. = Early English Text Society, original series

L. = Linnaeus (in scientific botanical names)

*Misc. Rem.* = *Miscellaneous Remedies*

*PD* = *Peri Didaxeon*

*Rem. An.* = *Old English Remedies from Animals*

# Note on the Texts

Each text in the present edition has been newly edited on the basis of either its manuscript source or the most reliable prior scholarly edition or editions, as set forth below. In the section headed "Notes to the Texts," notice is taken of any departures from the source that serves as our base text. Punctuation and capitalization are normalized in accord with modern conventions. Abbreviations are expanded, silently for the most part, though occasionally in a manner explained in the notes. Although most Old English compound words are left unhyphenated, hyphenation is added in order to highlight the meter in passages of verse or, occasionally, for the sake of greater lexical transparency.

For detailed information about matters of orthography, script, scribal corrections, word division, manuscript layout, manuscript foliation, the physical condition of the manuscript at points where it is imperfect or damaged, or the history of textual emendation of manuscript readings, readers are referred to the most reliable scholarly editions as summarized below. These editions are best used in conjunction with such facsimile editions as are available, also as noted below.

## The Old English Herbal and
## Old English Remedies from Animals

The first two texts in the present volume, *The Old English Herbal* and *Old English Remedies from Animals,* are preserved in four medieval manuscripts:

*V* = London, British Library, MS Cotton Vitellius C.iii, folios 11r–85r. Illuminated. Date: ca. 1000–1025.

*B* = Oxford, Bodleian Library, MS Hatton 76, folios 68r–130r. Date: 1060–1220.

*H* = London, British Library, MS Harley 585, folios 1r–129v. Date: 975–1025.

*O* = London, British Library, MS Harley 6258 B, folios 1r–51r (present foliation). Date: late twelfth century.

The chief scholarly edition of this pair of pharmacopeias is by Hubert Jan de Vriend, *The Old English "Herbarium" and "Medicina de quadrupedibus,"* EETS o.s. 286 (Oxford, 1984), which is based on manuscript *V.* De Vriend's edition serves as our base text, and to this edition we refer readers for a full description of the manuscripts. Notes to the Texts record emendations to the readings of that edition that have been made by the present editor, as supported by the readings of the manuscripts. Latin chapter titles from the late manuscript *O* are cited in the notes to the *Herbal* if they preserve important variant readings; they are included in the main text of *Old English Remedies from Animals* since the earlier English redaction of that treatise, represented by *VBH,* has no chapter titles. For a printed facsimile of *V,* the reader is referred to Maria Amalia D'Aronco and M. L. Cameron,

*The Old English Illustrated Pharmacopoeia: British Library Cotton Vitellius C.iii,* Early English Manuscripts in Facsimile 27 (Copenhagen, 1998). Electronic facsimiles of *V, H,* and *O* are available in the online British Library resource, Digitised Manuscripts (https://www.bl.uk/manuscripts). In addition, facsimiles of *V* and *H* are available in volume 1 of the series *Anglo-Saxon Manuscripts in Microfiche Facsimile,* ed. A. N. Doane and Phillip Pulsiano (Binghamton, NY, 1994–2021), while a facsimile of *B* is available in volume 6 of that same series.

Readers may appreciate knowing how we handle the plant names that occur in *The Old English Herbal* with such frequency. Peculiar to the style of the Old English translation of the Latin *Herbal* is its use, at the beginning of every chapter, of a Latin or Greek plant name followed usually, but not always, by a translation of that name into Old English. The Latin or Greek name is then repeated in the course of the chapter, even when the Old English term has been given. This usage is retained in the present edition, with the Latin/Greek forms presented in italics. The spelling of a name is reproduced as the scribe wrote it, whether or not the spelling is consistent throughout the recipe. A different practice is followed in the facing-page English translations. Here the Greek and Latin terms of the original text are replaced by the lemmas used by Jacques André in *Les Noms de plantes dans la Rome antique* (Paris, 1985). In chapter 7 of the *Herbal,* for example, Latin *veneriam* (in the accusative case) appears in our translation as *venerea.* Where an Old English plant name is provided, we adopt the equivalent modern English name throughout the facing-page translation. In that same chapter of the *Herbal,* for example, the Old English name of that plant, *beowyrt,* is translated as "yellow iris." If no Old English name is provided, then the Latin or Greek term is used in the form cited by André. In chapter titles, the plant's modern English name is given so that there is no doubt as to which plant is being discussed.

## LACNUNGA

The present edition of *Lacnunga* is based on the unique copy of that text that is preserved in London, British Library, MS Harley 585, folios 130r–93r, a manuscript of the late tenth or early eleventh century (manuscript *H* as listed shortly above). The standard critical edition and translation is by Edward Pettit, *Anglo-Saxon Remedies, Charms and Prayers from British Library MS Harley 585: The Lacnunga,* 2 vols. (Lewiston, 2001). In addition, five Old English metrical charms included in *Lacnunga* have been edited and translated by Robert E. Bjork in his edition *Old English Shorter Poems, Volume II: Wisdom and Lyric,* DOML 32 (Cambridge, MA, 2014). Bjork's handling of these texts differs somewhat from what is offered here. A facsimile of the manuscript is available on the British Library's Digitised Manuscripts website, as well as in volume 1 of Doane and Pulsiano, *Anglo-Saxon Manuscripts.*

Most of *Lacnunga* is recorded in Old English, while some parts are in Latin. The Old English text is emended only in cases of obvious and substantive error. Irregular scribal spellings are allowed to remain, since the spelling of the vernacular language was never thoroughly systematized during this early period. Latin passages and texts, which are often liturgical in nature, are emended with a freer hand so as to bring them into closer accord with standard medieval Latin grammar and orthography. Here too, however, a few peculiarities remain, given the informal nature of certain of these texts and their occasional use of vocabulary that is unattested elsewhere.

## PERI DIDAXEON

*Peri Didaxeon* is preserved in a single manuscript, London, British Library, MS Harley 6258 B, folios 51v–66v, dating from the late twelfth century (manuscript *O* as listed above). The present edition of that treatise is based on the manuscript text with reference to two prior scholarly editions in particular. One of these is Max Löweneck's minimal critical edition, published as a special issue of *Erlanger Beiträge zur englischen Philologie* 12 (1896), while the other is an unpublished PhD dissertation by Danielle Maion, "Edizione, traduzione e commento del *Peri Didaxeon*" (Università degli Studi Roma Tre, 1999). Consulted more selectively are Cockayne's initial edition in *Leechdoms, Wortcunning and Starcraft of Early England* (London, 1864–1866), vol. 3, pp. 81–145, and an edition featured in an unpublished PhD dissertation by Linda Sanborn, "An Edition of British Library MS. Harley 6258 B, *Peri Didaxeon*" (University of Ottawa, 1983). As is noted above, a facsimile of Harley 6258 B is available on the British Library's website, Digitised Manuscripts.

The Old English text of *Peri Didaxeon* presents special problems for its editors and readers since the unique extant version of this work is a late twelfth-century copy of a text that may date from a hundred years or so earlier. Determining what is a scribal mistake and what is a late variant spelling of an Old English word can be a delicate matter. For words beginning with the letters *A* through *I* (the range of coverage at the time of preparation of the present volume), readers are encouraged to make use of the online search engine of the *DOE* (https://tapor.library.utoronto.ca/doe/) in order to identify the headword under which a word with nonstandard spelling will be found. Use of that dictionary's "attested spelling" option will facilitate the search. Strategic use of Bosworth and Toller can likewise facilitate searches of this kind for words beginning with the remaining letters of the alphabet: James Bosworth and T. Northcote Toller, *An Anglo-Saxon Dictionary* (Oxford, 1892), with *Supplement* by T. Northcote Toller (1921), and with *Enlarged Addenda and Corrigenda* by Alistair Campbell (1972).

Two other features of the present edition of *Peri Didaxeon* call for attention. These are the handling of chapter divisions and the treatment of the rubrics by which, in the unique manuscript copy, a number of individual remedies are labeled, usually in Latin though sometimes in Old English. For ease in cross-reference, the numbering of chapter divisions that is adopted here corresponds to the system used by Löweneck on the basis of Cockayne's prior edition. Since this same system is used by the *DOE,* it is retained here despite its somewhat capricious nature. As for the rubrics that accompany the Old English text on an irregular basis, often at the foot of an item rather than at its head, they are omitted here. Many of them have the character of annotations made after the main text was written out, and their redundancy would be painfully evident if each were translated into modern English at the head of an item along with the remedy itself. In those relatively few instances where a rubric helps to clarify the nature of a remedy, it is cited in the Notes to the Translations.

## MISCELLANEOUS REMEDIES

The fourth and last part of the present volume is devoted to miscellaneous remedies that happen to be recorded in one or another manuscript of the early English period, often crowded into a margin or written out in a space that is otherwise blank. Many of these items have little or no discernible relation to the other contents of the manuscripts in which they occur. One hundred and one such remedies are assembled here. Their respective manuscript sources are listed below, along with information about facsimiles and the leading prior scholarly editions. Each text, where possible, has been checked against either the original manuscript or at least one facsimile.

Since this group of texts comprises a miscellany, and since many items are of late or uncertain date, they are organized here in a sequence based on the shelf marks of the manuscripts in which they occur. Such a system, though not ideal in every regard, will facilitate access to standard scholarly resources, including N. R. Ker, *Catalogue of Manuscripts Containing Anglo-Saxon* (Oxford, 1990); Helmut Gneuss and Michael Lapidge, *Anglo-Saxon Manuscripts: A Bibliographical Handlist of Manuscripts and Manuscript Fragments Written or Owned in England up to 1100* (Toronto, 2014); and, for the manuscripts of late date not included in either of those resources, the online catalogue Production and Use of Eng-

lish Manuscripts 1060 to 1220 (https://www.le.ac.uk/english/
em1060to1220).

1   Cambridge, Corpus Christi College, MS 41, pp. 206–8. One
    of a number of Old English entries added in the first half of
    the eleventh century to the margins of a manuscript consist-
    ing chiefly of the Old English version of Bede's *Historia eccle-
    siastica.* Text and marginalia date from about the same period.
    Facsimiles: Parker on the Web (https://parker.stanford.edu);
    Doane and Pulsiano, *Anglo-Saxon Manuscripts,* vol. 11, no. 25.
    Text based on the facsimiles, collated with R. J. S. Grant,
    *Cambridge, Corpus Christi College 41: The Loricas and the Missal*
    (Amsterdam, 1979), 5–6; and Karen L. Jolly, "On the Margins
    of Orthodoxy: Devotional Formulas and Protective Prayers
    in Cambridge, Corpus Christi College MS 41," in *Signs on the
    Edge: Space, Text and Margin in Medieval Manuscripts,* ed. Sarah
    Larratt Keefer and Rolf H. Bremmer, Jr. (Paris, 2007), 156–58.

2   Same source, p. 208, collated with Cockayne, *Leechdoms,* vol.
    1, p. 382.

3   Same source, p. 272, collated with Grant, *Loricas and Missal,*
    15–16, and Jolly, "On the Margins," 163.

4   Same source, p. 326, collated with Godfrid Storms, *Anglo-
    Saxon Magic* (The Hague, 1948), no. A4 (p. 314).

5   Same source, p. 326, collated with Storms, *Anglo-Saxon Magic,*
    no. A5 (p. 315), and Jolly, "On the Margins," 166n130.

6   Same source, p. 326, collated with Storms, *Anglo-Saxon Magic,*
    no. A6 (p. 315), and Jolly, "On the Margins," 166n131.

7   Same source, p. 329, collated with Grant, *Loricas and Missal,*
    18–19, and Lea Olsan, "The Marginality of Charms in Medi-
    eval England," in *The Power of Words: Studies on Charms and
    Charming in Europe,* ed. James Kapaló, Éva Pócs, and William
    Ryan (Budapest, 2013), 143.

8   Cambridge, Corpus Christi College, MS 367, part 2, folio 52r.
    A highly abbreviated addition made in the late eleventh or
    early twelfth century to an incomplete Latin life of Saint

Kenelm copied during that same period. Facsimile: Doane and Pulsiano, *Anglo-Saxon Manuscripts,* vol. 11, no. 54. Text based on the facsimile, collated with Storms, *Anglo-Saxon Magic,* no. 80 (p. 306), and with a transcription published in Doane and Pulsiano, *Anglo-Saxon Manuscripts,* vol. 11, p. 71.

9     Cambridge, Gonville and Caius College, MS 379, folio 49r. A twelfth-century compendium to which two remedies have been added. Text based on the manuscript, collated with Storms, *Anglo-Saxon Magic,* no. 72 (p. 302).

10     Same source, folio 49r. Text based on the manuscript, collated with Storms, *Anglo-Saxon Magic,* no. 62 (pp. 294–95).

11     Cambridge, Queen's College, MS 19 (formerly MS 7), folio 142v. A twelfth-century compendium. Text based on the manuscript, collated with Storms, *Anglo-Saxon Magic,* no. 64 (pp. 295–96).

12     London, British Library, MS Cotton Caligula A.xv, folio 129r. The first of four healing texts included in a miscellany of computistical and prognostic entries dating from the second half of the eleventh century. Facsimile: British Library Digitised Manuscripts. Text based on the facsimile, collated with Storms, *Anglo-Saxon Magic,* no. 68 (p. 300), and Karen L. Jolly, "Tapping the Power of the Cross: Who and for Whom?" in *The Place of the Cross in Anglo-Saxon England,* ed. Catherine E. Karkov, Sarah Larratt Keefer, and Karen L. Jolly (Woodbridge, 2006), 63.

13     Same source, folio 129r, collated with Bruce Dickins and R. M. Wilson, "Sent Kasi," *Leeds Studies in English* 6 (1937): 72.

14     Same source, folio 129r, collated with Storms, *Anglo-Saxon Magic,* no. A10 (p. 317).

15     Same source, folio 140r, collated with Storms, *Anglo-Saxon Magic,* no. 34 (p. 272).

16     London, British Library, MS Cotton Domitian A.i, folio 55v. A late tenth- or early eleventh-century addition to a mid-tenth-century ecclesiastical compilation. Facsimile: Doane

and Pulsiano, *Anglo-Saxon Manuscripts,* vol. 5, no. 187. Text based on the facsimile, collated with Pettit, *Anglo-Saxon Remedies,* vol. 2, p. 42.

17 London, British Library, MS Cotton Faustina A.x. Two groups of remedies (a total of ten, in two groups of five) added to a copy of the Old English version of the Benedictine Rule dating from the first half of the twelfth century. Facsimiles: British Library Digitised Manuscripts; Doane and Pulsiano, *Anglo-Saxon Manuscripts,* vol. 15, no. 193. This first text is from folios 115v–16r. Text based on the British Library facsimile, collated with Cockayne, *Leechdoms,* vol. 3, p. 292.

18 Same source, folio 116r, collated with Cockayne, *Leechdoms,* vol. 3, p. 292.

19 Same source, folio 116r, collated with Storms, *Anglo-Saxon Magic,* no. 82 (p. 307).

20 Same source, folio 116r, collated with Storms, *Anglo-Saxon Magic,* no. 40 (pp. 278–79).

21 Same source, folio 116r, collated with Storms, *Anglo-Saxon Magic,* no. 39 (p. 278).

22–26 Same source, folio 116r, written in the top and right margins in a poor, later twelfth-century hand; collated with Cockayne, *Leechdoms,* vol. 3, pp. 292–94.

27–28 London, British Library, MS Cotton Galba A.xiv, folio 118r. Two remedies included in a private devotional miscellany, badly damaged by fire in 1731, that dates from the early eleventh century. Facsimile: British Library Digitised Manuscripts. Base text: Bernard J. Muir, *A Pre-Conquest English Prayer-Book* (Woodbridge, 1988), 150, collated with the facsimile, which is legible only in part.

29 London, British Library MS Cotton Otho A.xiii. A remedy formerly to be found at the end of the now-burned eleventh-century manuscript; transcription by R. James in Oxford, Bodleian Library, MS James 27. Probably thirteenth century in date. Base text: Dickins and Wilson, "Sent Kasi," 72.

30–34 London, British Library, MS Cotton Otho B.xi (a tenth- and eleventh-century manuscript now largely destroyed by fire), together with London, British Library MS Add. 43703 (a sixteenth-century transcript made by Laurence Nowell), folios 261r–262v. Base text: Roland Torkar, "Zu den ae. Medizinaltexten in Otho B.xi und Royal 12.D.xvii, mit einer edition der Unica (Ker, no. 180 art. 11a–d)," *Anglia* 94 (1976): 330.

35   London, British Library, MS Cotton Titus D.xxvi, folios 16v and 17r. Added, in a hand of the first half of the eleventh century, to a copy, written out during that same period, of Bede's *De temporibus anni* and miscellaneous Latin prayers and devotions. Facsimile: British Library Digitised Manuscripts. Text based on the facsimile, collated with Beate Günzel, *Ælfwine's Prayerbook* (London, 1993), 157.

36   London, British Library, MS Cotton Vespasian D.xx, folio 93r. A mid-eleventh-century addition to a manuscript dating from the first half of the tenth century and consisting chiefly of a Latin penitential. Facsimile: British Library Digitised Manuscripts. Text based on the facsimile, collated with Storms, *Anglo-Saxon Magic*, no. 52 (pp. 289–90), and a transcription provided by Pettit, *Anglo Saxon Remedies*, vol. 2, p. 304.

37–38   London, British Library, MS Cotton Vitellius C.iii, folio 18v. A pair of remedies added to the early eleventh century *V* text of *The Old English Herbal* in a hand of the mid-eleventh century or a little earlier. Facsimiles: D'Aronco and Cameron, *The Old English Illustrated Pharmacopoeia;* Doane and Pulsiano, *Anglo-Saxon Manuscripts*, vol. 1, no. 253; British Library Digitised Manuscripts. Texts based on the British Library facsimile, collated with Cockayne, *Leechdoms*, vol. 1, p. 378.

39   The first of nine additional remedies added to the same manuscript at folios 82v–83v. This one is written out at folios 82v–83r, directly after the last item in *Old English Remedies from Animals*. A different hand of the early eleventh century. Text based on the British Library facsimile. This and the next two remedies are collated with Cockayne, *Leechdoms*, vol. 1, p. 374.

40   Same source, folio 83r, column 1.

41   Same source, folio 83r, column 1; a different, slightly later hand.

42      Same source, folio 83r, column 2; another hand of the late eleventh or twelfth century. This and the next five remedies are collated with Cockayne, *Leechdoms,* vol. 1, p. 376.

43      Same source, folio 83r, column 2. Yet another hand of the same approximate date.

44      Same source, folio 83r, column 2.

45–47  Same source, folio 83r, column 2; a different hand of the same approximate date.

48      London, British Library, MS Cotton Vitellius E.xviii, folio 15v. The first of four remedies included amid the matter prefacing the Vitellius Psalter, a manuscript of mid-eleventh-century date, now badly burned. Facsimiles: British Library Digitised Manuscripts; Doane and Pulsiano, *Anglo-Saxon Manuscripts,* vol. 2, no. 258. Text based on the British Library facsimile, collated with Phillip Pulsiano, "The Prefatory Matter of London, British Library, Cotton Vitellius E.xviii," in *Anglo-Saxon Manuscripts and their Heritage,* ed. Phillip Pulsiano and Elaine M. Treharne (Aldershot, 1998), 92; and Jolly, "Tapping the Power," 78.

49      Same source, folio 15v, collated with Pulsiano, "Prefatory Matter," 89, and Jolly, "Tapping the Power," 78.

50      Same source, folio 15v, collated with Pulsiano, "Prefatory Matter," 89, and Jolly, "Tapping the Power," 79.

51      Same source, starting at the foot of folio 15v and carrying over to the top of folio 16r, collated with Ker, *Catalogue of Manuscripts,* 300; Pulsiano, "Prefatory Matter," 90; and Jolly, "Tapping the Power," 79.

52–57  London, British Library, MS Harley 6258 B, folios 51r (for item 52) to 51v (for items 53–57), directly preceding the unique extant copy of *Peri Didaxeon.* Late twelfth century, in the same hand that wrote out the main texts of this anthology, including the *O* version of *The Old English Herbal.* Facsimiles: Doane and Pulsiano, *Anglo-Saxon Manuscripts,* vol. 1, no. 278; British Library Digitised Manuscripts. Text based on the British Library facsimile, collated with Cockayne, *Leechdoms,* vol. 1, p. 380.

58    Same source, folio 51v, collated with Cockayne, *Leechdoms,*
      vol. 1, pp. 380–82.

59–61  Same source, folio 51v, collated with Cockayne, *Leechdoms,*
      vol. 1, p. 382.

62    London, British Library, MS Royal 2.A.xx, folio 16v. The first
      of a number of additions made to a personal prayer book (the
      Royal Prayer Book) dating from the late eighth or early ninth
      century. Facsimiles: British Library Digitised Manuscripts;
      Doane and Pulsiano, *Anglo-Saxon Manuscripts,* vol. 1, no. 283.
      This entry is written in a mid-tenth-century hand. Text based
      on the British Library facsimile, collated with Alphons Au-
      gustinus Barb, "Die Blutsegen von Fulda und London," in
      *Fachliteratur des Mittelalters: Festschrift für Gerhard Eis,* ed.
      Gundolf Keil, Rainer Rudolf, Wolfram Schmitt, and Hans J.
      Vermeer (Stuttgart, 1968), 485–86 ("London A"); and Storms,
      *Anglo-Saxon Magic,* no. 57 (p. 292).

63    Same source, folio 16v, collated with Barb, "Die Blutsegen,"
      486–87.

64    Same source, folio 45v, in a later (twelfth-century) hand; col-
      lated with Storms, *Anglo-Saxon Magic,* no. A16 (p. 318).

65    Same source, folio 45v, collated with Storms, *Anglo-Saxon
      Magic,* no. 61 (p. 294).

66    Same source, folio 49r, collated with Barb, "Die Blutsegen,"
      487–88 ("London C"); and Storms, *Anglo-Saxon Magic,* no. 58
      (p. 292).

67    Same source, folio 49v, collated with Barb, "Die Blutsegen,"
      487–88 ("London D"); and Storms, *Anglo-Saxon Magic,* nos. 59
      and 60 (p. 293).

68    Same source, folio 52r, the first of five flyleaf remedies written
      in a twelfth-century hand. Text based on the British Library
      facsimile, collated with Storms, *Anglo-Saxon Magic,* no. 37
      (p. 278).

69    Same source, folio 52r, collated with Storms, *Anglo-Saxon
      Magic,* no. 38 (p. 278).

70    Same source, folio 52r, collated with Storms, *Anglo-Saxon
      Magic,* no. A11 (p. 317).

71      Same source, folio 52r, collated with Barb, "Die Blutsegen,"
        487 ("London B"); and Storms, *Anglo-Saxon Magic,* no. 55
        (p. 292).

72      Same source, folio 52r, collated with Dickins and Wilson,
        "Sent Kasi," 73.

73–75   London, British Library, MS Royal 12.E.xx, folio 162v. Three
        remedies included in a twelfth-century compendium of med-
        ical texts. Facsimile: British Library Digitised Manuscripts.
        Base text: the facsimile, collated with Edward Pettit, "Some
        Anglo-Saxon Charms," in *Essays on Anglo-Saxon and Related
        Themes in Memory of Lynne Grundy,* ed. Jane Roberts and Janet
        Nelson (London, 2000), 419.

76      London, Wellcome Historical Medical Library, MS 46, folio
        1r. Five remedies, dating from the eleventh century, found in
        the binding of a volume published in Antwerp in 1558. Fac-
        simile: Doane and Pulsiano, *Anglo-Saxon Manuscripts,* vol. 9,
        no. 320. Base text: a digital photograph of the manuscript,
        collated with A. Napier, "Altenglische Miscellen," *Archiv für
        das Studium der neueren Sprachen und Literaturen* 84 (1890): 325,
        item (a).

77      Same source, collated with Napier, "Altenglische Miscellen,"
        325, item (b).

78      Same source, collated with Napier, "Altenglische Miscellen,"
        325, item (c).

79      Same source, collated with Napier, "Altenglische Miscellen,"
        325–26, item (d).

80      Same source, collated with Napier, "Altenglische Miscellen,"
        326, item (e).

81      Oxford, Bodleian Library, MS Auctarium F.3.6, the recto of
        the folio preceding the one now numbered 1r (the same un-
        numbered prefatory folio is numbered 2r by both Ker and
        Pettit). A fragmentary remedy (or perhaps a pair of reme-
        dies), written in a hand dating from the mid-eleventh cen-
        tury, recorded in a manuscript of that same approximate date
        containing various glossed Prudentius texts and Prudentius's

*Psychomachia.* Base text: Pettit, "Some Anglo-Saxon Charms," 412.

82 Same source, the folio now numbered iv (and numbered 3v by Ker and 2v by Napier and Storms). Base text: a digital photograph of folio iv, collated with Napier, "Altenglische Miscellen," 323, item (b), and Storms, *Anglo-Saxon Magic,* no. 77 (p. 305).

83 Oxford, Bodleian Library, MS Barlow 35. A remedy (or a pair of remedies, one of them veterinary in nature) written in a hand dating from the beginning of the eleventh century; these were added to the end of the third part of a composite Latin manuscript of tenth-century date written by continental scribes. Facsimile: Doane and Pulsiano, *Anglo-Saxon Manuscripts,* vol. 15, no. 347. Base text: Edward Pettit, "Anglo-Saxon Charms in Oxford, Bodleian Library MS Barlow 35," *Nottingham Medieval Studies* 43 (1999): 34.

84 Oxford, Bodleian Library, MSS Junius 85 and 86, folio 17r–v. The first of four remedies written out as supplements to a mid-eleventh-century homiliary. Facsimile: Doane and Pulsiano, *Anglo-Saxon Manuscripts,* vol. 17, no. 390. Text based on the facsimile (where legible) and collated with Storms, *Anglo-Saxon Magic,* no. 45 (p. 283).

85 Same source, folio 17v, collated with Storms, *Anglo-Saxon Magic,* no. 49 (p. 286).

86 Same source, folio 17v, collated with Storms, *Anglo-Saxon Magic,* no. 41 (p. 279).

87 Same source, folio 17v, collated with Pettit, *Anglo-Saxon Remedies,* vol. 2, p. 305.

88 Oxford, Bodleian Library, MS Bodley 163, folio 227. Base text: Pettit, *Anglo-Saxon Remedies,* vol. 2, p. 23.

89 Oxford, Saint John's College, MS 17, folio 175r. One of a number of additions, made in an early twelfth-century hand, to a manuscript of about the same date (the Ramsay Scientific Compendium) that contains chiefly Latin computistical texts. Facsimile: McGill Library Digital Exhibitions and Col-

lections (https://digital.library.mcgill.ca/ms-17/). Text based on the facsimile, collated with Storms, *Anglo-Saxon Magic,* no. 54 (p. 291).

90      Worcester, Cathedral Library, MS Q 5. A remedy written on an otherwise blank leaf at the end of a manuscript dated to the early eleventh century. Base text: Napier, "Altenglische Miscellen," 324. The last ten words of the text, from *adiuro* on, are supplied from London, British Library, MS Harley 464 (a seventeenth-century transcript), folio 177, via Storms, *Anglo-Saxon Magic,* no. 36 (p. 276).

91–99    Louvain-la-Neuve, Université Catholique, Fragmenta H. Omont No. 3 recto. A stray parchment folio ("the Omont Fragment") dated variously to the mid-ninth century or to the beginning of the tenth. Base text: digital facsimile of the Omont Fragment available online courtesy of the North-vegr Foundation (https://jillian.rootaction.net/-jillian/world _faiths/www.northvegr.org/lore/omont/index.html), collated with Bella Schauman and Angus Cameron, "A Newly-Found Leaf of Old English from Louvain," *Anglia* 95 (1977): 291–93.

100     Vatican City, MS Reg. lat. 338, folio 91r. An Old English remedy, written in an early eleventh-century hand, added to a blank space in a tenth-century manuscript of continental origin. Base text: digital facsimile of the Vatican City manuscript available online at DigiVatLib (https://digi.vatlib.it/mss), collated with W. Stokes, "Glosses from Turin and Rome, IV: The Anglosaxon Prose and Glosses in Rome," *Beiträge zur Kunde der indogermanischen Sprachen* 17 (1891): 144, and with Ker, *Catalogue of Manuscripts,* p. 458 (no. 390a); Ker errs in identifying the manuscript's shelf mark as "Reg. lat. 388."

101     Same source, folio 111v, in a different English hand of about the same date, squeezed in at a ninety-degree angle to the main entry on the page. Base text: digital facsimile available online at DigiVatLib, partially collated with Ker, *Catalogue of Manuscripts,* p. 458 (no. 390b); Ker prints only the initial instruction, "Wið blodryne."

# Notes to the Texts

## The Old English Herbal

### Table of Contents

Incipiunt capitula: CAPIT^A V, CAPITA *de Vriend. V's reading is supported also by Humphrey Wanley, "Librorum veterum septentrionalium," vol. 2 of "Antiquae literaturae septentrionalis libri duo" (Oxford, 1705), 217: "Herbarium ... cui praemissa sunt Capitula Latine et Saxonice, cum hoc Titulo: Incipivnt Capitvla Libri Medicinalis" (A herbal ... preceded by a list of chapters in Latin and in English, with this title: Here begin the chapters of the book of medicine).*

54.1        Wið eagena sare: *omitted by VBH, supplied from main text, chapter 54*

55.2        hræce *BH*: ræce *V*

56          narcisus *BH*: na'r'cisus *V with interlined black* r *in modern hand*

73          Ulixe *BH*: Iulixe *V*

75.1        eagena *BH*: eagen *V*

87          savine *BH*: safinæ *V*

101         organe *BH*: organa *V*

103         salvia *BH*: salfia *V*

116         silvatica *BH*: silfatica *V*

135.1       ond wið banece *BH*: ond banece *V; see chapter135.1 of main text*

147.2       Wiþ heafodece: *omitted by VBH, supplied from chapter 147.2 of main text*

162.1       oððe: *omitted by VBH, supplied from chapter 162.1 of main text*

163.3       ymb: *emended from* ym

181.2       scurf *B*: scruf *V*

184.1       gedrecednesse *B*: gederednese *V*

## Main Text

1           Betonica: *title supplied from table of contents; omitted by VB*, De betonica *O*

1.5         eagena *B*: egena *V*

1.9         piporcorna *B*: piporcorn *V (de Vriend reads* piporcornu, *but the* u *was added by a later hand)*

1.13        hraðe: *emended from* raðe *V*

1.17        þreo: *de Vriend reads* þrec *(an apparent misprint)*

2           Wægbræde: De plantagine *O*

2.7         Nædre: *omitted by editor; see Notes to the Translations*

2.8         Scorpio: *omitted by editor; see chapter 2.7 above*

2.19        wes *B*: wesc *V, where* c *was inserted by the copyist above the line with a clear sign of insertion; B's reading is supported by the sense and by the Latin source. See Maria Amalia D'Aronco,"The Edition of the Old English Herbal and 'Medicina de quadrupedibus': Two Case Studies," in "Old Names—New Growth," ed. Peter Bierbaumer and*

*Helmut W. Klug (Frankfurt am Main, 2009), 42–51, and de Vriend, p. 288.*

| | |
|---|---|
| 3 | Fifleafe: Herba pentafilon v folia *O* |
| 3.9 | Nædre: *omitted by editor; see chapter 2.7 above* |
| 4 | Æscþrote: Vermenaca ascþrota *O* |
| 4.5 | þone *B*: þæne *V* |
| 4.6 | heafde *B*: hefde *V* |
| | Nædre: *omitted by editor; see chapter 2.7 above* |
| 4.8 | Attorcoppe: *omitted by editor; see chapter 2.7 above* |
| 4.11 | Nædre: *omitted by editor; see chapter 2.7 above* |
| 5 | Hennebelle: Symphoniaca hennebelle & belone *O* |
| 5.4 | sar: *supplied from table of contents* |
| 6 | Nædrewyrt: Viperina naddrawyrt *O* |
| | Nædre: *omitted by editor; see chapter 2.7 above* |
| 7 | Beowyrt: De veneria id est beowyrt *O* |
| 8 | Leonfot: Pes leonis *O* |
| 9 | Clufþunge: Scelerata clufþunca *O* |
| 10 | Clufwyrt: Baration clufwyrt *O* |
| | stowum *B*: landum *V* |
| 11 | Mugcwyrt: Arthemisia *O* |
| 12 | Herba artemesia tagantes: *supplied from table of contents; title omitted by VBHO, no separate chapter in VBHO* |
| 12.2 | gewes *BH*: gewæsc *V*, wes *O; see chapter 2.19 above* |
| 15 | Nædre: *omitted by editor; see chapter 2.7 above* |
| 15.1 | dracontean *BH*: dracontea *V* |
| 16 | Hreafnes leac: Saturion hrefnes leac *O* |
| | landum *BH*: landun *V* |
| 17 | Nædre: *omitted by editor; see chapter 2.7 above* |
| 18 | Slite: De orbiculari *O* |
| 19 | Unfortrædde: Proserpina fortredde *O* |
| 20 | Smerowyrt: De astrologya *O* |
| 20.1 | attres strengðe *B*: attres strenðe *V* |
| | ealle strengþe *B*: ealle strenðe *V* |
| 20.3 | afeormað *BH*: afcormcð *V* |
| 20.4 | Nædran: *omitted by editor; see chapter 2.7 above* |

20.5     gewes *BH*: gewesc *V; see chapter 2.19 above*

21     Cærse: De nasturcio *O*

23     Glofwyrt: De apollinaria *O*

24     Mageþe: De camemelon id est mægeðe *O*

25     Heortclæfre: De chamedris id est heortclæfe *O*

26     Wulfes camb: De cameelea Anglice wulfes camb *O*

27     Henep: Camepithis Anglice henep *O*

28     Hrefnes fot: Chamedafne id est hrefnes fot *O*

29     Lyðwyrt: De ostriago id est liþewyrt *O*

30     Hæwenhydele: Haewenhudela *O*

30.1     supan *BHO*: drincan *V*

32     Garclife: De agrimonia *O*

33     Wudurofe: astularegia *O*

35     Eorðgealla oððe curmelle: Centaurea maior *O*

36.2     beorhtnys B: beorhtnes H, beorˈtˈhnys *with* t *marked for insertion* V

37     Bete: Personaciam bete *O*

37.1     bete *BH*: boete *V*

37.3     Wið þæt cancor on wunde: þæt *supplied from table of contents; de Vriend emends from* O

38.1     fragam: *emended from* fragan

38.2     fremað *BHO*: hit fremað *V*

40     Ippirus: *supplied from table of contents; title omitted by VBH, chapter omitted by O*

42     Hundes tunge: De buglosa *O*

42.1.3     docce *BH*: doccoe *V*

43     Glædene: De bulbo scillitico Anglice gladene *O*

46     Harehune: Prassion marubium harehune *O*

47     Foxes fot: Xifion foxes fot *O*

49     Syngrene: Temolum singrene *O*

        Temolum: *for* te molum

50     Sigelhweorfa *H*: Sigelwearfa *V*

52     Hymele: Politricum *O*

53     Wuduhrofe: Item astularegia *O*

55     Oenantes: *supplied from table of contents; omitted by VHB,* Oenantes herba *O*

56 Halswyrt: De narciso id est halswyrt *O*

57 Brunewyrt: Splenion verio brunewyrt *O*

57.1 teucrion: teuerion *VBH*

58.1 þeah *BH*: þeh *V*

59 Cneowhole: De victoriola id est cneowholem Anglice *O*

59.1 þæs magan sare *BH*: þone magan *V*

61 Asterion: *supplied from table of contents; title omitted by VBH*, De asterion vel savina *O*

63 Dictamnus: *supplied from table of contents; title omitted by VBHO*

64 Solago maior: *supplied from table of contents; title omitted by VBH*, Solago id est solsequium *O*

64.1 Nædran: *omitted by editor; see chapter 2.7 above*

65 Solago minor: *supplied from table of contents; title omitted by VBH, no separate chapter in O*

66 Pionia: Pionia *O*
   peoniam *H*: peonian *VB*
   Creta: *emended from* Greca *V*, Creca *B*, Creaca *H; emendation of this shared error is supported by the Latin source*

67 Berbena: Peristerion id est vervena *O*

67.2 attru *BH*: atru *V*

69 Nymfete: *supplied from table of contents; title omitted by VBH*, Haec herba est nymfete *O*

70.1 crision: *for* cirsion

72 Scordea: *supplied from table of contents; title omitted by VHB*, Scordeon album . . . Anglice *O*

73 Feltwyrt: De verbascum id est feldwurt *O*
   Ulixe *BH*: Iulixe *V*, Uluxe *O*

74 Heraclea: *supplied from table of contents; title omitted by VBHO*

75.5 gebærned *BH*: forbærned *V*

76 Solsequia: Solosece solate *O*

77 Grundeswylige: Senecion grundesswulie *O*

78 Fern: De felice Anglice fern *O*

80 Glædene: *title omitted by O*

81 Boðen: De rosemarino *O*

82 Feldmoru: *title omitted by O*

83 Dolhrune: Perdiculus dolhrune *O*

|  | cænned *BH*: cened *V* |
|---|---|
| 83.1 | cneowa *BH*: cnewu *V* |
| 85 | Eforfearn: De radiolo id est pollipodio *O* |
| 86 | Wuduceruille: Sparagia wudecervilla *O* |
| 87 | Savine: Savina sive sabina *O* |
| 88 | Hundes heafod: *title omitted by O* |
| 89 | Bremel: *title omitted by O* |
| 90 | Gearwe: De millefolio . . . gearwe Anglice *O* |
| 90.13 | þam *BH*: þan *V* |
| 91 | Rude: De ruta *O* |
| 92 | Minte: *supplied from table of contents; title omitted by VHB,* De mentastro menstrastum *O* |
| 93 | Wælwyrt oððe ellenwyrt: Ebulum *O* |
| 94 | Dweorgedweosle: Pollegium dwerorgedwosle *O* |
| 94.2 | gewæs *BH*: wes *O*, gewæsc *V; see above, 2.19* |
| 94.10 | ond twegen *BHO*: twegen *V* |
| 95 | Nepte: De nepta *O* |
|  | Næddre: *omitted by editor; see above, chapter 2.7* |
| 95.1 | nemdon *B*: nendun *V; B's reading is supported by usage and grammar* |
|  | gecnucude *H*: gecnude *V* |
| 96 | Cammoc: Peucedanum cammoc *O* |
|  | Næddre: *omitted by editor; see above, chapter 2.7* |
| 98 | Nædre: *omitted by editor; see above, chapter 2.7* |
| 99 | Sundcorn: Saxifragia sundcorn *O* |
| 100.1 | þa stanas *BH*: heo stanas *V* |
| 100.3 | seofoðan *BH*: seofoþam *V* |
|  | seofontyne *B*: feofortyne *V* |
| 100.5 | wundum *BH*: wundun *V* |
| 101 | Organe: Serpillum organa *O* |
| 102 | Wermod: De absinthio *O* |
| 103 | Salvie: Savina *O* |
| 104 | Celendre: De coliandro *O* |
| 105 | Porclaca: *supplied from table of contents; title omitted by VBH, chapter omitted by O* |
| 107 | Brocminte: Brocminte sisimbrium *O* |
| 108 | Olisatra: De oleastro *O* |

| | |
|---|---|
| 109 | crinion: erinion *VB* |
| 110 | Lacterida: Titimallos calatites vel lacteridam *O* |
| 112 | Lupinum montanum: *supplied from table of contents; title omitted by* *VBH*, *chapter omitted by O* |
| 114 | heo ys lactuca leporina *BH*, heo ys lactuca leporinam *V* |
| 115 | Hwerhwette: De cucumere *O* |
| 116 | cannave *BH*: cannane *V* |
| 117.1 | nemdon *B*: nendun *V* |
| 118 | Seofenleafe: Eptafilon *O* |
| 119 | Mistel: De ocimo Anglice mistel *O* |
| 120 | Merce: De apio *O* |
| 122 | Minte: De menta *O* |
| 123 | Dile: De aneto Anglice dile *O* |
| 124 | Organe: De origano *O* |
| 125 | Sinfulle: Semperviva synfulle & singrene Iouis barba *O* |
| 127 | Liðwyrt: erifion *O* |
| 129 | Petersilie: Petrosillinum triannem *O* |
| 131 | Nædderwyrt: De basilica id est nædderwyrt *O* |
| 132 | Mandragora: De mandragora *O* |
| 134 | Action: *title supplied from table of contents; title omitted by VBH*, De action *O* |
| | gelice leaf: *the construction is uncommon, but it is shared by VBO; H emends* lang leaf gelice |
| 135 | Suþernewuda: De abrotano *O* |
| | suðernewuda: *inserted by a slightly later hand V, omitted by BH* |
| 136 | Laber: De sion id est lafere *O* |
| 137.3 | weartum *BH*: weartun *V* |
| 138 | Spreritis: *supplied from table of contents; title omitted by VBH*, Spreritis *O* |
| 139 | Aizos minor: *supplied from table of contents; title omitted by VBH*, Aizos minor *O* |
| 140 | Tunsingwyrt: *supplied from table of contents; title omitted by VBH*, Elleborum album *O* |
| 141 | Buoptalmon: *supplied from table of contents; title omitted by VBH*, De buoptalmon *O* |
| 142 | Gorst: Tribulus gorst *O* |

143    Coniza: *supplied from table of contents; title omitted by VBH, chapter omitted by O*

145    Glycyrida: *supplied from table of contents; title omitted by* VBHO

146    Strutius: *supplied from table of contents; title omitted by VBH,* Structium byscupwyrt *O*

147    Aizon: *supplied from table of contents; title omitted by VBH,* Item de aizon singrene Jovis barba *O*

148    Ellen: Samsuchon elle *O*

148.4    Hy *H: V unreadable; de Vriend emends* hit *from O*

149    Stecas: *supplied from table of contents; title omitted by VBH,* Stecas *O*

150    Thyaspis: *supplied from table of contents; title omitted by VBH,* Tiapis *O*

152    Hypericon: *supplied from table of contents; title omitted by VBH, chapter omitted by O*

153    Acantaleuca: *supplied from table of contents; title omitted by VBH,* De acantaleace *O*

153.3    afyrred *BH:* afyrmeþ *V; BH's reading is supported by the sense and by the Latin source; see Maria Amalia D'Aronco, "'Afyrman,' a Ghost Word in 'The Old English Herbarium and Medicina de quadrupedibus,'" in "Per Teresa: Dentro e oltre i confini," vol. 1, ed. Giampaolo Borghello (Udine, 2009), 247–52*

154    Beowyrt: *supplied from table of contents; title omitted by VBH,* De acanto beowurt *O*

155    Cymen: De cimino *O*

156    Wulfes tæsl: *supplied from table of contents; title omitted by VBH, chapter omitted by O*

157    Scolymbos: *supplied from table of contents; title omitted by VBH,* Scolinbos *O*

158    Iris Yllyrica: *supplied from table of contents; title omitted by VBH, chapter omitted by O*

159    Elleborus albus: *supplied from table of contents; title omitted by VBH,* Elleborum album tunsigwyrt *O*

160    Delfinion: *supplied from table of contents; title omitted by VBH, chapter omitted by O*

161    Acios: *supplied from table of contents; title omitted by VBHO*

162 Centimorbia: *supplied from table of contents; title omitted by VBH, chapter omitted by O*

163 Scordios: *supplied from table of contents; title omitted by VBH,* Scordios *O*

163.3 ymb *BH*: ym *V*

164 Ami: *supplied from table of contents; title omitted by VBH,* Aini alias milvium *O*

165 Banwyrt: De viola *O*

166 Viola purpurea: *supplied from table of contents; title omitted by VBH,* Viola purpurea *O*

167 Zamalentition: *supplied from table of contents; title omitted by VBH,* Zima lentition *O*

168 Ancusa: *supplied from table of contents; title omitted by VBH,* Ancura *O*

169 Psillios: *supplied from table of contents;* coliandre *V, title omitted by BH,* Psillios pulicaria *O*

170 Cynosbatus: *supplied from table of contents; title omitted by VH, chapter omitted by BO*

171 Aglaofotis: *supplied from table of contents, title omitted by VH,* De aglaofota Anglice foxes glofa *O*

173 Eryngius: *supplied from table of contents; title omitted by VBH*

174 Philantropos: *supplied from table of contents; title omitted by VBH,* Philantropos clate *O*

175 Achillea: *supplied from table of contents; title omitted by VBH,* De herba quae dicitur acylleia *O*

176 Ricinus: *supplied from table of contents; title omitted by VBH,* De ricino *O*

177 Polloten: *supplied from table of contents; title omitted by VBH,* De porro quod polloton dicitur Grece *O*

178 Netele: De vrtica *O*

179 Priapisci: *supplied from table of contents; title omitted by VBH,* De vica pervica quod priaprissi dicitur *O*

180 Litosperimon: *supplied from table of contents; title omitted by VBH, chapter omitted by O*

181 Stavisagria: *supplied from table of contents; title omitted by VBH,* Stavisagria *O*

181.2        scurf *BH*: scruf *V*

182          Gorgonion: *supplied from table of contents; title omitted by VBH,*
             *chapter omitted by O*

183          Milotis: *supplied from table of contents; title omitted by VBH,* De
             herba que dicitur melotis *O*

184          Bulbus: *supplied from table of contents; title omitted by VBH,* De
             bulbo *O*

185          Colocyntisagria: *supplied from table of contents; title omitted by*
             *VBH, chapter omitted by O*

## OLD ENGLISH MEDICINE FROM ANIMALS

3            Medicina de cervo: *supplied from O*
3.7          earfoðnyssum *BH*: earfodnyssum *V*
4            Medicina de vulpe: *supplied from O*
5            Medicina de lepore: *supplied from O*
6            Medicina de hirco: medicina de hirco et capra *O*
7            Medicina de capra: medicina de hirco et capra *O*
7.12         to drince: drince *V*, drincan *B*, þæt he drince *O; see Notes to the*
             *Translations*
8            Medicina de ariete: Medicina ariete *O*
9            Medicina de apro: *supplied from O*
10           Medicina de lupo et de canibus: Medina lupo *O; see also chapter*
             *14 below*
10.11        wedehundes *B*: wedes hundes *V*
14           Medicina de canibus: De canibus *O*

## LACNUNGA

1            Wið: wit
2            wætere: wæ wætere
8            Þis is: þiis
11           ðicce: ðicge
19           ceorf of: ceorof, *with the second* o *inserted above the line*
20           roseo: hroseo
             unum: hunum
             plurimas: *Pettit misreads as* plurimus

|        | emigraneum: emigranecum |
|--------|-------------------------|
| 21.2   | eced: ece |
|        | drince: brince *with the ascender of* b *erased* |
| 22.3   | fæder: fæd |
| 22.4   | finit, amen: fintamen |
| 23     | scearfa: searfa |
|        | heald: seald |
| 25.1   | syðan: syþðan |
| 29.2   | comprehenderunt: conprehenderunt |
| 29.4   | Ber þonne: ber þon |
|        | ond "Pater noster": ond *supplied by editor* |
| 30     | gemænged sy: sy *supplied by editor* |
|        | Þonne hæfst: þonne hæfs |
| 31.1   | tyddernysse: tyddernysse sceal |
| 31.3   | apoldre rinde: apoldrinde |
| 31.5   | brunne: brimne |
| 31.6   | Magnificat: mangnificað |
| 32     | mearhsapan: ar sapan |
| 34     | suþernewuda ond isopo . . . ond rude: *these ingredients are an inter-* |
|        | *linear addition, written in the same hand as the main text, starting* |
|        | *above the words* banwyrt ond brunwyrt *and carrying on above the* |
|        | *next line of text, with no indication as to just where the words should* |
|        | *go* |
| 36     | teoh: teon |
| 41     | ond fifleafan: ond leafan |
|        | hæferþon: eferþon |
| 42     | ifig: efic |
|        | cyslybbe: cysbybbe |
| 43     | ut-yrnendne: ut yrnnendne; *Pettit takes the manuscript reading to be* |
|        | ut yrnendne |
|        | ond fiftene: ond *supplied by editor* |
|        | wyrm-melo: wyrmelo |
| 44     | tu hamwyrte: ðu hamwyrte |
|        | Berend: beren |
|        | ond ado: ł ado *(that is,* vel ado*)* |
| 46     | lange: lacnge |

| | |
|---|---|
| 47 | pipercorn: pipercor |
| 50 | ompran: oppran |
| 56 | hwilum ðære: hwilum geðære |
| 57 | ond betonican: ond *not in manuscript* |
| 60 | mest: nest |
| 61 | snæda: sæda |
| 63.1 | agrimonia: agrimonis |
| 63.2 | heowes: heowe |
| 63.3 | . . . ælcne ðriwa: *although there is no gap in the manuscript, something seems to be left out* |
| | þæt is: þæt his |
| 63.5 | orationes: orationis |
| 63.7 | rogo: rigo |
| | Fili: filii |
| | deleas: delas |
| | adminiculum: adminilum |
| | hominis: homines |
| | confiteatur: confiteantur |
| | habet: abet |
| | et . . . voluntatem: *something seems to be left out here* |
| | sententiam: sentiam |
| | hoc: oc |
| 63.8 | sanctos: sancti |
| | et angelorum: et *supplied by editor* |
| | exercitus: excitus |
| | tu fecisti: defecisti |
| | omnia: omni |
| | suam: suuam |
| | id est: it est |
| 63.10 | *Substituted for the imperfect version of this prayer as written out in MS Harley 585 is the corresponding part of a superior copy recorded in a medieval sacramentary from Autun (Pettit, vol. 2, p. 73).* |
| | solemnitate: sollemnitate |
| 63.11 | utuntur: uiumur (?) *hard to decipher* |
| 64.2 | adiuva: adiu |
| | Filius: filili |

Sanctus: sancti

64.4 biberant: biberunt
accipiens calicem: calicem *supplied by editor*
signaculum: singnaculum

64.5 subiecta sunt: subiacta sunt
subiecta est: subiacta est
et expavescit: et *supplied by editor*
audito: auditu
fugit: fuit
silet: silit
extinguitur: extinguetur
regulus: regulas
et animalia: et *supplied by editor*
salutis: salutes
operationes: operationis
mortiferas: mortiverous *corrected from* mortiverus
quas in se: quas i se
habet: habent
evacua: vacua
cor ut: coruit
intelligant: intellegant

65.4 maris: maris sonum
velut: velet

65.6 huius: uius
anni: ani
vanitas: unitas

65.8 caelestis: celestis
militiae: militige

65.9 ne: he

65.12 caelestis: celestis
militiae: militie

65.15 archangelos: arhangelos

65.21 navis: naves

65.28 potentia: potentie
tua: tue

65.31 daemones: demones

| | |
|---|---|
| 65.32 | librent: liberantur |
| 65.33 | gigram: gigran |
| | cephale: chephalem |
| 65.35 | taleas: talias |
| 65.36 | atque: adque |
| | binas: bonis |
| 65.37 | capillis: scapulis |
| | vertici: vertice |
| 65.39 | triformi: triforme |
| 65.40 | labio: labie |
| | faciei: facie |
| | timpori: timpore |
| 65.41 | barbae: barbe |
| | supercilis: superciliis |
| 65.44 | gingis: iguis |
| | anele: anile |
| 65.45 | linguae: lingue |
| | ori: ore |
| | gutturi: guttore |
| 65.46 | gurgulioni: guguilione |
| | cervici: cervice |
| 65.47 | centro: ceotro |
| | cartilagini: cartilagine |
| 65.49 | lorica: lurica |
| 65.51 | invisibiles: invisibilis |
| 65.52 | clavos: clabos |
| 65.53 | tege: tegescyld |
| | ergo: erto |
| | forti: forte |
| | lorica: lurica |
| 65.55 | cubitis: cubiis *with a small, faint* t *inserted above the line* |
| 65.56 | palmas: palmos |
| 65.60 | catacrines: catacrinas |
| 65.61 | gambas: cambos |
| 65.62 | poplites: polites |
| 65.65 | talos: talas |

| 65.69 | genitalia: genetalia |
|---|---|
| 65.72 | fibrem: fithrem |
| | obligio: obligia |
| 65.74 | fibras: fivras |
| | bucliamine: buclimi amni |
| 65.75 | inguinem: inguinam |
| | medullis: medulis |
| 65.76 | splenem: *Pettit misreads as* splenum |
| | tortuosis: totuosis |
| 65.77 | vesicam: visicam |
| | pantes: pantas |
| 65.78 | compaginum: conpagnum |
| 65.79 | pilos: piclos |
| | reliqua: reliquia |
| 65.81 | cum quinque: cumque |
| 65.83 | verticem: vertice |
| 65.85 | possit: posit |
| 65.86 | langor: languor |
| 65.87 | iam: nam |
| 65.91 | aetheria: ethera |
| 65.92 | laetus: letus |
| 69 | ond æscðrote: ond *supplied by editor* |
| 75 | se fic: se uic |
| 76.1.6 | þam: þa |
| 76.1.8 | opene: opone |
| 76.1.10 | breodwedon: bryo dedon |
| 76.1.20 | ðam: ða |
| 76.1.30 | magon: ongan |
| 76.1.43 | ond wið færbregde: ond wið þæs hond wið frea begde |
| 76.1.44 | manra: minra |
| 76.1.53 | þystel-geblæd: þysgeblæd |
| 76.1.58 | adle: alde |
| 76.1.60 | þær: ond |
| 76.1.65 | mægðan: mageðan |
| 76.2 | mid æg-gemongc: mid aagemogc |
| | on do: on de |

76.3    on do: on de

77    oððe: oðð (?) *uncertain reading*

      upweardes: upweardnes

81    *The same recipe is written out twice in the manuscript, with slight variations, at lines 1–8 of folio 165r. Only the second version is included here. The first version is identical to it except in two regards: the use of ð instead of þ in* cyleþenigean, *and the inclusion of only the following string of symbols:*

      ✛ T ✛ ⱳ̄ Ā T

83.2    ðe: þe ðe

86.2    Martinianus: martimianus

      bufan: hufan

86.3    *line 1*: an spider-wiht: inspidenwiht

      *line 4*: legde þe: lege þe

      *line 13*: Fiat: fiað

87    ond bancoþum: ond *supplied by editor*

88    blegnum: blegnu

      Superare: suptare

93    mængc: mægc

104    *The text of entry 105 is erroneously written out here between the words* delf *and* þa moran.

      drincan: drican

105    *See note on 104.2.*

110    Genim: geni *plus an erased letter*

      ongemetlice: on megetlice *with* ge *above line*

113    ma gyf: ma gyf gif

116    piporcorna: piporcorn

120.1    þær seo adl: þær *supplied by editor*

121    drincan: drican

126.3    ducat: ducað

      conservat: conservæð

      liberet: liberat

      adiuvet: adiuvat

126.4    Contere: contrive

127.3.12    iserna: *a word may have gone missing before* iserna, *though there is no gap in the manuscript*

| | |
|---|---|
| | wundrum: wund |
| 127.3.25 | fleoh: fled |
| | fyrgenheafde: fyrgen hæfde |
| 128 | an sealf: an *supplied by editor* |
| 133.2 | þa wyrt þe weaxaþ: þe weaxaþ *supplied by editor* |
| | worðigum: wordigum |
| 136 | clitan: glidan |
| 142 | wærce: wyrce |
| 149.3 | oriente reducat: *oriente reducað* |
| 149.4 | *aquilone reducat: aquilone reducað* |
| | forhelan: ferholen |
| 150.1 | dolorem: dolorum |
| 150.2 | quos [. . .]: *some words have dropped out, though there is no gap in the manuscript* |
| | tu es oculus: tu es oculos |
| 150.3 | inluminas: inlumnas |
| 151 | constitue: constituam |
| 152 | mois: mor |
| 153.2 | neogone wæran: *these two words precede the title phrase* wið cyrnel *instead of following it* |
| 153.3 | cyrneles: cyrn neles, *divided over a line break* |
| 155 | extinguatur: extingunt |
| 156 | *After* catenis, *the words* contra dolorum dentium *follow; this heading properly pertains to entry 158* |
| 157 | Ad: ab |
| | dolorem constantem malignantem: dolorum constantium malignantium |
| | medicina: *transposed from later in the cure, where it is wrongly written after* in nomine |
| | ligavit: lignavit |
| | In nomine: in nomine medicina |
| 158.1 | Contra dolorem dentium: *heading transposed from entry 156* |
| | dolorem: dolorum |
| 158.2 | marmoreum: mamoreum |
| | tristis (2nd occurrence): tritis |
| 158.3 | palato: palpato |

frangere: fragere

190     *Text badly faded and mostly illegible; only a few words are reproduced here.*

191     puis: pius

## PERI DIDAXEON

1.1     Her: er, *with space for* H

gewitnesse: gewisnesse

gelæredustan: gelæredus

Þæt was: þæt wat

Esculafius: Esculapfius, *with a delete mark inserted below the* p *and with* f *added above the line*

feower: IIII^{or} *(for Latin* quattuor*)*

wintra a þusund: *the scribe first wrote* a sun þusund wintra, *then marked* sun *for deletion and marked* a þusund *and* wintra *for transposition*

cingc: cingi *(?)*; *Maion reads* cing

Scolafius *(?)*: *Maion reads* Sculafius; *Löweneck emends to* Escolafius

cmpiricam: empitricam

1.2     uþwytyna: aþwytyna

æfterfyligdun: þas æfterfyligdun

gehwylce: gehylce

and on þan dægen: and on þam dæge and on þan dægen, *with the first four words written at the end of folio 52r and then repeated at the start of folio 52v*

1.3     Drihtne: drihte

nymen: mymen

æt fruman: æt *supplied by editor*

2     gehæled: *thus also Cockayne; Löweneck and Maion read* gehaled

3     betwyx: becwyx

4     bynd: byd

6     wið: wd wið

8     heo: heo heo

forswylhþ: forswyhþ

9.1     earan: earam

9.2 no þustre; no *supplied by editor*

13 beþece *(?): Cockayne, Löweneck, and Maion read* beþete, *and Cockayne and Löweneck emend to* beþege

14 wyllð: wylld

15.1 bebind: bebin

þarof: þarto

15.2 ambrotena: ambrocena *(?)*

16.2 pina-hnutte: pinahnurtte

17.1 Þis sceal: þis scead

17.2 godes: gedes

mægn: mæng

forleosen: forleaosen

Wryng: wyng

18.1 oþerhwylum: oþer hylum

drincaþ: *after this word is written the redundant clause and* þa healsgunda syndan twa cunna

18.2 wyrce: wyrlce

bersten: besten

sona: þona

20 swyþe: syþe

21 cicles *(?): Cockayne, Löweneck, and Maion read* citles *and emend to* cicles

bynd: byð

22 croh and safran: croh safran

24 nenge: nengi *(?); Maion reads* neng

þe þærto: þe þe þærto

28 Þis ys: ys *supplied by editor*

29 hym byð: byð *supplied by editor*

31 bærne togædere: bærne *supplied by editor*

33.1 gesteal: *Maion reads* gesceal *and emends to* gesteal

33.2 fyrst: syst

þane mete underfoð: on gemete wisdom underfoð

þa yfel: þa ufe

wyrtrume: wyrtume

33.3 læcecræft: læcræft

35 Eftsona: ftsona

36      seoð: seod

37      deah: deaþ; *Löweneck and Maion read* deah

        meolce (?): *Löweneck and Maion read* meolte *and emend to* meolce

38      þat wos: wos *supplied by editor*

40      leohtne: leohne

41      spic: swic

42.1    manne: manme

        hnecca-sar: hnencca sar

43.1    untrumnesse: untwumnesse

        And þa: anð þa

        to þan sculdre: sculdre *supplied by editor*

43.2    læcecræft: læcræft

        se rusel: se ruse

        swylce: swyce

44      and galpanan: and galpania and anan

        hwile: wile

45      fingra: fringra

46      þanne streuw: þanne þanne streuw

48      Wyrc: wryc

50.1    untrumnyss: untrunyss

        þingum: þringum

        miclum æte: miclum hæte

50.2    of earfodnysse: of earferd earfodnysse

        in ane: in an ane

        gemenged wæs: wæs *supplied by editor*

51.1    hwylum mid: hylum mid

        wriþaþ: riþaþ

51.2    beo næþer: beo nærþer

        togædere: *Cockayne, Löweneck, and Maion read* togadere

        wætere betwex þan sculdrun: wætere betwex þan scaldrun

52.1    soþ: þoþ

        wætt (?): *Cockayne, Löweneck, and Maion read* wæte

        ædran: ærdran

        byþ fah: byþ faþ

        hrigge: hriggfe

        swilce: swice

| | |
|---|---|
| 52.2 | tacnunge: tacnuge |
| 52.3 | þanne ut: þanne uf |
| 52.4 | forhabban sceal: sceal *supplied by editor* |
| | flæsc þe: flæsc and þe |
| | fif-fingran: fiffringran |
| | ys þæt: hyt þæt |
| 52.6 | do hwyt cudu: hytt cudud; do *supplied by editor* |
| 53 | frencissen: fr frencissen |
| 54 | galpanan feorfer scellinga: galpanan feorfer scelliga |
| | pyne: wyne |
| 55 | niwes: nifes |
| 57 | hwit: hyit |
| 58.2 | and þa wurt: and þa wrut |
| 59.1 | læcecræfte: læcræfte |
| | and hy habbaþ swyþe: and hy habbaþ swyþe and hy habbaþ swyþe |
| | ne eglad: ne *supplied by editor* |
| | eglas of: eages hof |
| | gyf hy: of hy |
| 60 | gewyht: wewyht |
| 61.1 | þa-hwylc: and þa hylc |
| | betra: bera |
| 62.3 | eftsana: eftswana |
| | wurtrumen: wurtrumem |
| 63.1 | werig: wergi |
| | singalice: singanlice |
| 63.2 | ancleowe: ancweowe |
| | eallunga: ealluga |
| 63.4 | do wylle: do *supplied by editor* |
| | angen: aforenan renangen |
| | þa handa and smyre: þa hunda and smyre |
| | pin-hnutena: pinhutena |
| | þweal: þweald |
| | after spiwe: after and spiwe |
| 63.5 | clympran: clymppan |
| | swa litle: þa litle |

63.6    grind: gnind
63.8    Bynd: byð
64      miht gecnawen: milt gecwawen
65.1    Hit: his
        hwylum of þare lungone: hyylum of þare lungone
        hwylum of þan magen: hyylum of þan magen
65.2    of þan heafode: oþ þan heafode
        of þan uve: and of þan uve
        ingehwyrfþ: ingehwyfþ
        ut-hræcþ: uthræþ
        goman byþ: byþ *supplied by editor*
65.3    æt hys: gæð hys
        synd tobrocene: synd *supplied by editor*
65.4    þearmum: þearnum
        hwyt: hyt
66.1    smalran: smaran
        hym læten: hy læten
        of þan: oþan
        wæter: wætc
66.2    spongiam: spogiam
        hluttrun: hlultrun

## MISCELLANEOUS REMEDIES

1.3     querit: queri
1.4     silvas: silva
1.5     pessimum: persimum *with* per *abbreviated*
        nec celari: nec celare
2       in þa: *Cockayne misreads as* in þam
3.1     feonda: feoda
3.3     per verba: per verbo
        emendatio: amendatio
3.4     conpaginibus: *Jolly reads* compaginibus
        ab homine isto: *with superscript* a *over the* o *of* isto
3.5     nec tangendi . . . nec tacendi: nec tangendi . . . nec tangendi
        nec in fulgendo: nec in fulgendo ll effuie

regnas: *supplied where the reading is hard to make out*

secula: secula secula

4.3      constitutum: constraitum

munitas: munitur

consequatur: consequam *with a macron over the* m *(for* consequamur*)*

5      Fondorahel: fondorohel *with superscript* a *over the third* o

6      Wið: *Storms represents as* wiþ

sancto nomine: sancto nomine sanatione ad ad

7.3      adiutorem: adiutorium; *Olsan reads* adiuturium

famulae tuae: famulā tuā *(for* famulam tuam*)*

8.1      Medicina contra febres: *transposed from the foot of this entry, where it follows after* NOMEN

9.3      Thigat . thigat: *Storms incorrectly inserts a third* thigat

tedes: *Storms misreads as* sedes

herclenom: *Storms misreads as* hercleno

arcocugria: *Storms reads as* arcocugtia

10.2      ista orationem: *Storms misreads as* ista novem

11.5      per cherubin: *Storms misreads as* et cherubin

11.8      et "Salvum fac servum tuum, Domine": et ne nos salvum fac servum esto ei Domine

11.10      animae: *Storms reads as* animam

12.3      andies . mandies: *Storms reads as* audies . maudies

13.2      portaret: portare

15.5      prophetis: *Storms misreads as* profetis

eternis: eterilis

Christus: Christi *(abbreviated)*

ac fidelibus: *Storms misreads as* de fidelibus

16      cearfille: ceruille, *with* a *inserted above the initial* e, *and* f *inserted above* u

17.2      G̅T̅: *evidently an abbreviation of some kind*

ribbe ond gearwe: ribbe gearwe

swyþe lytlum ond litlum: *Cockayne omits* swyþe

19.1      an hrerenbræden æg: *Storms omits* an

19.3      huic: huic. ðæt is

21      famulum (vel famulam) tuum (vel tuam): famulum tuum *with a*

*crossed* l *(as an abbreviation for* vel*) and the alternative vowel* a
*written above the inflectional ending* um *in each case*
de omni populo: *Storms omits* de omni
24 cnuca: cunca
25 hraca: ranca, rauca, *or* raucka, *but scribal overwriting makes the
word hard to make out*
heortes: hortes
eara pið: hera wið *but with* wið *poorly written and hard to make out*
26 *Certain words are hard to make out because of the scribe's poor hand, a
wretched nib, and a densely abbreviated text.*
27 *Damage to the manuscript makes a number of readings uncertain.*
deah: ðeah
28 *Damage to the manuscript makes a number of readings uncertain.*
28.3 færlicum: fær
ge [. . .] ge: *this passage is scarcely legible; it may have been meant to be
cancelled*
29.1 *After* spiritus sancti *the letter* L *is written*
liberaretur: libaretur
33.2 wad-meolo: wahmealo
33.3 Meng eal tosomne: *some text is lost preceding these words*
35.2 smera: mera
eall swa fela dropena unhalgodes: swa *supplied by editor*
36.2 vermes . . . in me: *a blank space here leaves room for a dozen or so
characters*
de mortuis: de mortunrto *(?)*; *Storms reads as* demo rumto
36.3 Spiritus: spiritu
faucibus: facibus
36.4 perveniat: perveniad *(?) with the* d *or* t *miswritten*
37.3 ond bærn to ascen: *Cockayne omits by apparent oversight the words
that follow, from* ond nim þonne to ofergeot þa ascen
38.3 Nim þa: *Cockayne omits* þa
þærf: *Cockayne misreads as* þearf
39.1 wenne: *Cockayne misprints as* penne
39.2 hawenhydele: hæwene dile
salf: *corrected by a later hand to* sealf
helpð: help

41      Wið: Wð
        ealdes: ealdas
45      consolda: *Cockayne misreads as* consolida
        idem ius: ide iussum
46      menstruam: menstram
48      Gyf hryþeru: gyf *supplied where letters are lost to the burn*
        [. . .]ton hylle: *the beginning of the remedy is lost to the burn*
        mæssedæg: dæg *supplied where letters are lost*
        Do þærto: do *supplied where letters are lost*
        on middansumeres mæsse mergen: sumeres mæ *supplied where*
            *letters are lost*
        Miserere nostri: miserere *supplied where letters are lost*
        Exsurgat: exurgat
49      Genim lytel: genim *supplied where letters are lost*
        do þæt hi hraðor: hi hrað *supplied where letters are lost*
50      Sing ymb: sing *supplied where letters are lost*
51.1    Genim twegen: *supplied where letters are lost*
        on ælcere: on æ *supplied where letters are lost*
        lege þone forman þam berene: *multiple letters supplied where they*
            *are lost or illegible;* þone *and* þam *are the only whole words clearly*
            *visible*
        þone oðerne: oðer *supplied where letters are lost or illegible*
        on . . . ofer: *the part marked by an ellipsis is illegible*
        sticcan: an *supplied where letters are lost*
51.2    [. . .] þæt þær si: *folio 16a begins here; an unknown amount of text is*
            *lost before* þæt
        hyrnan: *only* hyrne *is visible because of a hole in the manuscript*
52      heafod-ece: eafod ece
        þæt is: is *supplied by editor*
53      wyrt: wurd
        hatað: hatad
55      ellan: eallan
60      *Between* optime *and* per *occur two marks whose significance is un-*
            *clear*
62.2    venas quae: venas *supplied by editor*
63      siderum: sidera

restringe trea: restrige trea
exiens restringe: *Barb misreads as* exiens restinge
flumen cruorem: *Barb omits* flumen
restrigentem: restrigantem
limitem: limentem

65.2 Deum vivum: *Storms misreads as* Deum unum
qui habeat: qui abeat

66.2 supplicis: supplices

66.3 labante: lavante
Deus in unitate: *Barb omits* Deus

67.3 salutis meae: *Storms misreads as* salutis mei

67.6 fluenta: *Storms interprets as* fluente

69 ut convalescat: ut convalescens
et tibi: et *supplied by editor*

70 Christe: *Storms misreads as* Christi
Blasii: *Storms misreads as* Blassi
NOMEN: *supplied by editor*

71.1 apud Deum: *Storms omits by apparent oversight the phrase that follows, from* et Deus erat *to* apud Deum

73 Finit: fi

74.4 *A cross is written above each instance of* AGYOS

76 feldwyrt: ferwyrt, *with the second letter corrected so as resemble a tailed* e, *so that the word is interpreted by the DOE as* færwyrt

78 elenan ond cyrfillan: elenan fillan

81.1 ... ond: *an unknown amount of text is lost before* ond

82.1 ierne: *the letter* r *is hard to discern*
sume: ssume; *the letter* e *is hard to discern*

82.2 grin: gerin (?) *there is a mark below the* e *that might be a mark of deletion*
grenn: *the first of the two letters* n *is hard to discern*
⫢: *construed by Storms as a series of three crosses*

82.3 horsse: *Storms reads* horse

83.1 Writ ðis: writ ð

83.3 protectione: protecte
restet: restat

84.3 ligatis: ligitas (?)

quatriduanus: *only* quatri . . . s *is clearly legible in the manuscript*

85.2 Writ: wrið

87 *The text breaks off abruptly at the end of the last folio of a quire.*

89.1 Wið blodrine: *Storms represents as* Wiþ blodrinu

89.2 Writ: wright; *Storms reads* wriht

90.4 et archangelos: et *supplied by editor*

From adiuro *to the end, the text is scarcely legible; the present construction of this part of the cure is supplied from a seventeenth-century transcript printed by Storms, no. 36, p. 276*

ullum potestatem: potestatem *supplied by editor*

93 wið smælum: wið smæng wið maelum

Clæm: clæn

geduinen: geðuinen

ðaeh ðe: ða ehde

94 Wyrc: wyrt

wæt ðu ða: wæt ða ðu, *with the two latter words reversed in order and with* wæt *miswritten; Schauman and Cameron read* weet

96 Wið lend-wærce: To ðon ilcan wið lend wærce

97 ond walwyrt: ond *supplied by editor*

ðorn: dorn

cuiicbeam: cuiicbean

baeð: *Schauman and Cameron read* bæð

98.2 secg: seecg

98.3 Waell: *all but the last two letters of the word have been trimmed away*

smeorue: smeome

Smire: *all letters but the final* e *have been trimmed away*

aslepen: alepen

99 Aec swilce: *all letters but the last three have been trimmed away*

swegles eapol: apol; swegles *restored on the basis of a parallel in Bald's Leechbook 1.23*

# Notes to the Translations

1 The contents of chapter 1, with its twenty-nine entries pertaining to the herb *betonica* (betony), circulated as a separate treatise in the earlier Latin herbal tradition.

1.2 *the weight of two tremisses*: The tremiss was a generic measure of weight, originally the equivalent of that of a small gold coin; see Hubert Jan de Vriend, *The Old English "Herbarium" and "Medicinia de quadrupedibus,"* EETS o.s. 286 (Oxford, 1984), lxxxiii.

*beer*: This is not the same beverage as goes by the name "beer" today; what Old English *beor* denotes was a more concentrated alcoholic drink concocted of honey and the juices of herbs.

1.8 For a variant of this remedy see *Bald's Leechbook* 1.6.1.

1.9 For variants of this remedy see *Lacnunga* 102 and 109. There is a similar remedy in *Bald's Leechbook* 1.21.3, and the present remedy may have the same Latin source as *Lacnunga* 116; see Edward Pettit, ed. and trans., *Anglo-Saxon Remedies, Charms, and Prayers from British Library MS Harley 585: The Lacnunga* (Lewiston, 2001), vol. 2, p. 208.

1.10 For a variant of this remedy see *Bald's Leechbook* 1.22.1.

1.12 *If a person is constipated*: Literally, "if a person's intestines are constipated." See the *DOE*, under *innoþ*, sense 4.b.ii.

*in three days' time*: Literally, "in three nights' time." The Old English idiom was to speak of the passing of time by the number of nights, not days.

1.13 *If a person spits blood*: Literally, "if blood flows through a person's mouth."

There is a different translation of this Latin remedy in *Bald's Leechbook* 1.7.1, while *Lacnunga* 102 presents another parallel; see Pettit, *Remedies,* vol. 2, p. 203.

1.14    *let him first of all take and eat*: That is, one should eat a bit of betony before starting to drink alcohol. "And" has been supplied editorially.

1.16    *ruptured internally*: Referring to an internal rupture, such as a hernia.

1.17    For a variant of this remedy see *Bald's Leechbook* 1.79.1.

1.20    *give him one to eat . . . drink three cupfuls of this liquid*: The translator seems to have had some problems with the Latin source, which calls for the patient to swallow one pill and then drink two cups of warm water: *ex quibus pastillum unum gluttiat et aquae calidae quiatos duo bibat* (de Vriend, *Herbarium,* 35). The remedy was then to be repeated, it seems.

1.25    For a variant of this remedy see *Bald's Leechbook* 1.69.1.

1.29    *Lacnunga* 119 is an analogous remedy.

      *Foot disease*: Old English *fotadle* (pain in the feet) can be a symptom of gout, but this is not always the case. Therefore the translation "foot disease" has been used throughout.

2.2    *until it becomes clear*: Interpreting *blacu* as "clear"; see the *DOE,* under *blac,* sense 2 (pale), and Anne van Arsdall, *Medieval Herbal Remedies: The Old English Herbarium and Anglo-Saxon Medicine* (New York, 2002), 142n82. De Vriend, *Herbarium,* 287, supplies *wlacu* (lukewarm), a reading that is supported by the Latin original, which has *tepefacito* (warmed).

2.11    *If there is a hardening on the body*: Translating Latin *duritia* (hardness). It is hard to say just what the ailment described here is.

2.12    *the fever that comes on a person every fourth day*: The quartan fever.

2.14    *the fever that comes on a person every third day*: The tertian fever.

2.15    *the fever that comes every second day*: The translator of the Latin text understood *secundarum* (the secundine, or afterbirth) in the indication *ad secundarum dolorem* as a third sort of fever.

2.19    *located on someone's face*: Old English *neb* could mean either the nose or the face; see James Bosworth and T. Northcote Toller, *An Anglo-Saxon Dictionary* (Oxford, 1898), with *Supplement* by

T. Northcote Toller (1921), and with *Enlarged Addenda and Corrigenda* by Alistair Campbell (1972), under *nebb*, sense III. The choice of "face" (here and elsewhere) is supported both by the Latin original and by the fact that Old English *nosu* denotes specifically the nose.

2.22 There are variants of this remedy in *Bald's Leechbook* 1.22.1 and in *Lacnunga* 103; see Pettit, *Remedies*, vol. 2, p. 204.

3.8 For a variant of this remedy see *Bald's Leechbook* 1.45.8.

3.9 *to heal an ulcerous sore*: See the *DOE*, under *abledan*, sense 3: "to close, to heal [an ulcer]."

*An old barrow pig*: That is, a gelded boar.

4.1 *hard swellings*: A number of different ailments, including tumors and boils, could be encompassed in Old English *cyrnel*, a term that comes up many times in these pages.

4.8 *spider bite*: Old English *attorcoppa* denotes a kind of poisonous spider.

4.9 *the poison has been drawn out*: The phrase "the poison" is supplied editorially.

5 *henbane*: White henbane (*Hyosciamus albus* L.), not *Hyosciamus niger* as is suggested by de Vriend, *Herbarium*, 289. The text distinguishes between the two species.

*Bellonaria*: A solanacea usually identified with poison gooseberry (*Withania somnifera* Dunal); see Jacques André, *Les Noms de plantes dans la Rome antique* (Paris, 1985), 35.

5.2 There is a similar remedy in *Leechbook 3* 50.1.

7.2 *If a person cannot urinate . . . stopped*: The reference is to dysury (pain and difficulty in discharging the urine) and strangury (painful passage of a few drops of urine instead of the normal flow).

8 *lion's foot*: Another name for the plant is "lady's-mantle."

8.1 *aristolochia*: "Birthwort"; see also *Herbal* 20, *smerowyrt*.

9 *he will take leave of his life laughing*: Crowfoots are known to be toxic and are said to cause convulsive laughter; see André, *Les Noms*, 20.

9.1 *the healthy flesh*: Literally, "the healthy body."

10.1 *a person suffering from lunacy*: Referring to persons considered mentally ill, afflicted by madness. The Old English term *monoðseoc* is a translation from Latin *lunaticus,* referring to a person under the dangerous influx of the moon.

11     *it prevents evil medicine*: That is, it averts "bad magic," or witchcraft.

      *It also turns away the evil eye*: Literally "it turns away the eyes of evil people."

12.4    *that it might have such great power*: That is, that tansy might have such great power.

13.2.2   *Diana*: Or Artemis, the goddess of the moon. Also known as the Great Mother, the healer and expert in plants.

      *The centaur Chiron*: The son of Chronos (Saturn) and of the nymph Philyra, the daughter of the Titan Ocean; he is the archetype of healers on account of his vast knowledge of herbs and healing. He was the teacher of Achilles, and another of his pupils was Esculapius (Asclepius), the god of medicine.

      *After the name of that same Diana*: That is, the plant is referred to by the Greek name for that goddess (Artemis).

14.1    *fumigate it*: Literally, "smoke it." See the *DOE,* under *berecan:* "to expose to smoke, to smoke."

15      *Apulia*: A region in southeastern Italy, forming the *tacco,* or heel, on the boot of Italy.

15.2    *a broken bone*: Old English *banbryce* could denote either a simple or compound fracture; see the *DOE,* under *banbryce.* In this case the remedy may be intended to extract bone fragments that protrude through the skin.

19.1    *you will perceive a wonderful change*: Literally, "you will perceive a wonderful thing," as in the original Latin phrase *miram rem experiris.*

19.6    *For diarrhea*: The remedy might also be for dysentery; the Old English term *utsihte* could denote either condition.

20.5    *sester*: A Roman measure of capacity corresponding to somewhat more than a pint. For discussion, see de Vriend, *Herbarium,* lxxxi–lxxxiv.

20.7    *an ulcer*: The apparent meaning of Old English *weargbræde.*

      *Cypress*: Probably a misinterpretation by the Anglo-Saxon translator, as the plant mentioned in the Latin source is not the cypress, but rather an aromatic plant, "English galingale," belonging to the *cyperus* family.

22.1 *oil of the cypress tree*: Here too the Anglo-Saxon translator misinterpreted his source. Latin *cyprinum oleum* is the ointment made from *cypros* (henna). See André, *Les Noms*, 85, and also *Herbal* 76.2 below.

22.2 *If spots*: Old English *nebcorn* is usually translated "pimple"; see Bosworth and Toller, *Dictionary*, under *nebcorn*. The translation "spots" is based on the fact that the word is documented only seven times in Old English, always translating Latin *lentigines* (freckles, lentil-shaped spots).

23 *Poison gooseberry*: The name *apollinaris*—the plant of Apollo—denotes several poisonous and hallucinogenic plants. *Apollinaris* grows in Europe, especially around the Mediterranean basin; it has been known since Dioscorides and Pliny. In late antique Latin texts it denotes "poison gooseberry"; see André, *Les Noms*, 21. De Vriend, *Herbarium*, 293, identifies this plant with the "lily of the valley," *Convallaria majalis* L. However, the use of this plant in medicine is fairly late. The *DOE*, under *glofwyrt*, accepts de Vriend's interpretation. There is full discussion in the Dictionary of Old English Plant Names (http://oldenglish-plantnames.org), under *glofwyrt*. *Glof* (glove) appears also in the compound *foxes glofa* that translates *strychnon manicon* (thorn apple) at *Herbal* 144.

*He thus gave it this name*: That is, Esculapius gave it this name.

24 *Chamomile*: See the Dictionary of Old English Plant Names, under *mageþe*.

24.1 *albugo*: A disease of the eye causing dimness of vision; perhaps cataracts. It is also treated at *Herbal* 75.1.

25 *Germander*: See the Dictionary of Old English Plant Names, under *heortclæfre*.

26 *Wild teasel*: *Chamelaea* normally denotes the *Daphne* species. It became a name for the thistle species because of a confusion with *chamaeleon*. See the Dictionary of Old English Plant Names, under *wild teasel*.

NOTES TO THE TRANSLATIONS

| | |
|---|---|
| 27 | *Ground pine*: This interpretation of Old English *henep* rests on the observation that the plant mentioned in the Latin text, *chamepithys*, is ground pine. However, *henep* usually denotes "hemp," as in *Herbal* 116; see the *DOE*, under *hænep*. |
| 28 | *Figwort*: Another name for the plant is "lesser celandine, pilewort." For discussion of the Old English equivalent name *hræfnes fot*, see the Dictionary of Old English Plant Names, under *hræfnes fot*, and the *DOE*, under *hræfn*, sense 3. |
| 29 | *Madder*: Latin *ostriago* is usually identified with madder; see André, *Les Noms*, 183. For discussion of the Old English equivalent name *lyðwyrt*, see the Dictionary of Old English Plant Names, under *lipwyrt*. |
| 29.1 | *to his annoyance*: Or, "to his harm." |
| 29.1.2 | *free of defilement*: Literally, "clean, purified," translating Latin *mundus*. |
| 30.3 | *it is not available at all times*: Literally, "it does not appear at all times," as the plant dies off in winter. |
| 31.2 | *this should be collected without smoke*: That is, the honey should have been collected without smoking the bees, a common practice of beekeepers when harvesting honey, intended to make the bees drowsy and less likely to swarm. |
| 32.1 | *the defect*: Old English *tal* is attested by Bosworth and Toller, *Dictionary*, only in the sense of "evil-speaking, calumny"; see, however, de Vriend, *Herbarium*, 296, who suggests "fault, defect." |
| 32.8 | *from an iron blade or a cudgel*: More generally, perhaps, from a metal tool or implement or from a wooden one. |
| 33.1 | *and make it very soft*: This must be the sense of the Old English phrase, though the grammar is less than clear. |
| 34.1 | *crumbs*: Literally, "the crumb." |
| 35 | *Yellowwort*: Another name for the plant is "greater celandine." |
| 36 | *febrifugia*: A common name for plants, such as the lesser centaury, that have the medicinal quality of driving away fevers. Today "feverfew" denotes another plant, bachelor's buttons (*Tanacetum parthenium* L.). |
| 36.2–3 | For variants of these remedies see *Bald's Leechbook* 1.2.20. |

36.3    *If someone is subject to that same danger*: Referring to the eye pain
        described in the previous remedy.

        *jugful*: Old English *amber* is a measure of capacity; here that
        word translates Latin *hemina*. See the *DOE*, under *amber,* sense
        2.a.i, and for discussion, see de Vriend, *Herbarium,* lxxxiii.

36.6    *If worms should be doing harm around the navel*: Alluding, per-
        haps, to wormlike movements in the area of the navel that can
        signify a parasitic infection in the intestines. A description
        of these symptoms—"painful cramps located around the na-
        vel"—appears in the Latin medical tract *Tereoperica,* part of
        which was translated into the English-language treatise *Peri
        Didaxeon*. See Laura López Figueroa, "Estudio y edición crítica
        de la compilación médica latina denominada *Tereoperica*" (PhD
        diss., Santiago de Compostela, 2011), 389.

37.6    *hawthorn leaves*: See the *DOE*, under *haguþorn, hægþorn.*

40.1    *If a person has a bloated stomach*: In the table of contents preced-
        ing the *Herbal,* this same remedy is listed as being *wiþ utsiht* (for
        diarrhea). The same is true of *Herbal* 53.1 and 69.1.

41.3    There is a parallel version of this remedy in *Lacnunga* 118; see
        Pettit, *Remedies,* vol. 2, p. 208.

42      *Bugloss*: Old English *glofwyrt* is used erroneously as a synonym
        for "hound's-tongue" or bugloss.

        *Grows*: In the Old English text, after the formulaic phrase that
        introduces the description of the plant, the personal pronoun
        *heo* (she or it) is redundant. See also *Herbal* 98, 109, 152.1, 185 for
        similarly redundant pronouns.

43.3    *paronykia*: Nail disease caused by bacterial or fungal infection, or
        by agnail.

46.1    There is a parallel version of this remedy in *Lacnunga* 174; see
        Pettit, *Remedies,* vol. 2, p. 347.

46.3    There is a similar version of this remedy in *Lacnunga* 130; see
        Pettit, *Remedies,* vol. 2, p. 262.

        *Harmful worms around the navel*: See note to 36.6 above.

        *And lupin*: For another interpretation of Old English *elehtre,* see
        Maria Amalia D'Aronco, "A Problematic Plant Name: *Elehtre,* a

Reconsideration," in *Herbs and Healers from the Ancient Mediterranean through the Medieval West,* ed. Anne van Arsdall and Timothy Graham (Farnham, 2012), 187–216.

46.6      *impetigo*: An infection of the skin, Old English *teter;* see also below, *Herbal* 122.1 and *Rem. An.* 3.9–10.

46.7      There is a similar version of this remedy in *Lacnunga* 54; see Pettit, *Remedies,* vol. 2, p. 58.

47      *Sword grass*: Also known as wild gladiolus; see André, *Les Noms,* 278, and the *DOE,* under *fox,* sense 3.b.

47.2      *It will draw out the fractured bone*: See above, 15.2.

     *poisonous bones*: That is, the "poisonous" fangs of reptiles and snakes; see the *DOE,* under *ban,* sense A.1.f.

49      *moly*: This is the magic plant that Mercury (Hermes) gives to Ulysses (Odysseus) in Homer's *Odyssey* in order to protect him from Circe's magic powers. The plant has not been identified; in some illustrated Latin pseudo-Apuleius manuscripts, it is identified with some species of garlic, perhaps because it is described as having a round root similar to a leek's. Old English *singrene* usually denotes either "houseleek" or "common periwinkle." See Hans Sauer and Elizabeth Kubaschewski, *Planting the Seeds of Knowledge: An Inventory of Old English Plant Names* (Munich, 2018), under *singrene.*

50      *Heliotrope*: The Latin name of the plant signifies "a plant that turns, following the sun," as does the Old English name *sigelhweorfa,* from *sigel* (sun) and *hweorfan* (to turn). The plant is usually identified with *Heliotropium europaeum* L. This same plant is mentioned in *Herbal* 137. The reason for this doubling of certain entries is that some of the plants translated from the two pseudo-Dioscoridean tracts, *Liber de herbis femininis* and *Curae herbarum,* are described independently (and sometimes with another name) in the pseudo-Apuleius *Herbarius.* Thus there are other doublets as well: *pionia* and *aglaofotis* (both peony), at *Herbal* 66 and 171; *scordea* and *scordios* (water germander), at *Herbal* 72 and 163; and *gearwe* and *achillea* (yarrow), at *Herbal* 90 and 175.

50.2      *flux*: The corresponding remedy in the Latin source is for a sore, not for some sort of flowing matter. The translator of the Old English text mistook Latin *luxum* (dislocation, luxation) for *fluxum* (flow).

51 *Madder*: The identification of the herb *grias* with *mædere* is the Old English translator's; the plant *grias* has not been identified; see André, *Les Noms,* 113.

*Lucania*: An ancient region of southern Italy, now mostly corresponding to Basilicata.

53 *Asphodel*: This is not the same plant that is found at *Herbal* 33, although they share the same Old English synonym *wudurofe.* The plant mentioned in the Latin source of *Herbal* 53 is "clary sage." See André, *Les Noms,* 117. De Vriend, *Herbarium,* 300, suggests "woodruff"; see the Dictionary of Old English Plant Names, under *wudurofe.*

54.1 *bleary eyes*: See Bosworth and Toller, *Dictionary,* under *torenige,* a term glossed there as "blear-eyed." Here the word translates Latin *epiphora,* a disease characterized by excessive lacrimation.

55.1 *and by another name [. . .]*: Here and elsewhere in passages similar to this one, the scribe left room for someone else to fill in the blank with an appropriate name.

56.1 *throatwort*: In this case the plant mentioned can be identified with "narcissus." See the *DOE,* under *healswyrt,* sense a.

57.1.2 *broad beans*: Old English *bean* translates Latin *faba,* denoting a plant that is a Mediterranean native, quite different from the common or French beans commonly used in today's cuisine, which are native to Central America.

*Cilicia and Pisidia*: Cilicia is the former Roman province on the southern Mediterranean coast of Turkey; Pisidia is a region of ancient Asia Minor corresponding to today's province of Antalya, Turkey.

58.1 *This will give proof of the same thing*: That is, "the second method will prove as efficacious as the first." See Bosworth and Toller, *Dictionary,* under *onfundelness.*

59.1 *rheum*: Old English *dropa* can denote humoral disorders, as it does here; see Lois Ayoub, "Old English *Wæta* and the Medical

Theory of the Humours," *Journal of English and Germanic Philology* 94 (1995): 342–43; and the *DOE,* under *dropa,* sense 4: "medical: gout, rheum."

60.1   *excessive flow of menstrual blood*: See the *DOE,* under *flewsa,* sense c.

60.2   *internal rupture*: Internal tearing of some organ, or a hernia.

63      *Dittany of Crete*: Another name for the plant is "white dittany." Crete is the largest of the Greek islands; Mount Ida is its highest peak.

63.2   *by iron or by a pole*: That is, by a metal weapon (such as a knife) or a wooden implement (such as a cudgel).

63.5   *stitchwort*: See the *DOE,* under *æpelfeorþingwyrt.*

*Hindheal*: See the Dictionary of Old English Plant Names, under *hind-heoloþe.*

64      *Heliotrope*: The Latin name *solago* means "plant of the sun." It is usually identified with plants that turn with the sun, usually some kind of heliotrope; see André, *Les Noms,* 242.

66      *Peony*: The same plant is mentioned in *Herbal* 171 under the name *aglaofotis.*

*Peonio*: That is, Paeos, physician to the gods, son of Esculapius (Asclepius), the god of medicine; see Homer, book 5 of the *Iliad.*

*Malum granatum*: Pomegranate.

*Its seeds are like cockles*: That is, they are like tares (vetch seeds); but the Old English word *coccele* represents a mistranslation of Latin *grano cocci* (cochineal grain).

66.2   *hip-bone ache*: Pain in the hip, probably sciatica; the Latin source has *ad sciaticos.*

67      *Gypsywort*: See the Dictionary of Old English Plant Names, under *berbene.*

68.1   *customary drinks*: The Old English translator seems to have simplified the difficult Latin text, which recommends mixing the plant *in tyriacis potionibus* (in drafts of theriac). Theriac was a medicinal compound made up from numerous ingredients, used as an antidote to snake venom or other poison.

70      *thistle*: This interpretation of Old English *clæfre* rests on the ob-

servation that the plant mentioned in the Latin text, *crision/ cirsion*, is a thistle. However, *clæfre* usually denotes "clover"; see the *DOE*, under *clæfre*, sense 1.c.i.

70.1 *For sore throat*: The Old English plural noun *goman* (genitive plural *gomena*), the etymon of modern English "gums," can refer to the whole inside of the mouth (including the jaws), not just the throat.

71 *ad serpentis morsum*: That is, "for snakebite." A curious synonym for this plant's name, one that is common to some manuscripts of the Latin pseudo-Apuleius *Herbarius*.

72 *Water germander*: The same plant is mentioned at *Herbal* 163 under the name *scordios*.

74 *Ironwort*: This interpretation of Latin *heraclea* rests on André, *Les Noms*, 120. De Vriend, *Herbarium*, 305, identifies the plant as "red-star thistle."

75.2 For variants of this remedy see *Bald's Leechbook* 1.2.1 and 1.2.17.

76.3 *berries*: Here, Old English *croppas* translates Latin *bacas* (berries); see the *DOE*, under *crop, croppa*, sense 2.b.

77.3 There is a variant of this remedy in *Lacnunga* 120; see Pettit, *Remedies*, vol. 2, p. 209.

78.2 *If a young man is ruptured*: The disease mentioned is hydrocele, a swelling of the testicles common in newborns.

80.2 For a variant of this remedy see *Bald's Leechbook* 2.41.4.

*in light wine*: The Latin text has *in vino lenissimo, id est inerticio*, referring to a type of mild, sweet wine.

82.1 For a variant of this remedy see *Leechbook 3* 37.1.

82.2 *For women's cleansing*: The remedy is for the purification of the woman after childbirth.

84.1 *For hardening of the abdomen*: Here and elsewhere, what this probably means is "for constipation."

86 *Asparagus*: Here, Latin *asparagus agrestis* is translated *wuducerville* or *wuducerfilla* (probably *Asparagus officinalis* L.); at *Herbal* 97.1 and 126.2, however, the same Latin plant name has been translated *eorðnafola* (asparagus), perhaps with reference to some sort of wild asparagus. See the *DOE*, under *eorþnafela*, sense 1.

86.4    He will be unbound: That is, "He will be freed from the enchantment."

87.1    *the king's disease*: According to antique medicine, a disease that can be cured by a king. The identity of the disease varied from period to period, ranging from jaundice to leprosy. For the interpretation of the disease as was current in Anglo-Saxon England, see Frank Barlow, *The Norman Conquest and Beyond* (London, 1983), 25–26. See also de Vriend, *Herbarium,* 307–8.

87.3    This remedy has a parallel in *Lacnunga* 93; see Pettit, *Remedies,* vol. 2, p. 199.

88.1    *the plant that is called canis caput*: The Latin name for the plant signifies "dog's head," as does the Old English name *hundes heafod.*

90      *Yarrow*: The same plant is mentioned at *Herbal* 175 under the name *achillea.*

90.11   *heartburn*: Either hiccups or heartburn (acid reflux) might be meant.

90.13   *venomous creature called spalangius*: The Old English term *næddercynn,* meaning, literally, "type of snake," can refer more generally to any poisonous creature. The *spalangius* was a kind of venomous spider. See Pettit, *Remedies,* vol. 2, p. 81.

        *if it is closing up too quickly*: In premodern times, if a wound closed up too quickly there would have been problems of sepsis. Until the nineteenth century, the techniques for wound healing were not much different from those given by the Roman physician Galen.

91.2–3  For a variant of these remedies see *Leechbook 3* 69.2–3.

91.5    *lethargus*: The normal meaning of this term is "drowsiness, lethargy"; see Bosworth and Toller, *Dictionary,* under *ofergitolness.*

91.7    For a variant of this remedy see *Bald's Leechbook* 1.1.2.

92.2    *skin disease*: Old English *hreofla* is a comprehensive term for various skin diseases, including leprosy. See the *DOE,* under *hreofla* 2, sense 1.

94.2    For variants of this remedy see *Bald's Leechbook* 2.6.3 and *Leechbook 3* 69.4.

94.3     *Lacnunga* 122 is a variant of this remedy; see Pettit, *Remedies,* vol. 2, p. 210.

94.4     For a variant of this remedy see *Leechbook 3* 70.1.

94.11    There is a parallel remedy in *Lacnunga* 142; see Pettit, *Remedies,* vol. 2, p. 270.

97.1     *asparagus*: Probably wild asparagus, as in *Herbal* 126.2. The plant mentioned in *Herbal* 86, *wuducerville,* seems to be another kind of asparagus.

97.3     *harmful worms around the navel*: Old English *rengwyrmas* are intestinal worms, corresponding to Latin *lumbrici* (worms) in the general sense.

98       *This plant is called*: Here, in the Old English formulaic phrase that introduces the description of the plant, the relative pronoun *þe,* "which, that," is redundant. See also *Herbal* 42, 109, 152.1, 185 for similarly redundant pronouns.

        *Cynoglossum* and *lingua canis*: The Latin names for the plant signify "dog's tongue."

100     *Ground ivy*: Perhaps a mistranslation on the part of the Old English translator. The plant mentioned in the Latin source is *hedera nigra,* the common or black ivy; see the *DOE,* under *eorþifig,* sense 1, and *ifig,* sense 1.b.

100.2    There is a parallel version in *Lacnunga* 130; see Pettit, *Remedies,* vol. 2, p. 262.

100.4    *venomous creatures . . . spalangiones*: Old English *wyrm* could denote a "creeping insect or worm." See Bosworth and Toller, *Dictionary,* under *wyrm,* sense II. For *spalangiones,* see above, note 90.13.

102.2    *horehound, and lupins*: See *Herbal* 46, *harehune,* and 112, *lupinum montanum.*

103.2    *anus*: The remedy is for perianal itching. For a variant of this remedy see *Bald's Leechbook* 1.29.1.

104.2    *delivery*: "The entire process of birth"; see the *DOE,* under *geeacnung,* sense 5.

110.3    *severe skin disease*: In this remedy Old English *hreofl* denotes leprosy.

*Resin*: Here and at *Herbal* 152.1, *tyrwe* translates Latin *resina;* in
*Rem. An.* 4.1 it translates Latin *bitumen,* "tar."

114 *which is called lactuca leporina*: That is, it is called "hare's-lettuce,"
from Latin *lepor* (hare).

115.2 *born prematurely*: See Bosworth and Toller, *Dictionary,* under *misboren*. The Old English translator did not understand that his
source treats of an abortion and that the woman should wash
herself with the potion rather than washing the child; see M. L.
Cameron, *Anglo-Saxon Medicine* (Cambridge, 1993), 180.

116 *Hemp*: Here the term correctly denotes *Cannabis sativa* L.

116.2 *For frostbite*: Literally, "for chill burns."

117.3 For a variant of this remedy see *Bald's Leechbook* 2.24.10.

118 *heptaphyllum*: Literally, "seven leaves," like Latin *septifolium*.

121 *because it bears seeds that are like grains of gold*: The etymon of *crisocantes* is Greek χρυσός (gold).

122.1 *impetigo and a pimply body*: In this remedy, *teter ond pypylgende lic*
correspond to Latin *ignis sacer* (holy fire), perhaps Saint Anthony's fire. See Alessandra Foscati, *"Ignis sacer": Una storia culturale
del 'fuoco sacro' dall'antichità al Settecento* (Florence, 2013).

126.1 There is an analogous remedy in *Lacnunga* 100; see Pettit, *Remedies,* vol. 2, p. 202.

127 *the mountain called Soractis*: There is no mountain of this name in
France (Gaul), though there is a Mount Soratte in Italy, not far
from Rome. This incorrect information is in the Latin source.

128 *Comfrey*: This interpretation of Old English *halswyrt* rests on
the observation that the plant mentioned in the Latin source is
comfrey; see André, *Les Noms,* 253, and the Dictionary of Old
English Plant Names, under *halswyrt*.

129.1 *one shilling-weight*: In medieval times coins could be used in the
system of weights; see also the note on "tremiss" at *Herbal* 1.2.

130.2 There is a variant of this remedy in *Bald's Leechbook* 1.21.6.

131 *Sweet basil*: Old English *nædderwyrt* translates Latin *basilisca* (the
plant of the snake *basiliscus*).

132.0.2 *will be deceived in that same way*: The allusion is to the centuries-
old legend about the mandrake's scream when pulled from the
ground, causing death to all living beings who hear it. A dog is

therefore to be deceived into pulling it up so that people in the area can cover their ears and be spared. See John M. Riddle, *Goddesses, Elixirs, and Witches: Plants and Sexuality Throughout Human History* (New York, 2010), 61–62.

132.3     *foot disease*: Here, as at *Herbal* 173.4, Old English *fotadle* probably refers specifically to gout.

135     *but this kind*: The Old English text has only the pronoun *heo*, "she" or "it," at this point. While this may refer back to the feminine noun *wyrt* with which the entry begins, the entry lacks clarity and is subject to different interpretations.

135.1     *pain in the thigh*: Perhaps sciatica, as the Latin source indicates that the remedy is useful for *sciaticis*.

135.3     *will drive out all evils*: In the Old English text there is no expressed object for the verb; the sense is that the plant is useful for a lot of problems.

135.5     *malum cydoneum*: "Quince"; see André, *Les Noms*, 152.

136.2     *for movement of the bowels*: According to the *DOE* entry for *astyrung*, sense 2, this is the meaning of *innopes astyrung* (or *innopa astyrung*); alternatively, the phrase might denote "intestinal disorders."

137     *Heliotrope*: The same plant is described in *Herbal* 50 above.

137.2     *niter*: Saltpeter, the mineral form of potassium nitrate.

137.3     *verrucaria*: That is, "the plant which cures warts," from Latin *verruca* (wart).

139     *is of no use*: Of no medical use, that is.

139.1     *erysipelas*: A severe skin disease characterized by burning pain, perhaps "Saint Anthony's fire," Latin *ignis sacer*. See above, 122.1.

139.4     *for intestinal flux*: What is meant very possibly is liquid discharge from the intestines.

140.1.2     *a light soup*: Or "clear soup." Usually Old English *blac briwe* is interpreted as "black soup," from Old English *blæc* (black). However, *blac* meaning "bright, pale" and *blæc* meaning "black" cannot easily be distinguished, and the context does not always make clear which word is involved. In this case it seems that the soup is to be light, in contrast to the strong medicine it conveys.

140.2     *oriza*: Rice.

141.1     *buphthalmon*: The etymological meaning of Latin *buoptalmon* is
          "ox eye"; an alternative modern English name for the same
          plant is "yellow oxeye." The plant mentioned here could be ei-
          ther "corn marigold" or "crown daisy"; see André, *Les Noms,* 41.
          *The city of Meonia*: Perhaps a mistranslation on the part of the
          Anglo-Saxon translator; the Latin source states that the plant
          grows *iuxta moenia civitatum* (near city walls). See de Vriend,
          *Herbarium,* 320.

142       *Land caltrop*: This interpretation of Old English *gorst* rests on
          the observation that the plant mentioned in the Latin source is
          *tribulus.* See André, *Les Noms,* 263. Usually, Old English *gorst* de-
          notes gorse or juniper.

143       *Fleabane*: Another name for the plant is "spikenard."

144       *Thorn apple*: The Old English name *foxes glofa* corresponds to
          Latin *trycnos manicos,* where *glof* seems to be the translation of
          *manicon,* interpreted as Latin *manica,* "sleeve," hence by exten-
          sion "glove." In the antique medical tradition, however, *mani-
          con,* "a plant whose juice maddens," is the name of very poison-
          ous plants such as poison gooseberry and thorn apple. See
          André, *Les Noms,* 154.

144.2     *pimply body*: Here, as distinguished from *Herbal* 122.1 above, the
          Old English phrase *pypelgende lic* translates Latin *herpes,* a con-
          dition characterized by painful red spots that appear on the
          skin, symptoms that could be interpreted as those of Saint An-
          thony's fire.

145.1     *simmer it; give to drink in warm water*: Literally, "simmer it in
          warm water; give to drink."

146.3     *the plant called capparis*: That is, "caper."

146.4     *severe skin disease*: In this remedy Old English *hreofl* denotes lep-
          rosy.

147.1     *gives the appearance of being always alive*: This observation is based
          on etymology: the meaning of Latin *semperviva* is "always alive."
          *An ell long*: Old English *eln* (ell) is an old linear measure corre-
          sponding to Latin *cubitus,* more or less the length of the fore-
          arm of an adult man.

148     *Sweet marjoram*: See André, *Les Noms*, 225. In this case the Old English term *ellen* may result from the translator's confusion of Latin *sampsuchum* (marjoram) with *sambucus* (dwarf elder).

148.1    *difficulty in urinating*: There is a discrepancy between Old English *unmihticlicnyss* (inability, impossibility) in the table of contents and *unmihticnyss* (weakness, difficulty) in the text. De Vriend, *Herbarium*, 321, suggests that the second of these forms is a corruption of *unmihticlicnyss*. However, since all four Old English manuscripts present the form *unmihticnyss*—attested in Old English only here, where it translates Latin *urinae difficultati* (a problem urinating)—this form has been preferred.

150     *it has whitish flowers*: The flowers of this plant, shepherd's purse (which is quite common almost everywhere), are whitish, *subalbidos* as the Latin source states. The *DOE*, under *hæwen*, offers the meaning "blue, gray-blue"; see, however, sense 1.h: "discolored?"

157     *Golden thistle*: See André, *Les Noms*, 231.

158     *iris Illyrica*: The plant derives its name from the region where it was originally found, that is, from Illyria, which, in classical antiquity, was a region in the northwestern part of the Balkan Peninsula.

       *Xiphion*: In *Herbal* 47, *xifion* is glossed as *foxes fot*, "fox's foot" or "sword grass."

159     *Smooth rupturewort*: This interpretation rests on Hofstetter's proposed identification of the Latin source of this chapter. See Walter Hofstetter, "Zur lateinischen Quelle des altenglischen Pseudo-Dioskurides," *Anglia* 101 (1983): 342–43.

163     *Water germander*: See André, *Les Noms*, 231. The same plant is described at *Herbal* 72.

       *And for this reason is called scordion*: The plant smells like garlic, σκόρδον in Greek.

163.2    *and, as we have just said*: "And" has been supplied editorially.

163.5    *the growth of flesh*: What may be meant is an excess of flesh caused by scar tissues in the healing of wounds.

165.1    *applied to the woman's lower parts*: Literally, "applied below."

165.2    *rhagades*: That is, "fissures."

166.2      *For hardening of the stomach*: The Latin source used by the translator seems to have read *arduram* (hardness), not *ardorem* (heat, burning), the reading published by de Vriend, *Herbarium*, 211.

167      *Zamalentition*: A plant that has not been identified.

167.1      *It will heal them all*: Or, with close to the same sense, "it will heal them entirely."

168      *Persia*: Present-day Iran.

169      *Fleawort*: Also known as "fleaseed."

     *psyllium . . . pulicaris*: The name *psyllium* derives from Greek φύλλα, "flea," and *pulicaris* is from Latin *pulex,* with the same meaning.

169.1      *oil jar*: A measure of capacity corresponding to something less than one pint; see de Vriend, *Herbarium*, lxxxiii.

170      *Dog rose*: Also known as "eglantine."

171      *Peony*: Peony has already been discussed in *Herbal* 66, where it is called *pionia;* here it is called *aglaofotis*.

173.2      *the plant called holus atrum*: The plant is alexander; see *Herbal* 108 above.

173.4      *foot disease*: Here, as at *Herbal* 132.3, Old English *fotadle* probably refers specifically to gout.

174      *Cleavers*: Also known as "stickyweed" (or "goose grass") on account of its sticky leaves and seeds.

175      *Yarrow*: The same plant is described in *Herbal* 90, where it is called *gearwe*.

175.2      *the woman while she is seated*: Literally, "the women while they are seated."

177      *its power is sharp*: That is, the plant is very powerful.

178.1      *wounds caused by cold*: Chilblains.

178.7      For a variant of this remedy see *Bald's Leechbook* 1.81.1.

180.1      *take five pennyweights of these stones*: The stones to be used are the *litospermon* stones mentioned in the previous entry, not the stones in the patient's bladder.

181      *Stavesacre*: Also known as "lousewort."

181.1      *corrupt humor*: Here, Old English *yfelan wætan* renders *humoribus* (humors) in the Latin source. A reference to humoral theory is sustained by the indication that the plant expels the corrupt humor through vomiting. For two other probable references to the doctrine of the humors, see *Rem. An.* 3.1 and 7.12, and note further Ayoub, "Old English *Wæta,*" 338–39.

184 *Tassel hyacinth*: See André, *Les Noms,* 40, sense 1; the second plant is "squill"; see also *Herbal* 43 above.

*It is called* scilliticus: Here *scilliticus* indicates *bulbus scilliticus,* corresponding to *scillodes* in the Old English text.

184.3 *hostopyturas*: A word perhaps formed on the basis of Greek πίτυρον (scurf, dandruff).

*Achoras*: From Greek ἀχώρ (scurf, dandruff); see de Vriend, *Herbarium,* 328.

184.4 *it is said that it is generated by dragon's blood*: There is no Latin source for this remedy. The reference to dragon's blood may derive from the plant name *dracontea;* see *Herbal* 15.

185 *frigilla*: This term is the result of a corruption of the Latin phrase *Afri gelilam;* the reading of the Latin source is *Afri gelilam vocant* (the Africans call it *gelila*).

## OLD ENGLISH REMEDIES FROM ANIMALS

Here and throughout this section (as is specified in Notes to the Texts), where chapter titles occur, they are supplied from manuscript *O*. That manuscript provides no titles, however, for chapter 1, consisting of the preface and medicine from the badger; chapter 2, for medicine from the mulberry tree; chapter 11, for medicine from the lion; chapter 12, for medicine from the bull; or chapter 13, for medicine from the elephant. Moreover, chapters 6 and 7 are treated in *O* under a single heading, for medicine from the wild goat and the domestic goat.

1.1 *Idpartus*: Hipparcus, the author of the *Epistula Hipparchi de taxone* (Letter of Hipparcus on the Badger).

*emperor Octavian*: The first Roman emperor, Augustus (born Gaius Octavius Thurinus), the adopted son and heir of Julius Caesar.

1.1.2 *that were given by Asclepius*: Here, Old English *ferdon* is used in the unusual sense "to obtain, derive (a remedy) from (someone)"; see the *DOE,* under *feran,* sense V. B. Asclepius is the Greco-Roman god of medicine.

*I am giving*: Supplied editorially; the sentence lacks a verb form.

1.2    *badger*: The European badger, *Meles meles.*

*The girdle of the prophet Obadiah*: This is a case of misinterpretation. The translator seems to have associated the verbal form *audias* (you hear) in his Latin original with *Abdias* — that is, the name of the prophet Obadiah. Then, since a few lines earlier, he had mentioned a linen cloth, he might have remembered an Old Testament account of a prophet and his linen girdle. However, he mistook Obadiah for Jeremiah (Jeremiah 13:4). For discussion, see de Vriend, *Herbarium,* 329.

1.7    *of the boundaries of your property*: Literally, "of your boundaries."

1.9    *in the days when you are abstinent*: Although the sense of the Old English phrase is somewhat opaque, the reference might be to days of ritual purification.

2.1.1    *For bleeding*: Literally, "for flow of blood," with probable reference to hemorrhage.

*For all people*: That is, for anyone regardless of sex; alternatively, "for all people of male sex" might be meant, as the Old English form *mannum* could have either meaning.

2.1.2    *Aps, aps, aps . . . æmesstanes*: An unintelligible magic formula.

3.1    *every humor*: See the note on *Herbal* 181.1.

3.2    There is a variant of this remedy in *Bald's Leechbook* 1.1.8.

*Relieves*: Literally, "restrains"; see Bosworth and Toller, *Dictionary,* under *gehaðerian,* "to restrain," and the *DOE,* under *geheaþorian,* sense b, "of a therapeutic drink: to reduce (pain)."

3.3    *For loose teeth*: Literally, "for looseness of teeth."

3.5    There is a variant of this remedy in *Bald's Leechbook* 1.48.4.

3.7    *hystem cepnizam*: A corruption of Greek ὑστερικὴ πνίξ (suffocation of the womb). For discussion, see de Vriend, *Herbarium,* 330–31.

3.8    *in a sweetened drink*: The Latin source prescribes oxymel, a mixture of honey and vinegar used in medicine.

3.11    *add to it*: The reference is evidently to the ointment specified in the preceding remedy.

3.12      *hard swellings*: Glandular swellings may be meant, though the Old English term *cyrnel* does not seem to be specific to any one disorder.

        *deer's cheek*: While Old English *heagospind* means "cheek," *patella* in the Latin original means "kneecap." The Old English translation may be the result of a misunderstanding of Latin *genu* (knee) as *gena* (cheek).

3.13      *To stimulate sexual arousal*: Literally, "to stimulate intercourse with a woman."

3.15      *on the same day*: Or, perhaps, "by day."

3.16      *For difficulty urinating*: That is, for strangury: see Bosworth and Toller, *Dictionary*, under *stede*, sense IV.c.

3.17      *and bind it*: "And" supplied by editor.

3.19      *deer marrow, burned . . . with you*: A conjectural translation of an obscure instruction. The Latin text reads *medulla cervina incensa et de ea suffumigabis aut tecum habeas* (burn the deer's marrow and fumigate the snakes' nest, or have it with you).

4.2      *For head sores*: Old English *heafodsar* can mean "rash, scurf, ulcer, tumor on the head"; see the *DOE*, under *heafodsar*, sense 2.

4.8      *For a sore throat*: The Old English plural noun *goman* (like the equivalent Latin plural noun *fauces*) refers to the internal tissues of the throat and the mouth.

4.10     *to make a mockery*: The meaning of this part of the remedy is obscure; probably the translator misunderstood Latin *irritamentum* (inducement, stimulus, incentive, spur) in his source as Latin *irritum* (something useless, vain).

4.11     *immerse frequently, and have him do the same thing*: The sense seems to be that the patient is to be immersed in the liquid in order to ease the ailing joints.

        *He should prepare this remedy for himself*: Or, alternatively, he (the physician) should provide this remedy for him (that is, for the patient). See the *DOE*, under *gearwian*, sense 5.a.ii.

4.13     There is a variant of this remedy in *Lacnunga* 8; see Pettit, *Remedies*, vol. 2, p. 7.

        *Honey from wild honeybees*: Here, Old English *doran hunig* translates *mel Atticum*, "honey from Attica" (the Attic peninsula, in central Greece). The translation "from wild honeybees" is based on *Rem. An.* 6.2 and 12.3, where Latin *mel Atticum* is translated by Old English *feldbeona hunig*, literally, "honey from field bees" (that is, wild bees).

4.15 *This will make walking easier for them*: Literally, "by this they will have lighter going."

5.2 *and fastened there*: That is, bandaged over the eyes.

5.3 *onto the top and bottom of the foot*: Literally, "over and under," without any more specific instruction.

5.4 *miscarry*: Literally, "lose their fetus."

5.7 There is a variant of this remedy in *Bald's Leechbook* 1.2.12. *The eyesight will improve*: Literally, "the eyes will brighten."

5.12 *let them both drink*: That is, both the husband and the wife.

5.13 *after her purification*: The Old English phrase refers to ritual purification after childbirth.

5.19 *after the tenth hour*: About two hours before sunset, depending on the season.

6.1 *a mountain buck—that is, a wild buck*: A male goat is meant.
*the liver of a mountain buck*: The Old English construction is rather unusual, literally reading "the liver of a mountain buck—that is, a wild buck—or of a goat, the liver."

6.8 There is a variant of this remedy in *Bald's Leechbook* 1.3.9.

7.5 *set over them along with the eyelids*: The Old English phrase is somewhat condensed; what is apparently meant is that the cheese is to cover the eyelids when the eyes are closed.

7.10 *for severe skin disease*: Old English *hreofla* can refer to leprosy or a similar skin disease.

7.17 There is a variant of this remedy in *Bald's Leechbook* 1.3.10.

7.22 There is a variant of this remedy in *Bald's Leechbook* 1.31.11.

7.23 There is a variant of this remedy in *Bald's Leechbook* 1.26.1.

8.3 *For ulcerous wounds on the face*: This interpretation is according to de Vriend's emendation of the Tironian abbreviation for *ond* in his base manuscript *V* to "on" based on the late manuscript *O*. However, the sense of the Latin original, *ad livores et suggellationes* (for bruises and contusions), would be maintained if Old English *anwlata* is interpreted "a livid bruise," as in the main volume of Bosworth and Toller, *Dictionary*, an entry deleted by Toller in his *Supplement*.

8.4 *and wipe it away . . . apply it*: That is, first wipe the skin with wool in order to remove the dead skin scales, then apply the salve.

9.12 There is a variant of this remedy in *Bald's Leechbook* 1.37.2.

10.1 *for an evil vision*: This affliction might include nightmares, which were often thought to result from demonic malice.

10.17 *a fever accompanied by delirium*: For *dweorg* in the sense "dwarf-driven fever," or "runaway fever," see the Introduction, in the section headed "Translating Words, Translating Cultures."

12.5 There is a variant of this remedy in *Bald's Leechbook* 1.3.11.

12.11 *For an injury*: Here, the Old English *bryce* renders Latin *alopecia* (hair loss), though normally, as in *Rem. An.* 14.9, *bryce* refers to a fracture or similar injury. It is likely that the translator did not understand the Latin term.

14.4 *king's disease*: See the note at *Herbal* 87.1.

14.5 There is a variant of this remedy in *Leechbook 3* 36.2.

## LACNUNGA

1 This cure is a variant of one recorded at an earlier date in *Bald's Leechbook* 1.1.5; for a transcription of that text, see Pettit, *Remedies*, vol. 2, p. 3. There it is *ham-wyrt* (houseleek) that is used rather than *hamor-wyrt*. In the present manuscript the *or* of *hamor-wyrt* is added above the line. Audrey L. Meaney, "Variant Versions of Old English Medical Remedies and the Compilation of Bald's Leechbook," *Anglo-Saxon England* 13 (1984): 255–64, identifies many parallels between cures included in *Lacnunga* and ones in *Bald's Leechbook*, with discussion of the complex nature of these relationships; not all the parallels to which she calls attention are mentioned here.

2 There is a nearly identical cure in *Bald's Leechbook* 1.1.6; see Pettit, *Remedies*, vol. 2, p. 3.

3 Compare the similar cure in *Bald's Leechbook* 1.1.4; see Pettit, *Remedies*, vol. 2, p. 3.

6      Compare the similar cure in *Bald's Leechbook* 1.2.41; see Pettit, *Remedies,* vol, 2, p. 6.

7      There is a similar cure in *Leechbook 3* 2.5; see Pettit, *Remedies,* vol. 2, p. 6. There the problem is "if mist forms before the eyes."

          *are stopped up*: The exact nature of the ocular ailment is unclear.

          *crab's gall . . . and salmon's gall*: Although the crab lacks a gallbladder, bitter extracts from either the crab or the salmon might have been thought to have curative properties.

          *cowslip*: A conjectural translation; mint or sweet basil are other possibilities.

          *a linen cloth*: The probable meaning of *lin-hæwenne cla\u00f0.* Perhaps a whitish or flax colored or blue-gray cloth is meant—or in effect, a clean, pure one.

          *will wake up*: That is, the patient's sight will be restored.

8      *honey from a wild honeybee*: See the note on *Rem. An.* 4.13.

9      See *Lacnunga* 32 for a variant of this cure.

10     A variant of this cure (from London, British Library, MS Cotton Vitellius C.iii) is included in this volume as *Misc. Rem.* 39.

10.1   *mites*: The Old English term *weormas* (a variant spelling of *wyrmas*), which can mean "worms," evidently refers here to small insects that can cause eye infections.

10.2   *pellitory-of-the-wall*: While this is the sense of *dolh-rune,* etymologically this Old English plant name is compounded of the simplex *dolh* (wound), plus *rune,* a noun that normally refers to "secret knowledge."

          *warm it at the fire*: Although Pettit translates "warm [the patient]," it is more likely the ointment that is to be warmed; on a hot day the patient might not need warming. It is curious that the instruction to warm the ointment follows after the instruction concerning how the ointment is to be applied, but this may be no more than a feature of the informal style of many of these remedies.

11     *Lacnunga* 107 closely resembles this remedy, though meant for lung disease and chest pain. In that cure, the chief ingredients are to be boiled.

*on an empty stomach*: Literally, "fasting for a night." See Bosworth and Toller, *Dictionary,* under *niht-nihstig.*

14      *acute inflammation on the neck*: That is, erysipelas of the neck.

*throatwort*: The Old English plant *healswyrt* (literally, "throatwort") may have been thought suitable for this remedy on account of its name.

*knapweed*: Here and at *Lacnunga* 63.1, "knapweed" is a somewhat arbitrary translation of the Old English plant name *isenhearde,* meaning, literally, "ironhard."

*the night before the first day of summer*: Literally, "the night before summer comes to town in the morning."

15      *green salve*: Probably "fresh salve" is meant; that is, a salve made of fresh ingredients rather than dried ones. The use of the salve is not specified.

*groundsel*: See the *DOE,* under *gunde-swelge, grunde-swelge, grundswelge.* The initial simplex of the Old English plant name *grundeswylie* (groundsel) probably derives etymologically from *gund,* denoting "pus," rather than *grund,* denoting "ground," while the second simplex derives from the verb *swelgan* (to swallow). The plant may thus once have been thought to "swallow" infections.

16      *For disease*: No particular disease is specified.

18      *for flying venom*: Most likely for infection, viewed as the result of airborne attack by malicious enemies. Compare *Lacnunga* 76, 126, and 127.

20      *enkausios, that is, migraine*: The hybrid Greek and Latin forms in this passage are difficult to construe and may not have made much sense to the writer. Greek *enkausios,* for example, means "heatstroke," not "migraine." The original sense of *emigraneum capitis* may have been "pain on one side of the head," from Greek *hemi* (half) plus Latin *caput, capitis* (head), although in practice, in the Anglo-Saxon context, the phrase might have meant little more than "a really bad headache."

22.2     *Caio laio, quaque voaque,* ofer *sæloficia*: A nonsensical string of short phrases whose force was enhanced by rhyme.

*the wyrm of men*: An apparent allusion to the belief that toothache results from the boring action of tiny creatures.

22.3      *name the person and his father*: That is, name the patient not just by his or her given name, but also by a patronymic that pinpoints that person's identity.

22.4      *Lilumenne*: Syllables of uncertain meaning.

23      *Jacob's ladder*: A somewhat speculative translation of Old English *hlæder-wyrt,* meaning, literally, "ladder plant." The translation "Solomon's seal" is another possibility.

24      *For a swelling*: The Old English word *geswel* might refer to a variety of ailments, including a tumor.

25.1      *the black boils*: That is, black pustules.

25.2      *Tigað . . . trauncula*: The word divisions adopted here differ slightly from those employed by the scribe. Pettit, *Remedies,* vol. 2, pp. 22–27, presents seven early medieval variants of this incantation, which, though nonsensical, incorporates both Irish and Latin verbal elements. For two of these variants see *Lacnunga* 63 and 83. For a third (from Cambridge, Gonville and Caius College, MS 379), see *Misc. Rem.* 9, and for a fourth (from Oxford, Bodleian Library, MS Bodley 163), see *Misc. Rem.* 88.

25.3      *Seek and you shall find*: Matthew 7:7.

            *Thou shalt walk upon the asp*: Psalms 90:13.

26.1      *drinks a wyrm*: Here Old English *wyrm* might refer to either a worm or a small insect.

26.3      *Gono mil . . . sir amum*: These word divisions reflect scholarly efforts to reconstruct elements of an Irish incantation that apparently underlies the present nonsensical text. The gist of that Irish charm is "I kill the beast." See Pettit, *Remedies,* vol. 2, pp. 31–34, for discussion.

27      *in case a wyrm has burrowed into the skin*: Literally, "against a burrowing wyrm."

29.1      *elf-wrought magic*: This is evidently the original sense of the word *ælf-siden,* and it is probably what is meant here, although either fever or demonic possession, or both, might be involved without conscious thought of elves as disease-causing agents. For wide-ranging discussion see Alaric Hall, *Elves in Anglo-Saxon England: Matters of Belief, Health, Gender and Identity* (Woodbridge, 2007), 119–56.

*all afflictions*: "All temptations" is an alternative translation.

29.2     *Write on a paten*: The words are to be written in Latin, the language of the liturgy.

*In the beginning was the word . . . understood it not*: John 1:1–5.

*and Jesus went all about . . . followed him*: Matthew 4:23–25.

*God, in thy name*: Psalms 53:3.

*God have mercy on us*: The beginning of the litanies of the saints for the week before Ascension Day.

*Lord God in our aid*: A slight variation on the words of Psalms 69:2.

29.3     *Take cristalle*: It is unclear what substance is meant. The *DOE* glosses this unique instance of the word as "a plant, perhaps fleabane or fleawort."

*a pitcher full*: Literally, "a sester-full," but the measure is likely to be an approximate one.

*a virgin*: Literally, "a spotless person," potentially of any age and of either sex, hence "virgin" in the older sense of that English word.

29.4     *Have Masses sung over it*: The three liturgical texts specified here are the three Latin collects generally known as *Ne despicas, Suscipe domine,* and *Tribulationem nostrum.*

*Sing these precatory psalms*: The five psalms to be sung are numbers 50, 53, 66, 87, and 85, respectively.

*and litanies*: The ones to be sung are not specified; they might vary according to the liturgical calendar.

30     *For a wen salve*: The Old English term *wenn* could refer to any type of small protuberance on the skin, such as a wart, cyst, tumor, or stye. A similar cure is recorded in London, British Library, MS Cotton Domitian A.i; see *Misc. Rem.* 16.

*cinnamon bark*: Old English *rind* means "bark." In this instance the word probably refers to the bark of the cinnamon plant.

*then cook it twice as much as before*: Evidently so as to make a reduction.

31.2     *limbwort*: A literal translation of Old English *liþ-wyrt*, a plant name of uncertain meaning, though perhaps to be identified with dwarf elder or common rue.

*greater stitchwort*: This tentative translation of Old English *hæferðe* is based on the surmise that that word, without initial *h*, is a shortened form of *æþel-ferþing-wyrt*, which in turn is thought to denote either greater stitchwort or chickenweed. See Sauer and Kubaschewski, *Planting the Seeds of Knowledge*, under *æferðe*.

31.4    *pour it into a mold*: The meaning of the Old English noun *trinda* is obscure; some kind of roundish mass may be meant.

31.6    *Blessed be the Lord, my God*: Psalm 143.

*Blessed be the Lord, the God of Israel*: The beginning of the prayer spoken by Zachary, the father of John the Baptist, at the time of John's birth, as recounted in Luke 1:67–75.

*the Magnificat*: The prayer spoken by Mary, the mother of Jesus, after the child Jesus leaps for joy in her womb, as recounted in Luke 1:46–55.

*the prayer "Matthew, Mark, Luke, John"*: Although this reference is not transparent, it might be to the prayer, involving the names of the four evangelists, that is featured in *Lacnunga* 126.3.

32    *bone-marrow soap*: The manuscript reading *ar* (bronze) makes little sense in this context and so is emended to *mærc* (bone marrow), also spelled *mearh* or *mærh*, on the basis of *Lacnunga* 9.

33    *simmer in beer*: See the note on *Herbal* 1.2.

*the pustules*: Although the purpose of this cure is not specified, like the preceding one it is evidently meant for pustules affecting the eyes. The word "pustules" is therefore supplied where the Old English text has only the pronoun *hy* (they).

34    *Wallflower*: Old English *banwyrt* refers either to this plant or to others, such as daisy, common centaury, or comfrey, that were thought to have bone-healing (or other medicinal) properties.

35    *the third kind of mint, the one with white blossoms*: Catnip is one possibility.

36    There is an abbreviated version of this cure in *Bald's Leechbook* 1.32.6; see Pettit, *Remedies*, vol. 2, p. 50.

*for a leprous body*: The ailment might be anything from a scabby body to leprosy.

*Give the person a steam bath*: I follow Pettit, *Remedies*, vol. 2, p. 50, in taking the corresponding Old English phrase *þweah . . . mid hate* (wash with heat) to refer to warming the patient with vapor in a stone-bath treatment, as in *Lacnunga* 73.

37 This cure is a variant of one in *Bald's Leechbook* 1.24.1; see Pettit, *Remedies*, vol. 2, p. 50. There the plants are put into ale, and the preparation serves as a bath and ointment as well as a drink.

*take weed runners*: The variant text from *Bald's Leechbook* 1.24.1 has *wudu-weax* (dyer's greenwood) rather than *weode-wisan* (weed runners).

*add meal*: Here, the text is almost certainly corrupt, since the plants are to be given to the patient to drink, yet no mention is made of a liquid ingredient. In the variant text, it is ale rather than meal that is to be added.

39 *For diarrhea*: The cure may be meant for either diarrhea or dysentery.

42 Although no purpose for this cure is specified, it too is likely to be for diarrhea or dysentery.

*shave ivy*: The manuscript reading *efu* looks to be corrupt and is emended to *ifig* (ivy).

*simmer all the shavings*: While the Old English text states only "simmer them all," the reference must be to the shavings.

43 *caper-spurge seeds*: The generally accepted meaning of Old English *lyb-corn*, though the identification is subject to some doubt. Caper-spurge seeds could make for a dangerously toxic emetic. It may be relevant that the related Old English noun *lyb-cræft* denotes "sorcery" or "witchcraft."

*meal made of small insects*: The possible meaning of *wyrm-melo*, if *wyrm* is indeed the correct reading here. Alternatively, meal made from dried earthworms might be meant.

44 *iris*: The plant name *glædene* can refer to any of several species of the genus Iris.

*a cubit*: Traditionally, a measure equal to the length of the arm from the elbow to the tip of the middle finger.

47 This drink too is evidently designed to induce vomiting.

49.1 *and for theor-disease*: A somewhat mysterious ailment. For discussion, see the Introduction, "Translating Words, Translating

Cultures." The term *peor* might refer to dermatitis or eczema, though other ailments too could be meant.

50    *water lilies*: Literally, "docks that swim."

56    Although no purpose for this cure is specified, it too seems to be meant for lung disease, like the five preceding ones, and, it seems, the five succeeding ones as well.

63.1    *For a holy salve*: The purpose of this salve is not specified; it is perhaps "the salve of salves," to be used for any ailments of the outer body.

63.2    *consecrate some water at the baptismal font*: In practice, what this may amount to is that one should obtain water that has been hallowed at the font.

63.3    *and square off its four sides*: A guess at the meaning of an obscure Old English phrase, one that is possibly a woodworkers' idiom.

*Sing these psalms . . . Blessed are the undefiled*: Since only one psalm is specified (Psalm 118), something seems to have dropped out of the text, hence the ellipsis; there is no gap in the manuscript.

*Glory to God in the highest*: The Gloria, the opening passage of the Mass, from Luke 2:14.

*I believe in God the Father*: The Creed, following soon after the Gloria in the canon of the Mass.

*My God and Father*: This prayer opens a later part of the Mass.

*In the beginning*: The text, from John 1:1, that initiates the closing part of the Mass.

*the wyrm charm*: An obscure allusion. It might be a reference to *Lacnunga* 26, with its garbled Irish incantation, or to *Lacnunga* 64, with its reference to Saint John's immunity to poison.

63.4    *Acre arcre . . . fidine*: A nonsensical incantation resembling the one in *Lacnunga* 25.

63.6    *Lord, holy Father*: For a close analogue to this prayer from the marginalia of CCCC MS 41, see *Misc. Rem.* 3, section 5.

63.7    *may come and . . . his will*: There is a breakdown in grammar and sense at this point. Some text may have been lost, though there is no gap in the manuscript.

63.8    *the mother of living creatures*: Possibly an error for "the mother of men," since the expected reading in prayers of this type is *virorum* (of men), not *vivorum* (of living creatures). The latter reading is retained, however, since Latin herbals of the early Middle Ages often include a hymn to the mother of living creatures; see Maria Amalia D'Aronco, "*Precatio terrae, Precatio omnium herbarum* in the Middle Ages," in *The Oxford Dictionary of the Middle Ages*, ed. Robert E. Bjork (Oxford, 2010), 1355.

*your servant* NAME: As is customary in remedies of this kind, Latin *nomen*, meaning "name," is written out where the practitioner is to utter the name of the person to be healed.

63.10  See Notes to the Texts. This paragraph consists of a superior version of the London, British Library, MS Harley 585 text as preserved in a medieval sacramentary from Autun. When read as a whole, the prayer to which the Autun passage pertains is a *benedictio pomorum*, that is, a blessing for the first harvest of apples in the fall.

63.11  *by your holy name*: A suitable closing formula; the Latin equivalent is *per nomen sanctum tuum*. I supply the phrase editorially in place of the scribe's abbreviation *per* (by). Other familiar liturgical formulas might have served just as well, for example *per Dominum nostrum* (by our Lord).

64  The prayers encompassed in this entry form a minianthology of Latin blessings to be used in conjunction with a consecrated drink.

64.1  *And Jesus went about all Galilee*: Matthew 4:23.

64.4–7  *Then when those who had drunk . . . that pertains to you*: The whole of this passage is based on part of chapter 8 of the apocryphal *Virtutes Johannes*. Pettit, *Remedies*, vol. 2, pp. 77–80, provides discussion and analogues.

64.7  *By our Lord Jesus Christ*: Another closing formula, supplied editorially to fill out the scribe's abbreviated phrase *per eundem* (by the same one). The equivalent Latin phrase is *per Dominum nostrum Iesum Christum*.

65  *The Lorica of Laidcenn*: Title supplied editorially. See the Introduction for discussion of this Latin poem and the tradition that it represents. Readers are referred to Michael W. Herren,

The *Hisperica Famina II: Related Poems* (Toronto, 1987), 23–31, 76–89, and 113–37; and to Pettit, *Remedies*, vol. 2, pp. 82–93, for discussion of its text and translation.

65.63    *my ten branches growing in unison*: That is, my ten toes.

65.82    *ten skillfully fashioned orifices*: The calculation of the number of the body's orifices varies in medieval tradition; see Pettit, *Remedies*, vol. 2, p. 92, for discussion.

66    *For the sudden onset of a disease*: While perhaps meant specifically for apoplexy or a seizure, this cure could well have had a wide application.

68    *For theor-disease*: See the note for *Lacnunga* 49.1, where the cure involves use of a salve rather than a drink.

   *a vessel full of ale made from bran*: The Old English term *cumb* (vessel) evidently refers to a measure of unknown capacity. The ale is to be brewed *mid gryt* (with grits, meal, or bran), a term that typically refers to barley bran.

69    This cure seems to be modeled on one in *Leechbook 3* 66.1; see Pettit, *Remedies*, vol. 2, p. 94, for that earlier version.

   *pour three pitchers of ale over them*: The Old English term *mæder*, here translated "pitcher," evidently denotes a measure larger than the sester.

70    *Take . . . centaury, sow thistle, common centaury*: There is possibility for confusion here on the part of the modern reader. "Centaury" translates Old English *felterre* (corresponding to Latin *centaurea*), while "common centaury" translates Old English *feferfuge* (corresponding to Latin *febrifuga* or *febrefugia*), a plant formerly known in English as "feverfew."

   *nothing fresh*: See the *DOE*, under *fersc*, sense 2: "of meat, fish, butter, cheese: fresh, not preserved with salt."

71    Earlier versions of this cure and the two succeeding ones are recorded in *Leechbook 3* 30.1–3; see Pettit, *Remedies*, vol. 2, pp. 94 and 96, for details.

   *eyebright*: While this is one proposed meaning of Old English *æg-wyrt*, the *DOE* considers this plant name to be of uncertain identification; "dandelion" is another possibility.

*plowman's-spikenard*: A tentative translation of Old English *peor-wyrt;* another possibility is great fleabane.

*good clear ale or good Welsh ale*: The Anglo-Saxons distinguished between ordinary ale and Welsh ale, also known as braggot (Welsh *bragawd*), a strong beverage made with herbs, spices, and honey.

74    *ash-tree bark taken from below ground level*: A conjectural translation. The ash-tree bark is to be found *in eorþan,* literally, "in the earth."

75    *If hemorrhoids persist*: The medical condition called *se fic,* or, literally, "the fig," is generally taken to be piles or hemorrhoids, with their fig-like protuberances.

*the hottest he can stand*: Literally, "as hot as you can stand."

76    *The Nine Herbs Charm*: Title supplied editorially; this text is also known in the scholarly literature as Metrical Charm 2. For an alternative handling of text and translation, see Robert E. Bjork, *Old English Shorter Poems, Volume II: Wisdom and Lyric,* DOML 32 (Cambridge, MA), 194–201. Although this remarkable cure begins without explanation, it is evidently meant to heal an infected wound or open sore. The cure has three components: a long incantation in alliterative verse, to be repeated many times, in the course of which nine plants with healing properties are directly addressed by name (lines 1–63); an act of exsufflation by which the disease-causing agent is blown away (line 63), an act that is apparently to be repeated with each reiteration of the charm; and the application of a salve, made up from those same nine plants, directly to the wound, which is likewise to be bathed twice with a special concoction (lines 64–68).

76.1.1   *Remember, Mugwort*: The plant is addressed directly by name and is attributed not just healing properties but also a prior history, as if in a psychodrama involving the healer, the patient, the herb, and the poison. Its name is therefore capitalized here, as are those of the other eight plants that figure in the incantation.

76.1.2    *what you determined at the Place of Proclamation*: An allusion to a
           mythological history that is not known from other sources and
           is not now recoverable except in its general outlines. The Old
           English phrase *æt Regenmelde* would seem to refer to a named
           but now unidentifiable place, "Regenmelde." The phrase is
           plausibly construed by Pettit to mean "at the Great (or Divine)
           Proclamation," from the prefix *regn-*, implying divine or inten-
           sive force, plus a nominal form related to the verbs *meldan* or
           *meldian*, meaning "to announce." The present translation "at
           the Place of Proclamation" combines these two possibilities.

76.1.3    *Una*: The Latin word for "one," in the feminine gender that suits
           both this plant and most of the others that are named.

76.1.5    *you have power against venom*: While here it is best to construe
           Old English *attor* as the common noun "venom," elsewhere in
           this charm there is reason to construe that word as signifying
           a personified Attor, the ancient enemy of Mugwort and the
           other curative plants. Compare capitalized "Attor" in 76.1.34.

76.1.6    *the enemy who journeys throughout the land*: A patristic topos; the
           devil of Christian belief is implied. Compare 76.1.43, *feondes
           hond* (the devil's hand).

76.1.7    *Waybroad*: The Old English name denotes either common plan-
           tain or dock; the latter is traditionally used as an antidote to
           nettle sting.

           *mother of plants*: The Old English phrase *wyrta modor* corre-
           sponds to Latin *mater herbarum* (mother of plants), a phrase at-
           tested in charms of the ancient Mediterranean tradition.

76.1.9    *over you chariots . . . have ridden*: While an alternative translation
           of this line (one that is adopted by both Pettit and Bjork) is
           "over you carts creaked, over you women rode," the Old Eng-
           lish noun *cræt* normally refers to chariots, while in this context
           the Old English form *cwene* can be construed as either the plu-
           ral of *cwene*, meaning "woman" or "wife," or the plural of *cwēn*,
           meaning "queen," a word used sometimes with reference to
           the queens of Old Testament history. The mythological con-
           text of this remedy favors the interpretation "queens," though
           "women" or "wives" in general may also be meant.

76.1.10    *over you brides have trampled*: The manuscript form *bryodedon* is here taken to be a scribal miswriting of *breodwedon,* the preterite plural form of *breodwian,* a verb whose probable meaning is "to strike down or trample." The reference might be to bridal parties. Other editors and translators retain the manuscript reading *bryodedon* as a preterite form of *breodian,* taking that verb to mean "to cry out aloud." No interpretation of this part of the charm is secure.

76.1.14    *Striker*: The plant's Old English byname, *Stune,* is closely related to the verbs *stunian* (line 15) and *wiþstunian* (lines 11 and 16), meaning "dash against" or "strike down." To judge from the list of ingredients of the healing salve that is introduced in the last paragraph of the cure (at 76.2), the plant's common name is *netele* (nettle). Nettles have a long history of medicinal use related to the stinging pain they can induce when grasped or brushed against.

   *she grew up on rocky ground*: Since modern English does not observe grammatical gender, the Old English feminine personal pronoun *heo* would normally be translated as "it" when used with reference to a plant. In the present mythological context, however, the nine healing plants are personified as if they were female warriors counteracting the effects of *attor* ("poison," a neuter noun), *onflyge* ("infection," a masculine noun), the Wyrm (a masculine noun), and the devil (*deofol,* a masculine noun). The translation "she" therefore seems justified.

76.1.16    *Unyielding*: The same plant's second byname, Old English *Stiðe,* represents the feminine form of the adjective *stiþ,* meaning "firm, hard, unyielding, resolute," often in a positive sense.

76.1.17    *the wrathful one*: Another reference to the devil, who is regularly characterized in medieval literature as being filled with fury.

76.1.18    *the plant that battled the Wyrm*: Another inscrutable mythological allusion. "Wyrm" is capitalized here, as befits an archetypal agent. The same creature is apparently referred to again at 76.1.31.

76.1.21    *Attorlathe*: The plant that fought the Wyrm is now revealed to be the one whose common name, *attor-laðe,* denotes cockspur grass or a similar plant. This name is compounded of *attor* (venom) plus the adjective *laþ* (inimical), in its feminine inflection. The name thus designates "she who is enemy to poison"— a sense that the term "cockspur grass" loses.

76.1.21–22   *the lesser the greater . . . the lesser*: An attempt to make sense of a difficult passage. The imperative form *fleoh*, from the verb *fleogan*, would normally mean "fly" or "flee." Here, *fleogan* may have been conflated with the weak class 1 verb *flygan* or *flyan*, meaning "put to flight, disperse." Perhaps what is meant is that Attorlathe will drive out both the venom (the greater power, because it is fast acting) and the infection (the lesser power, since it proceeds more slowly). But other interpretations are possible.

    *until he is healed of both*: "Both" plausibly refers to both venom and infection.

76.1.23   *Remember, Maythe*: The phrasing recalls that of 76.1.1, "Remember, Mugwort." The allusion is probably to the same mythological conflict that in that earlier passage is said to have been resolved with a truce at "the Place of Proclamation," a location that here is ascribed the name "Alorford." The literal meaning of the plant name *mægeþe* (spelled various ways) is evidently "maiden flower," while the plant that the name denotes—chamomile, or oxeye daisy—is known for its soothing properties.

76.1.24   *the pact you concluded*: The old English verbal form *ge-ændadest* is a variant spelling of the second-person preterite singular form of *ge-endian*, a verb that can denote "to bring (something) to completion" or "to determine, settle." See *DOE*, under *endian*, senses 1.c and 1.d.

76.1.26   *after chamomile was prepared*: The reference here is probably to the garden herb *mægðe*, or chamomile, rather than the mythic figure Maythe.

76.1.27   *Wergulu*: Although Pettit, *Remedies*, vol. 2, pp. 134–35, and others have held that this is a name for the stinging nettle that is an ingredient of the salve that is later introduced (see 76.2 below), it is more likely to be a byname for *lombes-cyrse* (shepherd's

purse), another of the salve's ingredients. In the reading adopted
here, the more aggressive "Striker" *(Stune)* of lines 76.1.14–17 is
the stinging nettle, while "Wergulu" is the soothing shepherd's
purse, which is known to have anti-inflammatory properties.

76.1.28   *Seal sent it over the sea's back*: "Seal," with that name capitalized, is
treated here as another character in the ancient mythological
drama. This aquatic animal's role, it seems, was to bring the
freshwater cress over the salt sea so as to aid the healing herbs
in their battle against the poisons.

76.1.29   *as a torment to venom, but a cure for someone else*: A difficult phrase
but not an impenetrable one. Shepherd's purse, here personi-
fied as "Wergulu," brings relief to the patient and thereby con-
founds poison. Bjork, *Old English Shorter Poems II*, 197, following Pettit, interprets the line differently, translating it "as a
remedy for the evil of another poison."

76.1.31   *The Wyrm*: Here capitalized, as suits an archetypal being.

    *it killed no one*: While other editors have emended the manu-
script reading *nan* (no one) to *man* (someone), no emendation is
called for: the Wyrm is said to have done no harm until Woden
struck it.

76.1.32   *nine wonder-twigs*: These mysterious "wonder-twigs" (or "twigs of
glory") are suggestive of a nine-branched mythological weapon
wielded by Woden. As in the Anglo-Saxon royal genealogies,
Woden may be conceived of here not as a deity but rather as a
powerful ancestral figure from some ancient time. Here he re-
leases poisons into the world that are resisted by the beneficent herbs, at least two of which, as we soon learn, were sent
into the world by Christ.

76.1.33   *it flew into nine parts*: While, in our own world, a nine-branched
weapon could be expected to strike an object into ten parts
rather than nine, the present passage rests on the incantatory
power of nines (nine snake parts, nine venoms, nine herbal an-
tidotes, a nine-branched weapon). Verisimilitude is not to be
looked for.

76.1.34   *Apple and Attor made a pact*: The reference may be to the pact or
pacts mentioned in lines 1–2 and 23–26. Here it seems reason-

able to capitalize "Attor" as denoting one of the mythological parties to that treaty.

76.1.35 *that it would never make its home in a house*: The implication is that poisonous snakes, with their venoms, are categorically excluded from human habitations, according to the myth presented here.

76.1.36 *Fille and Finule*: That is, chervil and fennel.

76.1.37 *plants that the wise Lord created*: Although there is no known source for the story that Christ sent beneficent herbs into the world at the time of his Crucifixion, the story has folkloric parallels. For discussion see Pettit, *Remedies,* vol. 2, pp. 149–50.

76.1.41 *She stands firm against pain*: The pronoun "she" (Old English *heo* or *seo*) that recurs here three times is of uncertain reference, as it might refer back to either of the two plant names that figure in the previous verse paragraph, namely *fille* and *finule*. Although *finol* is usually a masculine noun, it can also function as grammatically feminine, as it may do here. Perhaps the intended meaning of lines 41 and 42 is that one of these paired plants (*fille?*) stands firm against pain and strikes down venom, while the other one (*finule?*) has power against three and against thirty, and so forth. While such a suggestion is conjectural, a speaker's use of the voice for emphasis in the context of oral performance, or the use of gesture, could make such a distinction clear.

76.1.43 *sudden acts of cunning*: the Old English adjective *færlic* is most often used with reference to sudden events of a grievous kind, including pain, poison, disease, and death. The reference to "cunning" implies that the devil is involved.

76.1.47 *against foul venom*: The unique adjectival form *runlan* is tentatively construed as meaning "foul, dirty" on the basis of an apparent Old Icelandic cognate, *hrunull,* referring to a bad smell. If a specific type of venom is referred to, its identity is unknown.

76.1.50 *against deep-blue venom*: This second mention of deep-blue venom (see also line 48) may result from an error of transmission having to do with the need for an alliterating second half-line at this point. "Deep blue" is a plausible translation of Old English *weden*, an adjective that is likely to derive from the noun *wad* (woad), denoting a plant that produces a deep-blue dye.

76.1.55–57  *from the east . . . any from the west*: Although some editors have been tempted to emend this passage so as to add the phrase "or from the south" after the references to east, north, and west, such an intervention would work against the poem's numerology. Nine different venoms are named, followed by six different blisters plus three venoms flying from different directions. This makes for two nines, so that no emendation is called for despite the imperfect meter of line 56. Perhaps it was believed that flying venoms originated in only three points of the compass, while healing breezes came from the south.

76.1.60  *the nine serpents are on the watch*: The Old English verb *behealdan*, though normally transitive, here evidently functions without an object; see the *DOE,* under *behealdan,* sense A.2.a.

76.1.61  *May all woodlands now be filled*: This translation depends on taking the manuscript reading *weoda* as a variant spelling of *wuda* (woods) rather than of *weode* (weeds).

76.1.62  *may all seas disperse*: The verb *toslupan* can mean "to disperse," "to be released," or "to be rendered powerless." Perhaps the power of the herbs, which live on freshwater, stands against that of the venoms, associated with saltwater seas.

76.2  *shepherd's purse*: While the Old English plant name *cyrse* means "cress," probably watercress, the related name *lombes cyrse* is glossed by the *DOE* (under *cærse*) as "a cress-like plant, perhaps shepherd's purse or pennycress." See the note on 76.1.27, on the identity of the plant named "Wergulu."

*nettle*: See the note on 76.1.14, on the identity of the plant named "Striker" *(Stune)*.

*Make a cream*: The Old English noun *slype,* denoting a thin paste, has a modern English reflex, slip, that in pottery making refers to clay thinned to the consistency of cream. A paste with a similar consistency may be meant.

*bathe the person with an egg mixture*: This translation depends on emendation of the nonsensical manuscript reading *aagemogc* to *æg-gemongc* (egg mixture). The present reading follows Cockayne's basic understanding of the passage.

77  *If the wyrm has turned downward*: The reference might be to an anal fistula, spoken of figuratively as if it were a creature burrowing in the intestines.

*hold your hands up high*: The plural pronoun *hy*, which serves as the object of the verb, most likely refers to the hands or the arms, which are to be extended upright in a gesture of prayer; see Bosworth and Toller, *Dictionary*, under *up-weardes*, and the *DOE*, under *brædan* 2, sense 2: "to spread, extend."

*take a little cupful*: The shoots of greater celandine are apparently to be made into a drink, perhaps an herbal tea.

*have the person then drink it*: Here the feminine pronoun *hy* is taken to be an accusative singular form standing in for the feminine noun *cuppe* (cup).

*give him a bath*: The most likely meaning of the final part of a highly compressed set of instructions. The verb *beþian* (or *beþþan*), like *baþian*, can mean "to immerse in water," "to take or give a bath," or "to wash." The "tea" made from the shoots of greater celandine evidently also serves as a bath for the patient's sore tissues.

80  *For a distended body and for brinch-disease*: A distended body is suggestive of elephantitis, among other possible ailments. It is not known what "brinch-disease" is. The *DOE* discusses the word under the emended form "? *hring-adl*," tentatively glossing that hypothetical word as "a skin disease: ? ringworm . . . ? shingles."

*Then obtain salt*: Although its use is not specified, very likely the salt (like the sulfur mentioned in the next line) is to be added to the herbal preparation so as to make an astringent salve.

81  Two versions of this cure are written out in *Lacnunga*, one after the other. They scarcely differ from one another except as regards the inscription, which in the first version is given in an abbreviated form. Only the second version is given here; see Pettit, *Remedies*, vol. 1, p. 70, for the other one.

81.1  *for dwarf-driven fever*: Evidently a raging fever; for discussion see the Introduction, "Translating Words, Translating Cultures."

81.2 *T ✚ P . . . ✚ ō̄ Ā*: Although the meaning of this string of
symbols is unknown, *T* might stand for *Trinitas* (Trinity), *P* for
*Pater* (Father), *N* for *nomen* (the name of the patient), *UI* (or *VI*)
for *Victoricus*, and *M* for *Macutus*. *Ā* and *ō̄* are standard abbreviations for *alpha* and *omega*, respectively, denoting "the beginning" and "the end," a common way of calling on the name of
Christ.

81.3 *Saint Macutus, Saint Victoricus*: The former is a sixth- or seventh-
century bishop-saint, the latter a fourth-century martyr. It is
not clear why these particular saints are named or just how
their names are to be used.

83.1 *Tigað*: The scribe gives only the first word of the incantation; for
a full version see *Lacnunga* 25.2.

83.2 *for as long as is needed*: This is the apparent sense of the Old English phrase *ðe þearf sy*, taking *ðe* as a variant spelling of instru-
mental *ðy*.

86 This remedy is commonly known as "Metrical Charm 3." For
an alternative handling of text and translation, see Bjork, *Old
English Shorter Poems II*, 202–3; for a recent (but still not neces-
sarily conclusive) discussion of problematic details, see B. R.
Hutcheson, "*Wið dweorh*: An Anglo-Saxon Remedy for Fever in
Its Cultural and Manuscript Setting," in *Secular Learning in
Anglo-Saxon England: Exploring the Vernacular*, ed. László Sándor
Chardonnens and Bryan Carella (Amsterdam, 2012), 175–202.

86.1 *For dwarf-driven fever*: Here, the somewhat mysterious ailment
is counteracted by a spell meant to nullify the power of an inimical "rider" (see 86.3.3).

86.2 *Maximianus . . . Serafion*: The names of the legendary Seven
Sleepers of Ephesus, seven young men who entered a cave during the reign of the emperor Decius (d. 251 CE) so as to es-
cape the Roman persecution of Christians, fell asleep, and
emerged from the cave unharmed some three hundred years
later, thinking they had slept for only a day. The names of the
Sleepers vary somewhat from version to version. For medieval
and folkloric analogues to this feature of the remedy, see W.
Bonser, "The Seven Sleepers of Ephesus in Anglo-Saxon and

Later Recipes," *Folklore* 56 (1945): 254–56; and Pettit, *Remedies,* vol. 2, pp. 176–78.

*a virgin*: That is, a sexually pure person of either sex.

*hang it on his neck*: Although no explicit statement is made as to what is to be hung, it is very likely the set of seven wafers, enclosed, one imagines, in a small box.

86.3.1 *a spider-creature*: Since the present translation rests on the emendation of manuscript *spiden* to a form *(spider)* that is not otherwise attested before the Middle English period, not much weight should be put on it, though the association of dwarves and spiders in folkloric sources counts in favor of the emendation. For discussion see Pettit, *Remedies,* vol. 2, pp. 186–88.

86.3.2 *his skin-coat*: Elsewhere, particularly when used as the second element in compounds, the Old English noun *hama* denotes a coat or covering of some kind, such as the old skin that a snake might shed.

86.3.5 *They began to speed from the land*: the Old English verb *līðan* generally means "to go by sea, sail," rather than just "to travel," hence the present translation "to speed."

86.3.8 *the creature's sister*: Some editors would emend the Old English passage to read *dweorges sweoster* (the dwarf's sister), taking the dwarf to be the same as the spider-creature of line 1—a plausible suggestion, but one that reads much into the text.

86.3.9 *she made a pact and swore oaths*: Compare the mythological pacts referred to in the "Nine Herbs Charm" (*Lacnunga* 76.1, at lines 1–2, 23–26, 34–35).

86.3.10 *this could never harm*: That is, the creature's act of "riding" the sick person would do no lasting harm.

88 *Christ is born, a aius*. While this cure takes the form of a spell to be sung or chanted, it is not clear just how it is to be sounded out; personal instruction might have been required. *Aius* is probably a Latinized form of Greek *agios* (holy).

*holy*: The corresponding word *sanctus* (holy) may stand in for the hymn known as the Sanctus that forms part of the Ordinary of the Mass.

| | |
|---|---|
| 89 | For a variant of this cure see *Bald's Leechbook* 1.39.2. |
| 90 | For a variant text see *Bald's Leechbook* 1.39.3. |
| | *old draff*: That is, the spent malt that is the product of brewing. |
| 91 | For a variant see *Bald's Leechbook* 1.39.4. |
| | *Soak in a cold-water bath*: That is, soak the affected part of the body. |
| 92 | For a variant see *Bald's Leechbook* 1.39.5. |
| | *and break them up*: This is the apparent sense of the passage if the manuscript reading, *onbind,* is taken to be a variant of *unbind.* |
| 93–99 | For variants see *Bald's Leechbook* 1.39.5–11, respectively. |
| 97 | *Take the gall of a boar or other pig*: In the present instance the gall, or bile, is to be crushed before being applied. |
| 101 | *For morning nausea*: The reference might be to what is now called morning sickness. Alternatively, it might be to nausea brought on by causes other than pregnancy, including excessive drinking. |
| 102 | For a variant see *Herbal* 1.13, and for the Latin source and additional analogues, see Pettit, *Remedies,* vol. 2, p. 203. |
| | *the weight of three small coins*: The Old English text speaks of three *trymess,* using a term, borrowed from Latin *tremis,* that in England was used to denote a coin of the value of three pence, or the weight of such a coin; see Bosworth and Toller, *Dictionary,* under *trimes.* |
| 103 | For a variant see *Herbal* 2.22. |
| 105 | For a variant see *Bald's Leechbook* 1.2.33. |
| 107 | *Lacnunga* 11, for cough, is similar to this remedy. |
| 109 | For a variant see *Bald's Leechbook* 1.22.1; there, two pennyweights of betony, rather than ten pennyweights, are to be used. |
| 111 | For a variant see *Bald's Leechbook* 1.88.6. |
| | *If a horse is shot*: The reference here (as in *Lacnunga* 127 and 155) is to "elf-shot." For discussion see the Introduction, "Translating Words, Translating Cultures." |
| | *Irish wax*: Or "Scottish wax." It is not clear why wax of this kind is preferred except that ancient Ireland was famous for miracles. |

112      *If wens have settled*: As in *Lacnunga* 30, "wens" could refer to warts, cysts, tumors, or styes.

113      For a variant see *Bald's Leechbook* 1.83.1.

         *For hoarseness*: Literally, "for a person's voice."

116      The cure at *Herbal* 1.9 is similar to this one.

118      This cure too has a parallel in *Herbal* 41.1.

119      The cure at *Herbal* 1.29 is similar to this one.

120.1    *the ailment that doctors call "podagra"*: From the symptoms described here it is impossible to say just what ailment is meant; "podagra" can refer either to gout or to other afflictions.

121      *for pain from theor-disease*: The manuscript reading *þeor-werce* may be a mistake for *þeoh-werce* (pain in the thigh), a reading that makes better sense in this context since the preceding cures deal with the buttocks and the feet.

122      *For an itchy belly*: The Old English noun *wamb*, here translated "belly," can denote either the belly (its usual sense) or the female genitalia. The sexual sense of the word is surely implied here. A corresponding cure at *Herbal* 94.3 is for itching *þæra gesceapa*, meaning "the genitalia," whether male or female. See Pettit, *Remedies,* vol. 2, p. 210 for the Latin source of both these texts, and p. 211 for an additional parallel from *Bald's Leechbook* 2.34.9. The Latin source is directed *ad veretri prurigenem* (for itching of the genitalia).

         *Give it to the person*: The pronoun *he* in the Old English text could obviously refer to a man; in addition, it could stand in for *se wifmonn* (the woman), a noun whose grammatical gender is masculine. The remedy is meant for both men and women.

126.1    *for flying venom*: That is, for infection (most likely), evidently thought to result from an assault by maleficent agents.

126.2    *Cut four incisions on four sides*: Although the Old English text specifies only "on four sides," it is most likely the inflamed flesh that is to be marked out in this way.

         *an oaken implement*: What is meant is perhaps an oaken stick shaped to serve as a knife; iron is not to be used.

127      This remedy is commonly known as "Metrical Charm 4." For an alternative handling of text and translation, see Bjork, *Old English Shorter Poems II,* 204–7.

127.1   *For a stabbing pain*: That is, for sudden and sharp internal pain, evidently conceived of as the result of a dart having been lodged inside the body after being shot by an unknown assailant.

127.3.5   *I stood*: The Old English verb *stod*, with no expressed subject, can be construed as either a first-person- or a third-person-preterite form. It is best read here as first person, given the use of *ic* (I) three lines later.

*under the linden wood*: That is, under a linden-wood shield.

*a bright shield*: The adjective *leoht* could denote a shield that is both light-reflective and light in weight.

127.3.6   *the mighty women*: These uncanny figures are naturally to be associated with the riders of lines 1–2 and the witches of lines 17 and 22. Although some critics have associated these women with the Valkyries or *dísir* (minor deities) of Old Norse mythological belief, their identity is left open.

*gathered their powers*: A somewhat speculative translation of a difficult half-line.

127.3.7   *hurled screaming spears*: The Old English adjective *gyllende* (howling) might refer to the mighty women as well as to their spears.

127.3.11–12   *a little knife, out of iron bits*: Many readers have equated the "little knife" of line 11 with the "little spear" of lines 4 and 10. Alternatively, the healer's own knife, which is to be put to use as a surgical tool (see line 27), might well be meant. The Old English noun *isen* (iron, used here in the genitive plural case) can refer to a miscellany of objects made of iron. If a word or phrase went missing before *iserna*, as seems likely given the defective meter of line 12, then it would have alliterated on *w*.

127.3.17   *a witch's work*: Evidently, an allusion to the "mighty women" of line 6, as are the references to "witch's shot" at lines 22 and 24. We should not think in terms of the witches of witchcraft trials of the early modern period.

127.3.21   *shot by gods*: The reference is to powerful figures whom pagans regarded as gods but whom Christians considered to be worshipped falsely as such.

127.3.23–24   *This to cure you*: The editorial emphasis on "this" is meant to suggest that decisive action is called for, whether an apotropaic gesture or a surgical procedure.

127.3.25     *Fly away there*: The verb *fleoh,* if this emendation is accepted, functions as an imperative singular form (flee!). Compare the use of *fleoh* in *Lacnunga* 76.1.21.

127.4     *put it into the liquid*: A reference to the herbal bath mentioned at the start of the cure. If the knife is to be used for a medical procedure, then it is first placed in the herbal bath, perhaps preparatory to applying the herbs to the wound.

130     There are two parallel cures in the *Herbal,* 46.3 and 102.2; each addresses the problem of intestinal worms rather than lice. Reference to the navel makes better sense in that context than in the present one.

133.2     *The plants that grow in open spaces in settled areas*: Since the manuscript reading is defective at this point, the present translation depends on a pair of emendations that, though plausible, remain speculative.

    *the plant called "dog's piss"*: Perhaps the plant known today as "hound's-tongue," which is known for its strong odor.

    *peas of other kinds*: This interpretation depends on taking Old English *pys* as representing the noun *pise* (peas).

133.3     *and precious cloth*: The Old English noun *gode-web* (or *god-webb*) often denotes silk cloth.

133.4     *I will bless the Lord at all times*: The beginning of Psalm 33.

    *Benedicite*: The liturgical prayer, from Daniel 3:57–90, beginning *Benedicite omnia opera Domini Domino* (All ye works of the Lord, bless the Lord).

    *and litanies*: No instructions are given as to which litanies are to be used; the choice might depend on the cycle of the liturgical year.

    *Give the tenth penny to God*: That is, give a tenth part of the total value of the cattle to the Church, which has very likely provided the frankincense and the expensive cloth that are burned.

134     *afflicted with illness*: The verb *a-brecan* has medical reference to *broc,* an Old English noun denoting both sickness in general and certain specific conditions, including paralysis and dropsy.

136      *Sing four Masses over them*: Over the herbs, most likely, which would perhaps be brought to an altar, rather than over the pigs. The next clause refers to the pigs, however.

137      *Luben luben . . . delupih*: See Pettit, *Remedies*, vol. 2, pp. 267–68, for an account of Old Irish words and phrases on which certain parts of the present nonsensical incantation appear to be based.

138      This cure is almost identical to one in *Bald's Leechbook* 1.50.5; see Pettit, *Remedies*, vol. 2, p. 268.

139      There is a similar cure in *Bald's Leechbook* 1.50.6; see Pettit, *Remedies*, vol. 2, p. 269.

140      There is a similar cure in *Bald's Leechbook* 1.34.1; see Pettit, *Remedies*, vol. 2, p. 269.

144      *Rose and rue . . . carline thistle*: Although no instruction is given as to what to do with these herbal ingredients, they are perhaps to be made into an ale-based drink, as in the following entry.

149      This remedy is commonly known as "Metrical Charm 5." Bjork, *Old English Shorter Poems II*, 208–9, presents a slightly different handling of text and translation. At pp. 220–21, Bjork likewise edits and translates a close analogue, the text known as "Metrical Charm 10," from the marginalia of CCCC MS 41. For a looser analogue see *Misc. Rem.* 1, this one too from the marginalia of CCCC MS 41.

149.1    *your cattle are lost*: While here the cattle are said to be "lost," one infers from the following lines, with their reference to "this deed," that they have been stolen.

149.3    *May Christ's cross lead them back*: The speaker is to utter repeated versions of this key phrase in Latin, the language of the liturgy, though the rest of the charm is in the vernacular.

149.4    *Christ's cross was stolen away and was found*: A reference to the discovery of the True Cross by Saint Helena, the mother of the emperor Constantine I, after it had lain hidden in the ground for centuries.

150      For a variant text from CCCC MS 41, see *Misc. Rem.* 4.

150.2    *the many blind ones who . . .* : Some words have been lost at this point, though there is no lacuna in the manuscript.

150.3  *By the Father . . . amen*: Another closing formula, supplied editorially to fill out the scribe's abbreviation *per* (by).

152  *If a horse has sprained a leg*: The physical injury is evidently thought of as the work of a malicious being.

*Naborrede*: A pseudo-Latin nonsense word that is evidently directed against the power that has inflicted the injury, banishing it.

*I have believed, therefore*: The first words of Psalm 115, which continues: "I have believed; therefore have I spoken, but I have been humbled exceedingly."

*Alpha and Omega, the beginning and the end*: A common liturgical formula invoking the power of Christ.

*The cross is life to me and death to you, O enemy*: This same ancient Latin formula is used in *Lacnunga* 168.3 below, in a remedy that bears some resemblance to this one.

153.2  *Nine were they . . . became none*: A counting-down charm of a kind that has worldwide folkloric analogues, whether ancient or modern; see Pettit, *Remedies,* vol. 2, pp. 290–94, for examples, with discussion.

*the sisters of Noth*: The name "Noth" cannot reliably be connected to any known mythological figure; the "sisters" may personify the swollen tissues.

154  *Geneon genetron . . . nequetando*: An untranslatable incantation.

155  *If a horse is shot*: The problem is that of "elf-shot," as in *Lacnunga* 111 and 127, though here there is no reference to the power of elves. Whatever its ailment may be, the horse is to be healed through Latin prayers and the laying on of hands; hence the practitioner is likely to be a priest.

*May the animals in the world be healed*: A prayer with no identifiable source.

*They are troubled in health*: Also no known source.

*Which ones have we separated . . . of Christ?*: Since Latin *quas* (which ones) is feminine in gender, this pronoun might refer back to the preceding feminine noun *manus* (hand), though it is more likely that the phrase (here taken out of its original context) refers to a group of women.

156    *God . . . from chains*: This is a highly abbreviated allusion to a remedy calling upon Saint Peter to release, by God's command, the chains of the world. See Pettit, *Remedies,* vol. 2, p. 301, for that text, whose adaptation to the present context has been explicated by George Hardin Brown, "Solving the 'Solve' Riddle in B. L. Harley 585," *Viator* 18 (1987): 46.

       *from chains*: The unborn child needs to be "unchained" from the womb.

157    *In the name of the Father . . . Holy Spirit*: Expanded editorially from scribal *in nomine* (in the name).

158    A charm that is widely attested both in the Middle Ages and in more recent times; for variants and analogues see Pettit, *Remedies,* vol. 2, pp. 304–8; and Jonathan Roper, *English Verbal Charms* (Helsinki, 2005), 122–25. Two analogues that figure in the present edition are *Misc. Rem.* 36 (from London, British Library, MS Cotton Vespasian D.xx) and 87 (from Oxford, Bodleian Library, MSS Junius 85 and 86).

158.3    *malignant drop*: According to the medieval theory of the humors, headaches could be caused by an excess of fluid in the cranium. Here toothache is spoken of in analogous terms.

158.4    *Rex pax nax*: A garbled Latin invocation, perhaps deriving from a liturgical formula such as *rex pacis nascitur,* that is, "the king of peace is born." See Brown, "Solving the 'Solve' Riddle," 49n20.

159    *. . . God, who said*: The beginning of the prayer is lost.

       *Come to me . . . refresh you*: Matthew 11:28, Christ's words of comfort to the needy.

160    *For dysentery*: In other remedies, Old English *ut-siht* is best translated "diarrhea." Here, however, the word refers to an infectious disease.

160.2    *The angel brought this epistle to Rome*: What follows is an instance of the "Heavenly Letter" type of charm: a letter that is said to have been written by God or a divine agent and that offers the bearer protection from disease or other afflictions. For discussion of this example, see Rosanne Hebing, "The Textual Transmission of Heavenly Letter Charms in Anglo-Saxon Manuscripts," in Chardonnens and Carella, *Secular Learning,* 217–18.

160.3    *Ranmigan adonai . . . allelluia*: An untranslatable passage of
         Hebrew-like, Aramaic-like, Greek-like, and Latin-like words
         and phrases. *Adonai*, for example, is Hebrew for "the Lord"; *O
         ineffabile* is Latin for "O, ineffable one"; *Beronice* is "Veronica";
         *sother* is Latinized Greek for "savior"; *miserere mei Deus* is "have
         mercy upon me, Lord," the first words of Psalm 50.

161      For an alternative handling of the text and translation of this
         and the following two charms, see Bjork, *Old English Shorter
         Poems II*, 210–13. While Bjork treats *Lacnunga* 161, 162, and 163
         as three parts of a single cure (commonly known as "Metrical
         Charm 6"), they are treated here as three separate cures meant
         to address conditions that are different, though related.

161.1    *a woman who cannot nourish her child*: That is, apparently, one
         whose previous child (or children?) is believed to have died for
         lack of nourishment, whether during pregnancy or after she
         has given birth. While Bjork, *Old English Shorter Poems II*, 211,
         translates "who cannot bring forth her child from the womb,"
         the problem may be that the mother has been unable to breast-
         feed her child. But both this remedy and the two succeeding
         ones might be intended for use in a wide range of circum-
         stances having to do with a woman's present desire to give
         birth to a healthy, full-term child after a previous history of
         miscarriage or infant mortality.

161.2    *for grievous black birth*: Although the phrase does not admit of
         easy interpretation, it may have reference to a fetus that has
         died in the womb.

         *for hateful lame birth*: Although the reference could be to a child
         who is born lame or crippled, the Old English adjective *lama*
         has a broad enough semantic range to encompass the palsied,
         the paralytic, or the otherwise imperfect.

161.3    *when the woman is with child*: That is, apparently, when she be-
         comes pregnant again.

161.5    *when the mother feels that the child is alive*: That is, when she feels
         the fetus moving in the womb.

163      This third charm in the series seems to be intended either to
         avert miscarriage, or to prevent premature birth, or to address

problems relating to breastfeeding once the child is born; or it could be that all these potential sources of infant mortality are thought to be addressed.

163.2 *this splendid belly-strong one*: An evident allusion to the infant, addressed as if he or she were a great eater.

*this splendid food-strong one*: Probably the infant again, though conceivably the allusion is to the woman's uterus, spoken of as a strong and healthy source of nourishment.

*I will keep him as mine*: The Old English wording is not easy to construe. I take the word *þonne* (a common variant spelling of the demonstrative pronoun *þone*) as representing the masculine accusative singular pronoun *hine* (him). The pronoun *me* in the clause *ic me wille habban* is best taken to be reflexive (as is the pronoun *me* two lines earlier); hence the present translation, "I will keep him as mine," whether the reference is to the infant at the breast or the child in the womb.

164 For a variant of this cure see *Misc. Rem.* 19 (from London, British Library, MS Cotton Faustina A.x), a text that is directed against diarrhea or dysentery.

*Ecce dolgula . . . uhic*: Little sense can be made of this incantation, with its mixture of forms derived partly from Irish and partly from Latin. For discussion see Howard Meroney, "Irish in the Old English Charms," *Speculum* 20 (1945): 180; and Pettit, *Remedies*, vol. 2, pp. 331–32.

*For hard swellings*: This phrase might be meant to serve as the heading of the next cure rather than as the last words of the present one.

165 A veterinary cure for an unspecified ailment.

*Arcus supeð assedit . . . canabið*: There seems little point in attempting a translation of this passage of garbled Latin. See Pettit, *Remedies,* vol. 2, pp. 334–37, for discussion, citing later analogues where the remedy is for help with childbirth, not for a sick horse.

168.3 *Indomo mamosin . . . el marethin*: An untranslatable incantation.

*The cross is life to me*: The same formula is used in *Lacnunga* 152 above, a charm for an injured horse.

169.3    *O part, and O rillia pars*: Both garbled and coherent Latin are found in this passage.

*the beginning and the end*: An editorial expansion of the scribe's single word *initium* (beginning). Compare the use of the same phrase in *Lacnunga* 152.

170.1    *Arestolobius*: A king whose name is otherwise unattested.

170.2    *for jaundice*: Literally, "for the yellow sickness."

*for theor-riding*: A possible allusion to the belief that illnesses could result from inimical creatures "riding" a person. See the notes to *Lacnunga* 86, for dwarf-driven fever.

173    A pair of remedies written out in *Bald's Leechbook* 1.15.1–2 may be the direct source of this one; see Pettit, *Remedies*, vol. 2, p. 346. This is the first of a series of texts (extending through *Lacnunga* 180) that were copied either from that compendium or from a common source, with the remedies written out in the same order, though *Lacnunga* includes additional items as well.

173.3    *mashwort*: The infusion of malt in water before brewing.

*common polypody (the biggest part)*: What is apparently meant is that common polypody should be the chief ingredient in the recipe.

*houseleek*: The Old English plant name *singrene* (literally, "evergreen") is thought to denote either houseleek, barren privet, or common periwinkle.

174–75    For *Bald's Leechbook* 1.15.3–4, whose texts respectively correspond to these cures, see Pettit, *Remedies*, vol. 2, pp. 346–47.

175    *eyebright*: A conjectural translation based in part on the understanding that the prefix *ea-* in the Old English word *ea-wyrt* is a shortened form of *eag* (eye), rather than representing *ea* (water). Another possible translation is "dandelion."

*until . . .* : At least one folio is lost at this point. A comparison of this fragmentary remedy with the corresponding complete one in *Bald's Leechbook* helps to recover the substance of the lost text: the concoction is to be reduced by a third, and the patient is to drink it three times a day.

176    The word *gepigce* (consume, in the subjunctive mood), written out just before the start of this remedy, is evidently carried over from a lost folio.

177   For *Bald's Leechbook* 1.17.1, which corresponds to this text, see
      Pettit, *Remedies*, vol. 2, p. 350.
      *A handful*: See the *DOE*, under *gylm, gylma*: "handful / bundle (of
      something), sheaf (of wheat, etc.)." Bosworth and Toller, *Dic-*
      *tionary*, gloss the Old English word (under the spelling *gelm*) as
      "a yelm, a handful of reaped corn."

178   For *Bald's Leechbook* 1.17.2, corresponding to this text, see Pettit,
      *Remedies*, vol. 2, p. 350.
      *a stone bath*: See *Lacnunga* 73 for a brief account of this proce-
      dure.

179–80  For *Bald's Leechbook* 1.17.3–4, corresponding respectively to these
      texts, see Pettit, *Remedies*, vol. 2, pp. 350–51.

181   *and add to them*: That is, add the pottage to them.

182.1  *that we call "Egyptian"*: The "Egyptian" days, defined differently
      by different authorities, were unlucky ones. The tradition of
      identifying them goes back via Greek and Latin sources to
      Egyptian calendars.
      *either man or beast*: The Old English noun *nyten* refers especially
      to cattle or horses.

183   A prayer for deliverance from plague. This is the first of a series
      of eight healing texts or blessings that were added to the origi-
      nal compilation in the course of its transmission, thus making
      it even more of a miscellany. The first seven are in Latin, in a
      later hand.

183.2.1  *To the God of heaven*: The five stanzas that follow reproduce, with
      slight variations, a poem that is attributed to the sixth-century
      bishop Syagrius of Autun. See Pettit, *Remedies*, vol. 2, pp. 357–
      58, for a ninth-century copy of that earlier text, along with an
      attempt to reconstruct its original form.

183.3  *deliver this woman*: Unusually, this passage identifies the person
      to be healed as female.
      *upon me, NAME*: Unusually again, it is the name of the healer that
      is apparently to be supplied here, not the name of the patient,
      unless the two persons are the same. In that case the whole
      remedy is set in a woman's voice.

183.4  *Brigitarum dricillarum . . . rubebroht*: An incantation that resists
      translation, though certain of its elements derive from Latin

and Old Irish. For discussion see Meroney, "Irish in the Old English Charms," 174, and Pettit, *Remedies,* vol. 2, pp. 362–63. The feminine plural form *Brigitarum* is suggestive of a real or imagined group of nuns dedicated to Saint Brigid.

183.5      *Saint Rehhoc . . . Sigismund the king*: Saint Rehoc may be Saint Rihoch, an early Irish bishop. Saint Rehwald cannot be identified. Saint Cassian may be the fourth- and fifth-century founder of monasteries. Saint Germanus may be the fifth-century saint who traveled from Gaul to Britain, as is reported by Bede in his *Ecclesiastical History* 1.17–21. Saint Sigismund, who died in about 524, was king of Burgundy and a reputed martyr. It is not clear why this particular group of saints is invoked here, but certain of these names are suggestive of Irish and/or continental connections on the part of a contributor to the closing sections of *Lacnunga.*

185.2      *By the Father, the Son, and the Holy Spirit*: An editorial expansion of the scribe's abbreviation *per* (by).

188      The beginning of this blessing is lost because of damage to the manuscript. Basing his understanding on parallels recorded elsewhere (including *Lacnunga* 63.11), Pettit, *Remedies,* vol. 2, p. 367, offers the following reconstruction: *Benedic domine hunc fructum novarum arborum ut hi qui utuntur ex eo sint sanctificati. Per.* This can be translated: "Lord, bless this fruit of the new trees so that those who make use of it may be blessed, by your holy name."

189      This entry is so badly faded as not to be worth including, apart from three of its closing words. See Pettit, *Remedies,* vol. 1, p. 130, for an attempt to salvage more of the text.

190      Text not sufficiently legible to translate, apart from a few words. See Pettit, *Remedies,* vol. 1, p. 130, for a more detailed reconstruction of the badly faded passage.

     *wheat of holy . . .* : Since so little text is legible here, another possible translation is "wheat of Saint . . ."; the name of a particular saint would then have followed.

191     Written in a twelfth- or thirteenth-century hand in the Anglo-
        Norman dialect of French.

191.1   *garden herbs*: A specific herb may be meant, perhaps common
        purslane.


## PERI DIDAXEON

1.1     Latin rubric: *Incipit liber qui dicitur Peri Didaxeon,* or "here begins
        the book called *Peri Didaxeon.*"

        *Peri Didaxeon*: A title with the meaning "Concerning the Teach-
        ings" (from Greek δίδαξις, genitive plural διδάξεων), with ref-
        erence to the teachings, or schools, of the ancient founders of
        the science of medicine.

        *about its proper written expression*: Literally, "concerning its tes-
        timony," accepting the editorial emendation *gewitnesse* (for
        scribal *gewisnesse,* a reading that might be construed, though
        only doubtfully, as meaning "wise ways").

        *they brought forth*: The particular sense of the Old English verb
        *alysan* is "to deliver, release, liberate" (see the *DOE,* under *aly-
        san,* sense 1.e), thus the implication of this phrase is that the
        four ancient Greek physicians who are named in this para-
        graph set the science of medicine free from darkness into light.

        *Apollo*: That is, the figure whom the ancient Greeks worshipped
        as a god, but who in the Christian view was to be respected as
        an exceptional human being.

        *his sons Esculafius and Asclepius*: Max Löweneck, ed., "Peri Didax-
        eon," special issue, *Erlanger Beiträge zur englischen Philologie* 12
        (1896): 54, interprets the noun *suna* (sons), normally a nomina-
        tive plural form, as a singular form referring to Esculafius alone
        (as in the Latin source texts; see López Figueroa, "Estudio y
        edición crítica," 63). Yet the Old English text speaks in terms
        of two similarly named sons of the same father, the founding
        figure Apollo. The scribe initially wrote the name of the first
        son (Aesculapius, the Greco-Roman god of medicine) in ac-
        cord with its frequent medieval Latin spelling, *Esculapius* (see

Löweneck, "Peri Didaxeon," 2), then emended the spelling to *Esculafius* (see Notes to the Texts). In ancient Greco-Roman tradition, Aesculapius and Asclepius were alternative names for the same revered figure; here Asclepius is introduced as a separate person.

*Hippocrates*: The ancient Greek medical dynasty is carried on to a third generation by Hippocrates, whose name the scribe represents first as *Ypocrates* and then, later in this paragraph and at later points in this same treatise, as *Ypocras*. Hippocrates is regarded as a historical figure who was born on the Greek island of Kos around 460 BCE.

*Artaxis*: Artaxerxes I, the fifth king of Persia, who ruled from 465 to 424 BCE; he was the grandson of Darius I and the son of Xerxes I.

*Scolafius*: Evidently to be construed as the same person who, earlier in this same paragraph, is named "Escolafius" and whom we would call Aesculapius. The name is spelled variously as "Scolapius," "Scolafius," or "Escolapius" in the Latin texts cited by López Figueroa, "Estudio y edición crítica," 63.

*the preservation of law and life*: That is, the preservation of the natural order of things, unmarred by the intrusion of disease.

1.2    *philosophers*: This is the sense of the emended text *up-wytyna*, as opposed to scribal *ap-wytyna* (persons knowledgeable about the law).

*the circuit of earth*: Although Cockayne takes the unique phrase *middangeardes boga* (literally, "the bow of middle earth") to refer to the rainbow and is followed in this inference by the *DOE*, the phrase plausibly refers to the whole circuit of earth, with its groundwater and its freshwater streams.

*rough gall*: The Old English adjective *ruwa* (rough) is probably a mistranslation of Latin *rufus* (red, reddish). The same mistranslation is given a few lines down.

*from the eighteenth . . . April*: That is, from December 15 until March 25, according to Cockayne's reckoning of the old Roman calendar. Correspondingly, the remaining three quarters of the year run from the fifteenth to the twenty-fifth day of their respective months.

*the dog days*: The ancient Greeks referred to the unpleasantly hot days of late summer as the dog days, with reference to the return of the bright star Sirius to the night sky. Sirius was known as the Dog Star (from Greek κύων, "dog").

2    The Old English rubric reads *wið oman,* meaning perhaps "for erysipelas," a bacterial skin infection involving the upper dermis.

*litharge*: A yellow lead oxide, a derivative of the extraction of silver from its ore.

*fresh lime*: That is, the white mineral derived from limestone.

3    A Latin rubric reads *ad scabiosos,* or "for a scabious condition"; that is, for rough or pimply skin.

4    *take black beans*: The phrase denotes some type of dark or black beans other than what today are called "black beans," which are of Central American origin.

7    *Take some furze*: That is, take some shrubs growing on wasteland; other names for plants of the kind are "gorse" and "whins."

10    Latin rubric: *Ad dolorem capitis,* or "for headache."

*when insects have taken over the head*: The Old English noun *wurmas* (or *weormas* or *wyrmas*) could refer to any of a number of small creatures; here head lice are likely to be meant.

11    *the dye*: The meaning of the Old English noun *teafor* is unclear in this context; elsewhere the word denotes a pigment, one that is used, for example, to mark sheep with a red ocher color.

*galbanum*: An aromatic gum resin.

*take that salve*: The medicine must be thinned enough to be poured into the ears; perhaps "solution" is a better translation.

*until the ears have absorbed it all*: Literally, "until the ears have drunk it all in."

13    Latin rubric: *Ad tornionem capitis,* literally, "for turning-about of the head," or "for dizziness."

*his brains are turned about*: The malady must be dizziness, vertigo, or a similar ailment. The corresponding part of the Latin source (as printed by Löweneck, "Peri Didaxeon," 8) reads *versatum cerebrum habet* (he has a turned-about skull or brain).

*wool that has never been washed*: That is, wool that retains its natural oils.

16.1 Latin rubric: *De capitis purgatione,* or "for cleansing of the head."

16.2 *prepare a steam bath*: The Old English term *stuf-bæþ* refers to a vapor bath prepared with water thrown over red-hot stones.

17.1 Latin rubric: *Ad aures,* or "for the ears."

18.1 *neck swellings*: Evidently, glandular swellings are meant.

*the Greeks call kakote*: From the adjective κακός, "bad" or "evil."

19.1 Latin rubric: *Ad cecitatem oculorum,* or "for dimness of the eyes."

19.2 *place a cupping horn to the temples*: A traditional means of bloodletting, or of cupping the skin so as to draw the blood to the surface, as has been done using buffalo horns in parts of Asia in recent times.

*a purgative drink*: The Latin word *catarcum* is evidently a misspelling of *catharticum* (a purgative).

20 Rubric: *Wið totore egean,* or "for bleared eyes."

*or is bleared*: See Bosworth and Toller, *Dictionary,* under *toren-íge* (blear-eyed).

*black vitriol*: The apparent meaning of Old English *atrum* or *attrum,* from the medieval Latin *atramentum* (black pigment, or darkness).

22 Latin rubric: *Contra glaucomata,* or "for cataracts."

23 Latin rubric: *Item, contra cecitatem,* or "again, for blindness."

*this will cunningly cure it*: The Latin source (as printed by Löweneck, "Peri Didaxeon," 14) reads *postea prudenter sanat* (afterward this will cure it sagaciously). See also the *DOE,* under *gleawlice,* sense 3: "skillfully, cunningly."

24 *a buck's thigh joint*: That is, the ball part of a male goat's ball-and-socket joint.

25 *ordiolum*: The term has seen occasional use in the medical literature of England with reference to a stye; that is, a red, painful swelling near the edge of the eyelid. See in particular Juhani

Norri, *Dictionary of Medical Vocabulary in English, 1375–1550* (London, 2016), under *ordeolum*.

26 *bean meal*: Broad beans are meant (Latin *fabae*), not beans of the kind that are most commonly available today.

28 Latin rubric: *Ad sternutationem,* or "for sneezing."

*castoreum*: The yellowish secretion of a beaver's castor anal sac.

29 *this is a remedy for a person*: Both here and elsewhere in the treatise, the scribe's spellings vary unpredictably between what look to be grammatically singular and grammatically plural forms. The present translations aim for greater consistency in this regard.

31 *burn them together*: The main verb has dropped out of the text. While Löweneck, "Peri Didaxeon," 19, supplies *cnuca* (pound), I follow Danielle Maion, "Edizione, traduzione e commento del *Peri Didaxeon*" (PhD diss., Università degli Studi Roma Tre, 1999), at line 232 of her text, in supplying *bærne* (burn).

32 *that lifts up the teeth and loosens them*: The reference, though somewhat obscure, seems to be to decayed gums that no longer give proper support to the teeth as a result of untreated gingivitis.

33.2 *the most prominent teeth . . . first of all*: A difficult passage to construe in the original manuscript, which appears to be corrupt at this point. I follow Löweneck, "Peri Didaxeon," 19, in his reconstruction of the text into something more coherent, even if speculative.

*when it makes its way through*: My understanding of the Old English phrase *þan hyt vinð* rests on taking *vinð* to be a variant spelling of *winneþ* (it makes its way): see Bosworth and Toller, *Dictionary,* under *winnan,* sense III, and the *Middle English Dictionary,* ed. Hans Kurath and others (Ann Arbor, 1952–2001), under *winnen,* sense 11a (online at http://quod.lib.umich.edu/m/middle-english-dictionary).

36 Latin rubric: *Ad ufam,* or "for the uvula." The cure is directed against a tumor or swelling of the uvula (Old English *huf* or *uf*).

38 Latin rubric: *Ad strictum pectus,* or "for a constricted chest."

41 Latin rubric: *Ad inflationem guttaris,* or "For inflammation of the throat."

43.1 *from the jaw to the shoulder*: "Shoulder" (Old English *sculdre*) is a conjectural addition to the text; a word has apparently dropped out.

43.2 *using a cupping horn*: The reference is to a bloodletting procedure, as in *Peri Didaxeon* 19.2.

44 *and galbanum*: The Old English text is corrupt at this point. The present reconstruction of the sense of the passage relies on the *DOE* entry for *galban*, a loan word from Latin *galbanum*, denoting an aromatic gum resin.

49 *swail's apples*: It is unclear what kind of apple this is, if indeed apples are meant. The literal meaning of Old English *swegles æppel* is "apple of the sky," or perhaps "apple of the sun."

50.2 *castoreum*: The apparent meaning of *beferes herthan*, literally, "beaver's testicles"; see also the note to *Peri Didaxeon* 28. The equivalent Latin text as cited by Löweneck, "Peri Didaxeon," 30, reads *da ei castoreum et piper in vino calido* (give him castoreum and pepper in warm wine).

51.1 Latin rubric: *Ad strictum pectus, sive ad asmaticos,* meaning "for a constricted chest, or for shortness of breath."

 *accumulates in his throat and percolates*: The Old English verbs that occur in this passage are *weaxan*, literally meaning "to grow," and *seohhian*, literally meaning "to strain," respectively.

51.3 *Take ventuosa*: The Latin source reads *ventosas* (leeches).

 *in oil of cypress*: A misunderstanding of Latin *in oleo ciprino* (in oil of henna). See the note at *Herbal* 22.1.

51.4 *panax*: Any of a number of plants of the ginseng family, chiefly of East Asian distribution, valued for their medicinal properties.

 *three quinces*: Literally, "three apples of Cydonia," quince being a fruit associated with the ancient city-state of Cydonia on the island of Crete.

52.1 *someone suffering from a chest ailment*: Literally, a person "in whom there is illness in the breasts." Here (and elsewhere, for the most part), I adopt the translation "chest" where the Old English idiom involves the plural form "breasts."

*all the veins go lax*: Literally, "all the veins go to sleep."

52.5 *add some mast to it*: Here, Old English *æcern* probably refers to acorn mast, not acorn kernels.

 *wheat meal*: The likely sense of Old English *flysma*, a word that occurs nowhere else in the Old English corpus and that is glossed by the *DOE* as "? bran" or "? flour."

52.6 *eat the bread that you have warmed*: The probable sense of a passage that is grammatically condensed.

54 *the breastbone*: Literally, "the sharp bone located between the breasts."

56 Latin rubric: *Ad umbilicum*, or "for the navel."

58.2 *stir it up with*: Translating Old English *mæcige mid*, from *mecgan*, "to stir, mix."

59.1 Latin rubric: *Ad eos qui nimis salivam conspuunt*, or "for those who spit up too much spittle."

61.1 Latin rubric: *Potus provocans vomitum*, or "a drink to induce vomiting."

62.1 Latin rubric: *Potus levior ad vomitum*, or "a lighter drink for vomiting."

62.2 *asarabacca*: That is, *Asarum europaeum* L., a plant also known by the Latin name *vulgago* or, in English, as "European wild ginger," among other names.

62.3 *dwarf-elder juice*: This is the sense of the passage if Old English *eallan-wyrte* refers to the elder tree (usually spelled *ellen-wyrt*). But perhaps the text should read *eolone-wyrte* (elecampane juice)?

63.1 Latin rubric: *Contra nimium vomitum*, or "for excessive vomiting."

63.3 *resin*: Literally, "clear pitch."

63.4 *take some tow*: That is, take the coarse part of some flax.

63.7 *if a person is accustomed that* . . . : Since the verb *pyrete* that occurs in the Old English text makes no clear sense, an ellipsis is inserted here. There is no gap in the manuscript.

64 *Galen the physician*: That is, Galen of Pergamon (here called by the Latinate form *Galwenus*), the ancient Greek physician who practiced medicine in Rome and whose writings had a dominant place in medieval European schools.

65.1     Latin rubric: *Si dolor et infimitas sit in visceribus,* or "if there is pain and disease in the intestines" (with *infirmitas,* "disease," miswritten as *infimitas*).

*the two small tubes . . . inside the gullet*: That is, the esophagus and the windpipe.

66.1     *if he is of age*: The probable meaning of Old English *gif he para hulde habban.* This interpretation requires taking *hulde* as a late variant spelling of *alde* (age), which in turn is a derivative of the adjectival form *eald* (old). Compare the *DOE* entry for *eald,* sense I.A.2.b: of animals, "mature, full-grown, adult."

66.2     *tie the ointment to his chest*: That is, tie the poultice to his chest. The main text of the treatise stops here, at the foot of folio 66v. In the lower margin is written *gif þæt blod of þan innoþe cump* (if the blood comes from the inner organs), followed by a full stop and by what looks to be the Roman numeral *vii.* It is unclear if more text was to follow, or if this note refers back to the contents of section 65.

## MISCELLANEOUS REMEDIES

1     Although this charm is not medical in nature, it is included here since it is an analogue to *Lacnunga* 149, where a charm against cattle theft is interspersed among veterinary cures. Preceding the present entry in CCCC 41 are two other closely related texts: these are "Metrical Charm 9" and "Metrical Charm 10," both of which are for lost or stolen cattle. For those two texts, with translations, see Bjork, *Old English Shorter Poems II,* 218–20. For discussion of these and other variants, see Stephanie Hollis, "Old English 'Cattle-Theft Charms': Manuscript Context and Social Uses," *Anglia* 115 (1997): 139–64, and Peter Dendle, "Textual Transmission of the Old English 'Loss of Cattle' Charm," *Journal of English and Germanic Philology* 105 (2006): 514–39.

1.2     *sing this on its shackle or on its bridle*: Horses were often tethered by means of a fetter or shackle attached to one foot. "Sing this on" might loosely mean "sing this over," but the phrasing is more forceful than that: the shackle and the bridle are addressed directly.

1.3    *Peter, Paul . . . Felix*: It is not clear why this particular set of holy figures is named. References to Saint Patrick ("the Apostle of the Irish") and to Saint Brigid (the second patron saint of Ireland) are suggestive of an Irish connection on the part of whoever devised this remedy. "Philip" is probably Philip the Apostle, mentioned several times in the Gospel of John. Felix is probably Saint Felix, the seventh-century bishop who introduced Christianity to the people of East Anglia.

*Cyriacus*: One of the Seven Sleepers of Ephesus; see the note to *Lacnunga* 86.2. The obscure word *chiric* in the Latin text is taken by some, including Godfrid Storms, *Anglo-Saxon Magic* (The Hague, 1948), 207, to be a common noun meaning "the church."

*He who seeks will find*: After these words the scribe has written out an eighteen-line Latin hymn, omitted here as having no direct relation to the present charm. For that interpolated text see R. J. S. Grant, *Cambridge, Corpus Christi College 41: The Loricas and the Missal,* Costerus Essays in English and American Language and Literature, n.s. 17 (Amsterdam, 1979), pp. 5–6, lines 9 26; or Karen L. Jolly, "On the Margins of Orthodoxy: Devotional Formulas and Protective Prayers in Cambridge, Corpus Christi College MS 41," in *Signs on the Edge: Space, Text and Margin in Medieval Manuscripts,* ed. Sarah Larratt Keefer and Rolf H. Bremmer, Jr. (Paris, 2007), 157. With the words "May the cross of Christ lead them back," the charm against cattle theft resumes.

1.4    *houses*: The evident meaning of Latin *andronas,* from the Greek plural form ἀνδρῶνες, referring to the part of a house occupied by men, hence by extension the domicile itself.

3.3    *you will be a most unclean spirit*: Something appears to have been lost from the text at this point. The original sense of the passage is likely to have been "you, most unclean spirit, will be consigned to hellfire."

*The weeping of eyes*: Although this passage too is hard to construe, its sense may be "your actions in causing many people to weep" will result in your destruction; the water of tears contrasts with the hellfire *(gehenna)* suffered by devils.

3.4–5    *Go out from his head . . . forever and ever*: For a close parallel to this exorcism, see *Lacnunga* 63.6.

3.4    *so that devils may have no power*: Alternatively, "so that the devil may have no power"; there is grammatical confusion here between singular and plural forms.

   *over this man* NAME: The scribe's insertion of an alternate feminine inflectional ending for the Latin pronoun *isto* (see Notes to the Texts) permits the alternative translation "over this woman NAME."

3.5    *or while remaining silent*: The manuscript reading *tangendi* (touching), which is repeated from earlier in the sentence (where it seems to refer to devils harassing human beings), is probably an error for *tacendi* (remaining silent), to judge from an analogous passage in *Lacungna* 63.6.

   *or while shining*: This phrase makes little sense, and the text that immediately follows it is corrupt (see Notes to the Texts). In the same analogous passage from *Lacnunga,* the corresponding phrase is *nec in legendo* (or while reading).

4    There is a parallel to this text in *Lacnunga* 150.

4.2    *the son of Tobit*: In the biblical book of Tobit, it was actually Tobit's son, Tobiah, who was given the power to heal his father's blindness.

   *You are the hand*: In the Latin text, the phrase "you are" is left understood.

4.3    *By the Father . . . amen*: An editorial expansion of scribal *per* (by).

5    *Fandorahel*: Evidently, one of Satan's minions.

   *send him back from the torturing of his ears*: While this is the literal meaning of this passage, Jolly, "On the Margins," 166n130, offers the alternative translation "recede soon, torturing [pain], from his ears."

   *By the Father . . . amen*: An editorial expansion of scribal *per.*

6    *Laniel . . . Dormiel*: These names are of uncertain origin and signification.

*By the Father . . . amen*: An editorial expansion of scribal *per.*

7 Probably a tripartite prayer to accompany an offering for a successful childbirth, though it is possible that the three sections of the cure were meant to function separately.

7.1 *By the Father . . . amen*: An editorial expansion of scribal *per.*

7.2 The line consists of the *sator* formula, a palindrome popular in Europe from the sixth century on. The same formula has a long history as a word square, a form that resembles a square shield and hence is suggestive of prophylactic qualities. See Grant, *Cambridge, Corpus Christi College 41,* 19–22; Storms, *Anglo-Saxon Magic,* 281–82; and Jolly, "On the Margins," 169–70, for discussion of its history and its frequent cross-like visual deployment. Devised as an all-purpose lorica (or shield), it is often employed in medical contexts as an aid to childbirth. Although the five Latin words that constitute the charm do not make good grammatical sense, their gist is "the sower *(sator)* Arepo holds *(tenet)* the wheels *(rotas)* through his effort *(opera),*" a reference to steady plowing. Users of the charm would not necessarily have known this.

7.3 *so that they might grow and multiply*: The corresponding Latin clause, *ut crescere et multiplicare,* is fraught with grammatical difficulties, though the sense of the passage remains clear, with its echoes of Genesis 8:17, 9:1, and 9:7. For detailed but inconclusive discussion of the grammar, with observations on scribal interventions meant to clarify it, see Grant, *Cambridge, Corpus Christi College 41,* 18–19, his footnote to line 10 of the text (in his line numbering).

8.3 *Ire . . . arafax*: Although several words of this invocation are recognizably Latin or Latin-like, there is no point in offering a translation.

9.1 *For a stomach ailment*: The probable meaning of Latin *contra felon.* The term *felon* could also refer to gall, bile, or poison.

9.3 *Thigat . . . cunhunelaya*: An incantation that has no translatable meaning; for other versions see the note to *Lacnunga* 25.2.

9.4 *Seek and you shall find*: Matthew 7:7.

11.4 *Elyon, Elyo . . . Us, Ne, Te*: Many of the names of God that are listed in this part of the remedy are of no clear meaning and are left untranslated.

*Tetragrammaton*: The Greek term for the four-letter Hebrew name for God ("YHWH" in Latinized form, or "Yahweh"). In Jewish tradition, the tetragrammaton is considered unpronounceable.

*Sös*: The central letter in this word is surmounted by what might be a mark of abbreviation, but it is unclear what word is abbreviated, if indeed any is.

11.8    *these three psalms*: The first of these is Psalm 122, beginning "To thee have I lifted up my eyes"; the second is Psalm 66, beginning "May God have mercy on us"; and the third item to be intoned is the Athanasian Creed, beginning "Whosoever wishes to be saved."

*Glory to the Father*: The doxology, a succinct declaration of faith.

*Lord have mercy . . . Lord have mercy*: The Greek invocation *Kyrieleison* is from the first part of the Mass.

*Make your servant well, O Lord*: This is the gist of the garbled Latin passage. Compare Psalms 11:1: *Salvum me fac, Domine* (Save me, O Lord).

11.11    This set of instructions is written in a later hand.

*This prayer should be said*: The Latin term *carmen* (poem, song) was sometimes used to mean "prayer," particularly when an element of magic was involved, as it is here. This sense of the word is thought to be primary, other senses having accrued to it in the course of time.

12.3    An incantation without translatable meaning.

12.5    Another untranslatable incantation.

13.2    *Saint Nicasius*: The fourth- or fifth-century bishop of Rheims, the patron saint of smallpox victims. For variants of the first part of this prayer, see *Misc. Rem.* 29 and 72.

*his name written out on him . . .* : The passage breaks off here without any gap in the manuscript. A phrase such as *hoc malum non haberet* (would not suffer this evil) is to be understood after *scriptum* (written).

14    *Saint Blasius*: The early Christian Armenian monk Saint Blasius,

also known as Saint Blaise, was a physician before he became a monk. He is associated with the healing of sore throats.

15.1 *An angel brought this inscription from heaven*: Another instance of the "Heavenly Letter" type of remedy, like *Lacnunga* 160; for discussion see Hebing, "The Textual Transmission," 218–21.

15.2 *whether flying or earth borne*: This passage rests on a belief in flying venoms and other venoms that attack the body from without.

15.4 *Me abdicamus . . . pisticus*: No translation is offered for the pseudo-Latin passage written out at this point.

15.6 *and John the king*: The reference is apparently to John the Evangelist; the Greek term for "king" *(basileus)* is used.

15.7 No translation is offered for the italicized passages of garbled Latin that occur in this paragraph.

15.8 *The saints rejoice in glory*: Psalms 149:5.
*and swords*: Again, the Latin text is garbled; Latin *gladii* (swords) is very likely a mistake for *gaudia* (joys).

15.9 *Praise God in his saints*: Psalms 150:1 (Septuagint version); an alternative translation is "Praise God in his holy places."

17.1 *and for mites*: See the note to *Lacnunga* 10.1.

17.2 *on top of them put . $\overline{GT}$ .* : A word or phrase is wanted here corresponding to $\overline{GT}$ in the manuscript; it is unclear what word or words that symbol is meant to stand for.

19.1 *for diarrhea*: The Old English term *utsiht* might refer to either diarrhea or dysentery. The latter disease could be a mortal one.

19.2 *Ecce . . . dop*: An incantation without translatable meaning.

20.4 *Eugenius . . . Quiriacus*: An alternative version of the names of the Seven Sleepers of Ephesus; see the note to *Lacnunga* 86.2.

20.5 *Write these names*: Probably just the names are to be written, not the exorcism.

21 *the seven sleeping saints*: Another reference to the legend of the Seven Sleepers; see the note to *Lacnunga* 86.2.

25 *sour cream*: Cockayne's conjectural translation of Old English *unflid*, a word of uncertain meaning. The text may be corrupt at this point.
*a good oak drink*: Probably one made from oak bark.

*oat bran*: A conjectural translation of Old English *etriman dust*.

*the pith of the ears of grain*: That is, the pith of the oat kernels.

27 Variant texts of this cure and the next one are recorded in *Bald's Leechbook* 2.65.10 and 2.65.11.

28.2 *three days before midwinter*: It is unclear just when the people of early medieval England celebrated midwinter's day; customs might have varied somewhat from place to place and from century to century.

*Saint Stephen's feast day . . . John the Evangelist*: The feast commemorating Saint Stephen, the first Christian martyr, is celebrated on December 26 in the Gregorian calendar. The feast celebrating the birth of Saint John the Evangelist is on December 27.

28.3 *or Lenten illness or . . .* : There is a short blank space at this point in the manuscript. Either some words are left out, or the scribe was confused into thinking that they were.

29.2 *this man Roger*: The patient, evidently.

30–34 Meaney, "Variant Versions," 246–50 and 265–68, calls attention to the complex relations between this set of five cures and variant texts recorded in *Bald's Leechbook* and *Leechbook 3*.

30.3 *swail's apple*: See the note to *Peri Didaxeon* 49.

33.2 *meal made of woad*: A conjectural translation based on emendation of the nonsensical manuscript reading *wah mealo*.

33.3 *Mix it all together*: Because of a break in the manuscript, it is unclear if this passage pertains to this same cure or not.

35 *smear the yolks*: A conjectural translation based on emendation of the unsatisfactory manuscript reading.

36.2 *worms . . . in me*: What is meant here must be something like "worms are causing pain in me."

36.4 *and ambrosiana*: *Ambrosiana* in the Latin text is evidently a mistake for *ambrosia,* here probably denoting woodland germander, also known as wood sage.

37.3 *make this into a bed*: An obscure passage; perhaps a thin paste is meant.

38.2 *the rind that has grown back after having been cut*: Again, the translation is a tentative one.

39      For a variant of this cure, see *Lacnunga* 10.

39.2    *warm it at the fire*: That is, warm the ointment.

40      *yellow stone*: Probably ocher.

41      *daisy*: Old English *brysewyrt* might denote either this plant, or comfrey, or perhaps soapwort.

45      *For flow of blood*: Evidently, a reference to menstruation.

        *Take some comfrey—that is, consolda*: The Latin text provides two synonymous names for this plant. The second name appears only here and in Metrical Charm 7 (see Bjork, *Old English Shorter Poems II*, p. 214, line 6), but it has reflexes in Middle English *consolde* and modern French *consoude*. The corresponding Latin name, *consolida*, is suggestive of the traditional use of this plant to bind (or consolidate) materials, as in bonesetting or, here, the staunching of a flow of blood.

46      *leaves from an ash tree*: That is, make a broth from ash leaves.

48      *. . . and burn to ashes*: No translation can be offered for the imperfect text that precedes these words.

        *Midsummer's Eve*: The eve of the feast of Saint John the Baptist on June 24. This was a time of communal festivities, including bonfires.

        *sing these three psalms*: The songs or incantations to follow are Psalm 122, the third verse of which begins "Have mercy on us, O Lord"; Psalm 67, beginning "Let God arise, and let his enemies be scattered"; and the Athanasian Creed, beginning "Whosoever wishes to be saved."

49      *are diseased*: The apparent meaning of *on ylon*, a phrase that is perhaps equivalent to *on yfelum* (in a bad way), hence "diseased."

50      *Holy, holy, holy*: The first words of the Sanctus of the Mass; the Greek equivalent is *agios*.

51      This remedy might have served as another veterinary cure; it is followed by two blessings, the first of which is intended to control the ravages of mice. For those texts see Phillip Pulsiano, "The Prefatory Matter of London, British Library, Cotton Vitellius E.xviii," in *Anglo-Saxon Manuscripts and their Heritage*, ed. Phillip Pulsiano and Elaine M. Treharne (Aldershot, 1998), 90;

or Karen L. Jolly, "Tapping the Power of the Cross: Who and for Whom?" in *The Place of the Cross in Anglo-Saxon England,* ed. Catherine E. Karkov, Sarah Larratt Keefer, and Karen Louise Jolly (Woodbridge, 2006), 79.

51.1     *the other one* . . . : The meaning here is unclear. The two sticks when put together, perhaps, would make the sign of the cross.

51.2     *. . . that the sign of the cross is on*: This fragmentary paragraph most likely forms the concluding part of the same cure, though there can be no certainty in the matter.

       *Lammas Day*: A late summer holiday celebrating the making of the first bread from the year's harvest; the name comes from Old English *hlaf* (bread) plus *mæsse* (the Mass).

56     There is a pair of similar cures in *Bald's Leechbook* 1.26.3–4.

56.2     *Once it has settled*: The probable meaning in this context of Old English *gelogode,* which would normally mean "arranged" or "disposed."

57     There is a very similar cure in *Bald's Leechbook* 1.1.3.

58     There is a similar cure in *Bald's Leechbook* 1.27.1.

62     An incantation to staunch the flow of blood, similar to *Misc. Rem.* 66.2 and 67.6. The initial stanza, beginning *rivos cruoris torridi,* is the *R*-initial stanza from an alphabetical hymn by Sedulius *(A solis ortus cardine)* that alludes to the incident recounted in the Gospels where Jesus staunches a woman's flow of blood (Matthew 9:20–22; Mark 5:25–29; Luke 8:43–44). For discussion, see Alphons Augustinus Barb, "Die Blutsegen von Fulda und London," in *Fachliteratur des Mittelalters: Festschrift fur Gerhard Eis,* ed. Gundolf Keil, Rainer Rudolf, Wolfram Schmitt, and Hans J. Vermeer (Stuttgart, 1968), 485.

63     Much of the translation of this entry is conjectural, since the Latin text requires frequent emendation.

64     A short lorica, or prayer for health in all parts of the body.

65     A charm against demoniacal possession.

65.1     *Eulogumen . patera . . . nenonamini*: An untranslatable invocation, evidently meant to serve as an inscription to be written on a scrap of parchment.

65.2     *devil of Satan's elf*: "Elf" may stand in for any demonic creature, in accord with the belief that dementia is caused by the malice of elf, dwarf, or devil.

*this inscription written by Christ*: The reference is apparently to the preceding text beginning *Eulogumen*.

66     An incantation, similar to *Misc. Rem.* 62, meant to staunch the flow of blood.

67.2     *ϹΟΜΑΡΤΑ . . . ΕΚΤΥΤΟΠΟ*: A pseudo-Greek inscription; see Barb, "Die Blutsegen," 488–89, for discussion. Here the letter Ϲ corresponds to Greek *sigma* (or Latin *S*), while *P* corresponds to Greek *rho* (not to Latin *P*). If normalized along the lines of current Greek orthography, the passage would read ΣΟΜΑΡΤΑ ΟΣΟΓΜΑ ΣΤΥ ΓΟΝΤΟΕΜΑ ΕΚΤΥ-ΤΟΠΟ.

67.3     *Veronica*: The first-century saint, known in the East as "Beronike," identified with a woman whose long-lasting flow of blood was staunched by Jesus (Matthew 9:20–22; Mark 5:25–29; Luke 8:43–44).

67.4     *ϹΑϹΙΝΑϹΟ ΥϹΑΡΤΕΤΕ*: Another untranslatable passage of pseudo-Greek. If normalized, the passage would read ΣΑΣΙΝ-ΣΑΣΟ ΥΣΑΡΤΕΤΕ.

67.8     *ΑΜΙϹΟ . . . ΙΔΡΑϹΑϹΙΜΟ*: See Barb, "Die Blutsegen," 489–90, for discussion of this as a garbled Greek palindrome. If normalized, the passage would read ΑΜΙΣΟ ΣΑΡΔΙΝΟΡΟ ΦΙΦΙΡΟΝ ΙΔΡΑΣΑΣΙΜΟ.

67.9     *Digging out*: It is hard to make sense of this passage, though the theme of resurrection can be discerned.

       *usugma*: A word with no discernible meaning.

68     A version of the Seven Sleepers prayer; see the note to *Lacnunga* 86.2.

69     Although the first part of this prayer might be meant to arouse a person in a coma, its second part is designed to help someone who has been unable to sleep soundly, perhaps because of a feverish condition.

70     It is not clear if this prayer was meant to address any particular ailment.

       *Saint Blasius*: See the note to *Misc. Rem.* 14.

71 Another charm to staunch bleeding.

71.1 *your servant (or maidservant)*: The scribe has inserted alternative feminine grammatical inflections *(peccatrici, famulae, tuae)* directly above three words with masculine inflections *(peccatori, famulo, tuo)* so as to make the prayer suitable for persons of either sex.

*and from his wound (or body, or from mine)*: Just above *de eius* (from his [wound]), the same scribe, or perhaps another one, has written *vel meis* (or from mine), while just above *plaga* (wound) is written *vel corpore* (or body). The cure is thus made suitable for use either by a physician for a patient (whether male or female), or by a physician for himself.

72 A short version of a charm similar to *Misc. Rem.* 13 and 29.

72.1 *Saint Cassius*: The same person as Saint Nicasius. See note to *Misc. Rem.* 13.2.

73 *Cheilei cecce becce . . . noli et coli*: I offer no translation for these mixtures of Latin and nonsensical language.

*Strike the wyrm of men*: Compare *Lacnunga* 22.2, where the same phrase is employed. Edward Pettit, "Some Anglo-Saxon Charms," in *Essays on Anglo-Saxon and Related Themes in Memory of Lynne Grundy*, ed. Jane Roberts and Janet Nelson (London, 2000), 420–22, presents a helpful collation of these two cures.

*When it burns hottest on earth*: The underlying idea may be that fever cools down after it has peaked; again see *Lacnunga* 22.

*The servant of God*: The Latin noun *famulum* (servant) is in the accusative case, hence a prayer such as "liberate NAME the servant of God from evil" must be understood.

74.1 *In the name of the Father*: This sentence is written in a different ink, as an addition to the line in which it occurs; it may or may not be intended to serve as the start of this cure, though I take it as such.

*Whosoever wishes*: The first words of the Athanasian Creed.

74.2 *a good headache*: The word "good" stands for its opposite; the headache is of course a bad one.

*Obsorbis . . . edifa*: A passage of pseudo-Latin for which no translation can be offered.

74.3     *malignant drop, migraine*: Compare *Lacnunga* 158.3, where Latin *gutta* (drop), used with reference to migraine, is called "malignant." There is an allusion here to the theory of the humors, according to which headaches can result from an excess of fluid in the brain cavity.

74.4     *HOLY . . . HOLY . . . HOLY*: Translating Greek *agyos* (customarily spelled *agios*), from the Sanctus of the Mass. A cross is written above each instance of *agyos*.

75     *the Seven Sleepers*: See the note on *Lacnunga* 86.2. Above each Sleeper's name is written a cross and a word: *clarus* (illustrious) above Malcus, *probus* (virtuous) above Martinianus, *clemens* (merciful) above Dionisius, *gaudens* (rejoicing) above Constantinus, *sumens* (taking as his own) above Serapion, and *libens* (cheerful) above Iohannes (John).

76     *autumn crocus*: While this is the probable meaning of Old English *greate wyrt*, no secure identification of this plant can be offered; see the *DOE*, under *great*, sense 2.e.1: "the name of some bulbous plant."

*gentian*: A conjectural translation, as it depends on emending Old English *ferwyrt* (or *færwyrt*), a word that is not otherwise attested and has no certain meaning, to *feldwyrt*, the common name for gentian.

79.2     *clean lenten-berries*: It is uncertain what kind of berries these are other than that they evidently ripen during the spring (Old English *lencten-time*).

79.3     *for hemorrhoids*: The apparent meaning of the Old English phrase *wiþ þone flowendan fic* (literally, "for the bleeding fig"), with reference to either hemorrhoids or a comparable eruption elsewhere on the body.

80     *under the knee*: An idiom that seemingly denotes "close to the body."

81.1–2     *and thebal guttatim aurum . . . galabra Ihesus*: Little sense can be made of this passage, though *aurum* is gold and *thus* (for *tus*) is frankincense, while the words *thebal guttatim* are commonly found in other charms, with variations; for discussion with examples, see Pettit, "Some Anglo-Saxon Charms," 412–18. The syllable *de* may have become detached from Latin *deferentes* (carrying away) in Isaiah 60:6, as Pettit points out (pp. 416–17).

81.3     *For high fever*: Literally, "for a dwarf," although belief in dwarves as agents of disease is likely to have been no more than vestigial in this ecclesiastical context.

82     For a variant of this remedy see *Bald's Leechbook* 1.9.4. Jolly, "Tapping the Power," 59–60, offers a commentary on the two variants.

82.2     *Ær grin thonn . . . aær leno*: This is another inscription without translatable content. Several of its words are of Irish origin (for example, *struht fola,* "stream of blood"). In it, the symbol *ꛂ* (consisting of three crossed ascenders) is set off by raised points in the same manner that special symbols, including runes and cardinal numbers, are customarily set off, but its meaning here is unknown, unless three crosses are meant. See Jolly, "Tapping the Power," 60, for discussion.

83.2     *the sign of the Lord*: That is, the cross. Edward Pettit, "Anglo-Saxon Charms in Oxford, Bodleian Library MS Barlow 35," *Nottingham Medieval Studies* 43 (1999): 33–46, argues that this paragraph, when written out on a scrap of parchment, was meant to serve as an amulet, perhaps written in the shape of a cross, as with *Misc. Rem.* 89 below.

    *While . . . parchment remains . . .* : No translation is offered for the garbled Latin words *(maritum, obrido arigetus)* written out in this passage.

    *lose its strength, harden*: Evidently, an injunction meant to slow or stop the flow of blood.

83.3     *For the protection of a good cow*: While the corresponding Latin passage is ungrammatical, it makes reasonably good sense. Pettit, "Anglo-Saxon Charms in Oxford," 41–45, calls attention to a late fifteenth-century Icelandic parallel and suggests that the paragraph once served as an independent charm.

84.3    *When the Lord saw the sisters of Lazarus*: The allusion is to chapter
        11 of the Gospel of John.

86      *on your leech-finger*: It is the ring finger that is meant; its tip was
        evidently used to help measure out small amounts of a sub-
        stance, as in our phrase "a pinch."

        *inscribe around the sore and say*: The words of the Lord's Prayer (or
        else the words of the present charm that immediately follow?)
        are evidently to be inscribed at the sore spot while also being
        voiced aloud.

87      See *Misc. Rem.* 36 for the probable content of this fragmentary
        charm.

88      This charm is for black boils, to judge from its analogues. See
        the note on *Lacnunga* 25.2.

88.2    *Seek and you shall find*: Matthew 7:7.

89.2    *Write on his forehead*: The scribe wrote out the Latin phrase (de-
        rived from Greek) *stomen calcos* horizontally while writing *sto-
        men metafofu* vertically across its middle, thus creating a large
        cross-like formation on the page. After each of the two phrases,
        in addition, is written a smaller sign of the cross. Storms, *Anglo-
        Saxon Magic,* 291, points out that more accurate representa-
        tions of the original Greek phrases would have been *stomen
        calos* (let us stand respectfully) and *stomen meta fobou* (let us
        stand in awe).

90.4    *that you have no power*: The Latin word for "power" (*potestatem*) is
        supplied editorially in order to provide an object for the verb.
        The complete ending to the prayer is likely to have been words
        along the lines "no power to hurt this servant of God."

91      Meaney, "Variant Versions," 243–45, calls attention to parallels
        between cures in the Omont fragment (*Misc. Rem.* 91 through
        99) and ones in *Bald's Leechbook,* posing the possibility of a di-
        rect relationship between these two texts.

98.1    *For a paralyzed body*: Literally, "for a body that has fallen asleep."

100     *the person's name*: That is, the patient's.

        *Eugenius, Stephanus . . . Cyriacus*: Another invocation of the names
        of the Seven Sleepers; see the note to *Lacnunga* 86.2.

101.2     It is hard to determine what use was made of this untranslatable string of special characters; perhaps it was meant to be written out on parchment or another surface so as to serve as an amulet. Seven of the characters (*p, c, o, x, ẏ, z, B*) look like standard letters of the Old English alphabet, though *p* might be a miswriting of *p* (or *wynn,* the insular form of *w*) or might represent the Greek capital letter *P* (rho), while *c* might stand for *s* (see note to *Misc. Rem.* 67.2). Roman *o* is indistinguishable from Greek *o* (omicron) and Roman capital *B* from Greek capital *B* (beta), while Roman *x* is very similar to Greek $\chi$ (chi). Greek $\zeta$ (zeta) also sometimes resembles Roman *z*. Two other characters have the appearance of Greek letters: these are $\lambda$ (lambda) and $\varphi$ (phi). The symbol ᛗ is the *D*-rune, which is usually associated with the word *dæg* (day).

# Bibliography

## EDITIONS AND TRANSLATIONS

Barb, Alphons Augustinus. "Die Blutsegen von Fulda und London." In *Fachliteratur des Mittelalters: Festschrift fur Gerhard Eis,* edited by Gundolf Keil, Rainer Rudolf, Wolfram Schmitt, and Hans J. Vermeer, 485–93. Stuttgart, 1968.

Bjork, Robert E., ed. and trans. *Old English Shorter Poems, Volume II: Wisdom and Lyric.* DOML 32. Cambridge, MA, 2014.

Cockayne, Thomas Oswald. *Leechdoms, Wortcunning and Starcraft of Early England.* 3 vols. London, 1864–1866.

D'Aronco, Maria Amalia, and M. L. Cameron. *The Old English Illustrated Pharmacopoeia: British Library Cotton Vitellius C.iii.* Early English Manuscripts in Facsimile 27. Copenhagen, 1998.

de Vriend, Hubert Jan, ed. *The Old English "Herbarium" and "Medicina de Quadrupedibus."* EETS o.s. 286. Oxford, 1984.

Dickins, Bruce, and R. M. Wilson. "Sent Kasi." *Leeds Studies in English* 6 (1937): 67–73.

Doane, A. N., and Phillip Pulsiano, eds. *Anglo-Saxon Manuscripts in Microfiche Facsimile.* Multiple volumes. Binghamton, NY, 1994–2021.

Dobbie, Elliott Van Kirk, ed. *The Anglo-Saxon Minor Poems.* ASPR 6. New York, 1942.

Grant, R. J. S. *Cambridge, Corpus Christi College 41: The Loricas and the Missal.* Costerus Essays in English and American Language and Literature, n.s. 17. Amsterdam, 1979.

Grattan, J. H. G., and Charles Singer. *Anglo-Saxon Magic and Medicine.* London, 1952.

Grendon, Felix, ed. and trans. *The Anglo-Saxon Charms.* New York, 1909.

Günzel, Beate, ed. *Ælfwine's Prayerbook*. London, 1993.

Jolly, Karen L. "Magic, Miracle, and Popular Practice in the Early Medieval West: Anglo-Saxon England." In *Religion, Science, and Magic in Concert and in Conflict,* edited by Jacob Neusner, Ernest Frerichs, and Paul V. M. Flesher, 166–82. New York, 1989.

——. "On the Margins of Orthodoxy: Devotional Formulas and Protective Prayers in Cambridge, Corpus Christi College MS 41." In *Signs on the Edge: Space, Text and Margin in Medieval Manuscripts,* edited by Sarah Larratt Keefer and Rolf H. Bremmer, Jr., 135–80 and plates 1–4. Paris, 2007.

——. "Tapping the Power of the Cross: Who and for Whom?" In *The Place of the Cross in Anglo-Saxon England,* edited by Catherine E. Karkov, Sarah Larratt Keefer, and Karen Louise Jolly, 58–79. Woodbridge, 2006.

Ker, N. R. *Catalogue of Manuscripts Containing Anglo-Saxon.* Reissued with a supplement. Oxford, 1990.

Leonhardi, Günther, ed. *Kleinere angelsächsische Denkmäler 1.* Bibliothek der angelsächsischen Prosa 6. Hamburg, 1905.

Löweneck, Max, ed. "Peri Didaxeon." Special issue, *Erlanger Beiträge zur englischen Philologie* 12 (1896). Reprint, Amsterdam, 1970.

Maion, Danielle. "Edizione, traduzione e commento del *Peri Didaxeon.*" PhD diss., Università degli Studi Roma Tre, 1999.

Muir, Bernard J. *A Pre-Conquest English Prayer-Book.* Woodbridge, 1988.

Napier, A. "Altenglische Miscellen." *Archiv für das Studium der neueren Sprachen und Literaturen* 84 (1890): 322–27.

Olsan, Lea. "The Marginality of Charms in Medieval England." In *The Power of Words: Studies on Charms and Charming in Europe,* edited by James Kapaló, Éva Pócs, and William Ryan, 135–64. Budapest, 2013.

Pettit, Edward. "Anglo-Saxon Charms in Oxford, Bodleian Library MS Barlow 35." *Nottingham Medieval Studies* 43 (1999): 33–46.

——, ed. and trans. *Anglo-Saxon Remedies, Charms, and Prayers from British Library MS Harley 585: The Lacnunga.* 2 vols. Lewiston, 2001.

——. "Some Anglo-Saxon Charms." In *Essays on Anglo-Saxon and Related Themes in Memory of Lynne Grundy,* edited by Jane Roberts and Janet Nelson, 411–43. London, 2000.

Pollington, Stephen. *Leechcraft: Old English Charms, Plantlore, and Healing.* Norfolk, 2001.

Pulsiano, Phillip. "The Prefatory Matter of London, British Library, Cotton Vitellius E.xviii." In *Anglo-Saxon Manuscripts and their Heritage,* edited by Phillip Pulsiano and Elaine M. Treharne, 85–116. Aldershot, 1998.

Sanborn, Linda. "An Edition of British Library MS. Harley 6258 B, *Peri Didaxeon.*" PhD diss., University of Ottawa, 1983.

Schauman, Bella, and Angus Cameron. "A Newly-Found Leaf of Old English from Louvain." *Anglia* 95 (1977): 289–312.

Stokes, W. "Glosses from Turin and Rome, IV: The Anglosaxon Prose and Glosses in Rome." *Beiträge zur Kunde der indogermanischen Sprachen* 17 (1891): 144–45.

Storms, Godfrid. *Anglo-Saxon Magic.* The Hague, 1948.

Torkar, Roland. "Zu den ae. Medizinaltexten in Otho B.xi und Royal 12.D.xvii, mit einer edition der Unica (Ker, no. 180 art. 11a–d)." *Anglia* 94 (1976): 319–38.

Van Arsdall, Anne. *Medieval Herbal Remedies: The Old English Herbarium and Anglo Saxon Medicine.* New York, 2002.

## FURTHER READING

Arthur, Ciaran. *"Charms," Liturgies, and Secret Rites in Early Medieval England.* Woodbridge, 2018.

Barley, Nigel. "Anglo-Saxon Magico-Medicine." *Journal of the Anthropological Society of Oxford* 3 (1972): 67–77.

Bonser, Wilfrid. *The Medical Background of Anglo-Saxon England: A Study in History, Psychology and Folklore.* London, 1963.

Cameron, M. L. *Anglo-Saxon Medicine.* Cambridge, 1993.

Crawford, Sally. "The Nadir of Western Medicine? Texts, Contexts and Practice in Anglo-Saxon England." In *Bodies of Knowledge: Cultural Interpretations of Illness and Medicine in Medieval Europe,* edited by Sally Crawford and Christina Lee, 41–51. BAR International Series 2170. Oxford, 2010.

Dendle, Peter. "Plants in the Early Medieval Cosmos: Herbs, Divine Potency, and the *Scala Natura.*" In *Health and Healing from the Medieval Garden,* edited by Peter Dendle and Alain Touwaide, 47–59. Woodbridge, 2008.

Fisher, Rebecca M. C. "Genre, Prayers and the Anglo-Saxon Charms." In *Genre—Text—Interpretation: Multidisciplinary Perspectives on Folklore and Beyond,* edited by Kaarina Koski and Frog with Ulla Savolainen, 137–51. Helsinki, 2016.

Glosecki, Stephen O. *Shamanism and Old English Poetry.* New York, 1989.

Kesling, Emily. *Medical Texts in Anglo-Saxon Literary Culture.* Woodbridge, 2020.

Lee, Christina. "Body Talks: Disease and Disability in Anglo-Saxon England." In *Anglo-Saxon Traces,* edited by Jane Roberts and Leslie Webster, 145–64. Tempe, 2011.

Meaney, Audrey L. "Extra-Medical Elements in Anglo-Saxon Medicine." *Social History of Medicine* 24 (2011): 1–56.

———. "The Practice of Medicine in England about the Year 1000." *Social History of Medicine* 13 (2000): 332–37.

Rawcliffe, Carole. "Medical Practice and Theory." In *A Social History of England 900–1200,* edited by Julia Crick and Elisabeth Van Houts, 391–401. Cambridge, 2011.

Riddle, John M. "Theory and Practice in Medieval Medicine." *Viator* 5 (1974): 157–84.

Rubin, Stanley, "The Anglo-Saxon Physician." In *Medicine in Early Medieval England: Four Papers,* edited by Marilyn Deegan and D. G. Scragg, 7–15. Manchester, 1987; corrected reissue, 1989.

Van Arsdall, Anne. "Medical Training in Anglo-Saxon England: An Evaluation of the Evidence." In *Form and Content of Instruction in Anglo-Saxon England in the Light of Contemporary Manuscript Evidence,* edited by Patrizia Lendinara, Loredana Lazzari, and Maria Amalia D'Aronco, 415–34. Turnhout, 2007.

Voigts, Linda E. "Anglo-Saxon Plant Remedies and the Anglo-Saxons." *Isis* 70 (1979): 250–68.

Weston, Lisa M. C. "Women's Medicine, Women's Magic: The Old English Metrical Childbirth Charms." *Modern Philology* 92 (1995): 279–93.

# Index of Modern English
# Plant Names

# Index of Old English and Latin
# Plant Names

Old English names are in roman font, while Latin names are in italics.

# General Index

swelling(s) *(continued)*
breasts; eye ailments; foot ailments; genitalia; head ailments;
neck ailments; sinews; stomach
ailments

tar, *Rem. An.* 4.1; *Lacnunga* 138; *Misc.
Rem.* 98.3. *See also* pitch
teeth, *PD* 33–35; growth of in children, *Rem. An.* 5.16, 10.8, 14.13;
loose, *Herbal* 97.2; *Rem. An.* 3.3.
*See also* toothache
Telephos, *Herbal* 90.1
theor-disease, *Lacnunga* 49.1, 68–
74, 121, 144, 170.2
thread, *Herbal* 10.1, 104.2, 183.1
tiredness, *Herbal* 1.17
Tobit (biblical character), *Lacnunga*
150.2; *Misc. Rem.* 4.2
toothache, *Herbal* 1.8, 5.3, 30.3, 76.3,
81.1, 86.2, 90.2, 97.2, 153.3, 181.3;
*Rem. An.* 6.9, 14.11; *Lacnunga* 22,
158; *Misc. Rem.* 36, 73, 87; canker,
*Herbal* 165.3; tooth decay, *PD* 33–
35. *See also* teeth
tow, *PD* 63.4
tremors, *Herbal* 13.2, 171.3

Ulysses, *Herbal* 73
urinary distress, *Herbal* 4.5, 7.2, 12.1,
55.1, 80.1, 90.5, 107.1, 108.1, 117.4,
135.1, 143.1, 146.1, 148.1, 152.1,
153.2, 154.1, 156.2, 157.2, 163.1,
163.2, 164.1, 173.1, 180.1; *Rem. An.*
3.16, 9.11, 9.12; *Lacnunga* 170.2;
*PD* 63.1, 65.4

veins, *Herbal* 4.3, 86.3; *PD* 52.1–2,
64–66; hardened, *Herbal* 90.9
venoms (infections), *Lacnunga* 28,
64.5, 76.1, 170.2; *Misc. Rem.* 9.5,
15, 88.3; flying, *Lacnunga* 18, 126.1;
nine mythological, *Lacnunga*
76.1. *See also* poison
vertigo. *See* head ailments: giddiness
vessels, *Rem. An.* 1.6, 7.1; *PD* 14, 36;
brass/electrum, *Herbal* 75.2; *Misc.
Rem.* 18; bronze, *Lacnunga* 38,
49.2; *Misc. Rem.* 17.2, 18; copper,
*Lacnunga* 173.3; *Misc. Rem.* 18;
earthenware, *Herbal* 126.2; *Rem.
An.* 1.6; *PD* 16.2, 17.2, 33.3, 52.5,
61.3; *Misc. Rem.* 17.2; glass, *Herbal*
31.2, 117.1, 132; *PD* 16.2, 17.2; horn,
*Lacnunga* 5; wood, *Herbal* 25.1,
117.2; *PD* 62.2
veterinary remedies: for cattle/live-
stock, *Lacnunga* 26.1, 111, 132, 133;
*Misc. Rem.* 50, 51, 83.3; for horses,
*Lacnunga* 111, 152, 154, 155, 165,
168; *Misc. Rem.* 82; for pigs, *Lacnunga* 136; for sheep, *Lacnunga*
134, 135; *Misc. Rem.* 49
vinegar, *Herbal* 1.8, 2.7, 2.17, 10.2,
12.2, 18.3, 32.5, 36.5, 37.3, 43.1, 43.3,
47.1, 54.2, 58.1, 60.3, 75.4, 85.1,
87.2, 90.5, 91.3, 91.5, 91.7, 94.2,
94.7, 94.10, 94.12, 96.1, 96.2,
100.8, 101.2, 112.1, 116.2, 117.4,
119.1, 122.1, 143.1, 146.4, 148.3,
151.3, 155.2, 155.4, 158.2, 158.4,
163.4, 165.5, 181.3, 184.3, 184.4;